Molecular Characterization of Gram-Negative Bacteria: Antimicrobial Resistance, Virulence and Epidemiology

Molecular Characterization of Gram-Negative Bacteria: Antimicrobial Resistance, Virulence and Epidemiology

Editor

Theodoros Karampatakis

MDPI

Basel • Beijing • Wuhan • Barcelona • Belgrade • Novi Sad • Cluj • Manchester

Theodoros Karampatakis
Microbiology Department
Papanikolaou General Hospital
Thessaloniki
Greece

Editorial Office
MDPI AG
Grosspeteranlage 5
4052 Basel, Switzerland

This is a reprint of the Special Issue, published open access by the journal *Antibiotics* (ISSN 2079-6382), freely accessible at: www.mdpi.com/journal/antibiotics/special_issues/Gram_Negative_AMR.

For citation purposes, cite each article independently as indicated on the article page online and using the guide below:

Lastname, A.A.; Lastname, B.B. Article Title. *Journal Name* **Year**, *Volume Number*, Page Range.

ISBN 978-3-7258-1418-3 (Hbk)
ISBN 978-3-7258-1417-6 (PDF)
https://doi.org/10.3390/books978-3-7258-1417-6

Contents

antibiotics

Editorial

Molecular Characterization of Gram-Negative Bacteria: Antimicrobial Resistance, Virulence and Epidemiology

Theodoros Karampatakis

Microbiology Department, Papanikolaou General Hospital, 57010 Thessaloniki, Greece; tkarampatakis@yahoo.com

Multidrug-resistant (MDR), extensively drug-resistant (XDR) and pan-drug-resistant (PDR) Gram-negative bacteria constitute a huge public health problem. According to the European Centre for Disease Prevention and Control (ECDC), MDR is defined as 'acquired nonsusceptibility to at least one agent in three or more antimicrobial categories' and extensively drug-resistant is defined as 'nonsusceptibility to at least one agent in all but two or fewer antimicrobial categories (i.e., bacterial isolates remain susceptible to only one or two categories)'. PDR is defined as 'nonsusceptibility to all agents in all antimicrobial categories' [1]. Among them, MDR Gram-negative bacteria and especially *Klebsiella pneumoniae*, *Pseudomonas aeruginosa*, and *Acinetobacter baumannii* are the main bacteria leading to increased rates of antimicrobial resistance [2]. Infections caused by these MDR bacteria adversely affect the outcome of patients' hospitalization and increase mortality rates [3]. In addition, they can increase healthcare costs significantly [4]. The increased consumption of antimicrobials and the poor implementation of infection control measures in the hospital setting are the two main causative factors for their emergence. The deep knowledge of the virulent factors that these bacteria produce could provide useful information regarding their spread. Whole-genome sequencing (WGS) has contributed to this aim over recent years [5].

The main problem concerning infections caused by these bacteria is that treatment options are extremely limited, as only a limited number of novel antimicrobials have been launched over the last few years [6]. The molecular epidemiology of infections caused by Gram-negative bacteria is significant as it can determine which of these few new antimicrobial agents could be effective for their treatment [7]. This Special Issue entitled 'Molecular Characterization of Gram-Negative Bacteria: Antimicrobial Resistance, Virulence and Epidemiology' sought manuscript submissions that could expand our knowledge of antimicrobial resistance in Gram-negative bacteria. Submissions on the mechanisms of antimicrobial resistance, the presence and function of virulent factors, as well as the molecular epidemiology of Gram-negative bacterial infections were especially encouraged. Ultimately, 12 manuscripts were submitted for consideration for the Special Issue, and they were all accepted for publication. The contributions are briefly analysed below.

In the first contribution, Braspenning et al. examine the frequency and incidence of colistin heteroresistance among a huge number of MDR *K. pneumoniae* strains. The authors have performed WGS and examined the in vitro susceptibility to colistin. Colistin-heteroresistant isolates are shown to have increased sequence type (ST) diversity, while colistin-resistant *K. pneumoniae* isolates are associated with particular STs, such as ST101, ST147 and ST258/ST512.

In the second contribution, Papa et al. attempt to elucidate the differences in virulence features in normoxia and anoxia between MDR and PDR *P. aeruginosa* clinical strains and sensitive strains to antimicrobials, isolated in cystic fibrosis patients. The authors have performed pyocyanin, pyoverdine, protease, zymography and motility assays, as well as biofilm quantification, and they reveal that these features are highly diversified in anaerobiosis, which depicts the condition of cystic fibrosis chronic infection.

check for updates

Citation: Karampatakis, T. Molecular Characterization of Gram-Negative Bacteria: Antimicrobial Resistance, Virulence and Epidemiology. *Antibiotics* **2024**, *13*, 402. https://doi.org/10.3390/antibiotics13050402

Received: 14 April 2024
Accepted: 23 April 2024
Published: 28 April 2024

In the third contribution, Meletis et al. report the isolation of bla_{NDM-1} in three different bacterial species (*K. pneumoniae*, *Proteus mirabilis* and *Morganella morganii*) isolated from a single patient. The authors have used WGS and bioinformatic tools to detect antimicrobial resistance genes and plasmids. They disclose that bla_{NDM-1}, along with its neighbouring genes, belongs to the same cluster, implying the in vivo transfer of the bla_{NDM-1}-containing cluster through three different species.

In the fourth contribution, Khalid et al. present the in vitro synergistic effectiveness of 19 different combinations of antimicrobials in 31 New Delhi metallo β-lactamase (NDM) producing MDR isolates of *Enterobacterales*. The authors have used the 2D (two-dimensional) checkerboard method. They have revealed that three combinations of antimicrobials, namely doripenem with cefoxitin, doripenem with streptomycin, and imipenem with cefoxitin, are effective against these MDR isolates, suggesting further in vivo and pharmaceutical studies need to be conducted.

In the fifth contribution, Zarras et al. study 24 carbapenem-resistant *K. pneumoniae* strains, isolated from a single healthcare institution. The authors have applied WGS and various pieces of bioinformatics software to identify antimicrobial resistance genes, plasmids and STs. In addition, they have compared multiple genome alignments and identify core genome single-nucleotide polymorphism sites (SNPs). The authors disclose that the isolates are assigned to seven different phylogenetic branches, with the phylogeny changing every seven SNPs.

In the sixth contribution, Zhang et al. determine the prevalence and molecular characteristics of $bla_{CTX-M-55}$-positive *Escherichia coli* isolated from duck–fish polyculture farms in Guangzhou, China. The authors have used WGS, southern hybridization and pulse field gel electrophoresis (PFGE) to detect the possible horizontal transfer and clonal dissemination of $bla_{CTX-M-55}$. The authors have revealed that the F18:A-:B1 plasmid might have major significance in the transmission of $bla_{CTX-M-55}$ in *E. coli*.

In the seventh contribution, Gupta et al. report the cloning and analysis of *phl*D (involved in the biosynthesis of non-volatile metabolite phloroglucinol) from soil-borne Gram-negative bacteria *Pseudomonas* spp. In addition, the authors have analysed the structural details of the PHLD protein, providing novel strategies for the combinatorial biosynthesis of natural but pharmaceutically important metabolites with enhanced antibacterial and biocontrol effects.

In the eighth contribution, Osman et al. present the molecular analysis of 86 *K. pneumoniae* strains, with most of them being MDR, isolated from different hospitals of Khartoum, Sudan. The authors have performed WGS to detect virulence and resistome profiles. *Ybt9* is the most common virulence gene detected, while the authors report the detection of various antimicrobial resistance genes. In addition, transmissions between patients are found to be rare.

In the ninth contribution, Hao et al. present the differences in the molecular characteristics and the proteomes of sensitive, MDR and XDR *K. pneumoniae* strains. The authors' enrichment analysis has revealed that a majority of differentially expressed proteins are involved in various metabolic pathways which are beneficial to the evolution of antimicrobial resistance in *K. pneumoniae*.

In the tenth contribution, Karruli et al. present a literature review to assess the antimicrobial activity of cefiderocol and sulbactam-durlobactam against carbapenem-resistant *A. baumannii* (CRAB). The authors have concluded that cefiderocol is non-inferior to colistin/other treatments for CRAB infections and displays a better safety profile. Combination treatment is not correlated with improved outcomes compared to monotherapy. Sulbactam-durlobactam could also be another valuable option against CRAB infections.

In the eleventh contribution, Karampatakis et al. review the dynamic evolution of the molecular epidemiology of carbapenem-resistant *K. pneumoniae* (CRKP), its virulence factors and the latest updates for the treatment of CRKP infections. In addition, they report the latest guidelines for treating these infections, as proposed by international organisations.

In the twelfth contribution, Li et al. describe the genetic and structural compositions of the five secretion systems that exist in *A. baumannii*. The authors have underlined the importance of these systems in the fitness and pathogenesis of *A. baumannii*, and their contribution to the emergence of antimicrobial resistance.

In conclusion, the manuscripts published in this Special Issue reveal recent data which contribute to the better understanding of the virulence of MDR Gram-negative bacteria, and they provide important updates on the evolution of their molecular epidemiology at a global level.

List of Contributions

1. Braspenning, A.J.M.M.; Rajakani, S.G.; Sey, A.; El Bounja, M.; Lammens, C.; Glupczynski, Y.; Malhotra-Kumar, S. Assessment of Colistin Heteroresistance among Multidrug-Resistant Klebsiella pneumoniae Isolated from Intensive Care Patients in Europe. *Antibiotics* **2024**, *13*, 281. https://doi.org/10.3390/antibiotics13030281.
2. Papa, R.; Imperlini, E.; Trecca, M.; Paris, I.; Vrenna, G.; Artini, M.; Selan, L. Virulence of *Pseudomonas aeruginosa* in Cystic Fibrosis: Relationships between Normoxia and Anoxia Lifestyle. *Antibiotics* **2023**, *13*, 1. https://doi.org/10.3390/antibiotics13010001.
3. Meletis, G.; Malousi, A.; Tychala, A.; Kassomenaki, A.; Vlachodimou, N.; Mantzana, P.; Metallidis, S.; Skoura, L.; Protonotariou, E. Probable Three-Species In Vivo Transfer of *bla(NDM-1)* in a Single Patient in Greece: Occurrence of NDM-1-Producing *Klebsiella pneumoniae, Proteus mirabilis*, and *Morganella morganii*. *Antibiotics* **2023**, *12*, 1206. https://doi.org/10.3390/antibiotics12071206.
4. Khalid, S.; Migliaccio, A.; Zarrilli, R.; Khan, A.U. Efficacy of Novel Combinations of Antibiotics against Multidrug-Resistant-New Delhi Metallo-Beta-Lactamase-Producing Strains of Enterobacterales. *Antibiotics* **2023**, *12*, 1134. https://doi.org/10.3390/antibiotics12071134.
5. Zarras, C.; Karampatakis, T.; Pappa, S.; Iosifidis, E.; Vagdatli, E.; Roilides, E.; Papa, A. Genetic Characterization of Carbapenem-Resistant *Klebsiella pneumoniae* Clinical Isolates in a Tertiary Hospital in Greece, 2018–2022. *Antibiotics* **2023**, *12*, 976. https://doi.org/10.3390/antibiotics12060976.
6. Zhang, L.J.; Yang, J.T.; Chen, H.X.; Liu, W.Z.; Ding, Y.L.; Chen, R.A.; Zhang, R.M.; Jiang, H.-X. F18:A-:B1 Plasmids Carrying bla(CTX-M-55) Are Prevalent among *Escherichia coli* Isolated from Duck-Fish Polyculture Farms. *Antibiotics* **2023**, *12*, 961. https://doi.org/10.3390/antibiotics12060961.
7. Gupta, P.; Dash, P.K.; Sanjay, T.D.; Pradhan, S.K.; Sreevathsa, R.; Rai, R. Cloning and Molecular Characterization of the *phlD* Gene Involved in the Biosynthesis of "Phloroglucinol", a Compound with Antibiotic Properties from Plant Growth Promoting *Bacteria Pseudomonas* spp. *Antibiotics* **2023**, *12*, 260. https://doi.org/10.3390/antibiotics12020260.
8. Osman, E.A.; Yokoyama, M.; Altayb, H.N.; Cantillon, D.; Wille, J.; Seifert, H.; Higgins, P.G.; Al-Hassan, L. Klebsiella pneumonia in Sudan: Multidrug Resistance, Polyclonal Dissemination, and Virulence. *Antibiotics* **2023**, *12*, 233. https://doi.org/10.3390/antibiotics12020233.
9. Hao, L.; Yang, X.; Chen, H.; Mo, Z.; Li, Y.; Wei, S.; Zhao, Z. Molecular Characteristics and Quantitative Proteomic Analysis of Klebsiella pneumoniae Strains with Carbapenem and Colistin Resistance. *Antibiotics* **2022**, *11*, 1341. https://doi.org/10.3390/antibiotics11101341.
10. Karruli, A.; Migliaccio, A.; Pournaras, S.; Durante-Mangoni, E.; Zarrilli, R. Cefiderocol and Sulbactam-Durlobactam against Carbapenem-Resistant *Acinetobacter baumannii*. *Antibiotics* **2023**, *12*, 1729. https://doi.org/10.3390/antibiotics12121729.
11. Karampatakis, T.; Tsergouli, K.; Behzadi, P. Carbapenem-Resistant *Klebsiella pneumoniae*: Virulence Factors, Molecular Epidemiology and Latest Updates in Treatment Options. *Antibiotics* **2023**, *12*, 234. https://doi.org/10.3390/antibiotics12020234.

12. Li, P.; Zhang, S.; Wang, J.; Al-Shamiri, M.M.; Han, B.; Chen, Y.; Han, S.; Han, L. Uncovering the Secretion Systems of *Acinetobacter baumannii*: Structures and Functions in Pathogenicity and Antibiotic Resistance. *Antibiotics* **2023**, *12*, 195. https://doi.org/10.3390/antibiotics12020195.

Conflicts of Interest: The authors declare no conflicts of interest.

References

1. Magiorakos, A.-P.; Srinivasan, A.; Carey, R.B.; Carmeli, Y.; Falagas, M.E.; Giske, C.G.; Harbarth, S.; Hindler, J.F.; Kahlmeter, G.; Olsson-Liljequist, B.; et al. Multidrug-resistant, extensively drug-resistant and pandrug-resistant bacteria: An international expert proposal for interim standard definitions for acquired resistance. *Clin. Microbiol. Infect.* **2012**, *18*, 268–281. [CrossRef]
2. Jean, S.S.; Harnod, D.; Hsueh, P.R. Global Threat of Carbapenem-Resistant Gram-Negative Bacteria. *Front. Cell. Infect. Microbiol.* **2022**, *12*, 823684. [CrossRef]
3. Lodise, T.P.; Bassetti, M.; Ferrer, R.; Naas, T.; Niki, Y.; Paterson, D.L.; Zeitlinger, M.; Echols, R. All-cause mortality rates in adults with carbapenem-resistant Gram-negative bacterial infections: A comprehensive review of pathogen-focused, prospective, randomized, interventional clinical studies. *Expert Rev. Anti. Infect. Ther.* **2022**, *20*, 707–719. [CrossRef]
4. Nelson, R.E.; Hatfield, K.M.; Wolford, H.; Samore, M.H.; Scott, R.D.; Reddy, S.C.; Olubajo, B.; Paul, P.; Jernigan, J.A.; Baggs, J. National Estimates of Healthcare Costs Associated with Multidrug-Resistant Bacterial Infections among Hospitalized Patients in the United States. *Clin. Infect. Dis.* **2021**, *72* (Suppl. S1), S17–S26. [CrossRef] [PubMed]
5. Ellington, M.J.; Ekelund, O.; Aarestrup, F.M.; Canton, R.; Doumith, M.; Giske, C.; Grundman, H.; Hasman, H.; Holden, M.T.G.; Hopkins, K.L.; et al. The role of whole genome sequencing in antimicrobial susceptibility testing of bacteria: Report from the EUCAST Subcommittee. *Clin. Microbiol. Infect.* **2017**, *23*, 2–22. [CrossRef]
6. Yusuf, E.; Bax, H.I.; Verkaik, N.J.; van Westreenen, M. An Update on Eight "New" Antibiotics against Multidrug-Resistant Gram-Negative Bacteria. *J. Clin. Med.* **2021**, *10*, 1068. [CrossRef] [PubMed]
7. Bassetti, M.; Garau, J. Current and future perspectives in the treatment of multidrug-resistant Gram-negative infections. *J. Antimicrob Chemother.* **2021**, *76*, iv23–iv37. [CrossRef] [PubMed]

antibiotics

Review

Carbapenem-Resistant *Klebsiella pneumoniae*: Virulence Factors, Molecular Epidemiology and Latest Updates in Treatment Options

Theodoros Karampatakis [1], Katerina Tsergouli [2] and Payam Behzadi [3,*]

1 Microbiology Department, Papanikolaou General Hospital, 57010 Thessaloniki, Greece
2 Microbiology Department, Agios Pavlos General Hospital, 55134 Thessaloniki, Greece
3 Department of Microbiology, Shahr-e-Qods Branch, Islamic Azad University, Tehran 37541-374, Iran
* Correspondence: behzadipayam@yahoo.com

Abstract: *Klebsiella pneumoniae* is a Gram-negative opportunistic pathogen responsible for a variety of community and hospital infections. Infections caused by carbapenem-resistant *K. pneumoniae* (CRKP) constitute a major threat for public health and are strongly associated with high rates of mortality, especially in immunocompromised and critically ill patients. Adhesive fimbriae, capsule, lipopolysaccharide (LPS), and siderophores or iron carriers constitute the main virulence factors which contribute to the pathogenicity of *K. pneumoniae*. Colistin and tigecycline constitute some of the last resorts for the treatment of CRKP infections. Carbapenemase production, especially *K. pneumoniae* carbapenemase (KPC) and metallo-β-lactamase (MBL), constitutes the basic molecular mechanism of CRKP emergence. Knowledge of the mechanism of CRKP appearance is crucial, as it can determine the selection of the most suitable antimicrobial agent among those most recently launched. Plazomicin, eravacycline, cefiderocol, temocillin, ceftolozane–tazobactam, imipenem–cilastatin/relebactam, meropenem–vaborbactam, ceftazidime–avibactam and aztreonam–avibactam constitute potent alternatives for treating CRKP infections. The aim of the current review is to highlight the virulence factors and molecular pathogenesis of CRKP and provide recent updates on the molecular epidemiology and antimicrobial treatment options.

Keywords: carbapenem-resistant *Klebsiella pneumoniae*; molecular epidemiology; antimicrobial agents; virulence factors

Citation: Karampatakis, T.; Tsergouli, K.; Behzadi, P. Carbapenem-Resistant *Klebsiella pneumoniae*: Virulence Factors, Molecular Epidemiology and Latest Updates in Treatment Options. *Antibiotics* **2023**, *12*, 234. https://doi.org/10.3390/antibiotics12020234

Academic Editor: Michael J. McConnell

Received: 26 December 2022
Revised: 13 January 2023
Accepted: 16 January 2023
Published: 21 January 2023

1. Introduction

Klebsiella pneumoniae is a non-motile Gram-negative opportunistic pathogen responsible for approximately 10% of nosocomial bacterial infections. Infections caused by carbapenem-resistant *K. pneumoniae* (CRKP) isolates are a major threat for public health. Such infections can increase the mortality rates of critically ill and debilitated patients hospitalised in intensive care units (ICUs) and can have a negative impact on the financial costs of their hospitalisation all over the world [1–4]. Remarkably, the mortality rate among patients with pneumonia caused by *K. pneumoniae* is about 50% [5]. Another major topic for public health is the effect of CRKP infections in disability-adjusted-life-years (DALYs) per 100,000 population, with a median of 11.5 in the European Union, and Greece being among the countries with the highest numbers [6]. The rate of carbapenem resistance for *K. pneumoniae* isolates reached 66.3% in 2020 in Greece [7]. A recent meta-analysis shows that the prevalence of CRKP colonisation ranges worldwide from 0.13 to 22% with a pooled prevalence of 5.43%, while the incidence of CRKP colonisation ranges from 2% to 73% with a pooled incidence of 22.3% [8]. CRKP isolates are usually classified as multidrug-resistant (MDR), extensively drug-resistant (XDR) and pandrug-resistant (PDR), which cause even more difficulty in treating infections. According to the European Center for Disease Prevention and Control (ECDC), MDR is defined as 'acquired non-susceptibility to at least one agent in $\geq$ three antimicrobial categories, XDR is defined as 'non-susceptibility to at least one agent in all but $\leq$ two antimicrobial categories (i.e., bacterial isolates remain susceptible to only one or two categories)' and PDR is defined as 'non-susceptibility to all agents in all antimicrobial categories' [9]. The molecular epidemiology of CRKP isolates is significant as it can determine potential treatment options [10].

The aim of the current review is to highlight the virulence factors and molecular pathogenesis of CRKP and provide recent updates on the molecular epidemiology and antimicrobial treatment options.

2. Genomic Pool

Despite the unclear reasons for the high frequency of infections caused by *K. pneumoniae* compared to other Gram-negative opportunistic bacterial pathogens, there are some suggestions comprising genetic element exchanges with human microbiome populations through DNA molecules, mobile genetic element exchanges bearing genes associated with virulence enhancers and antimicrobial resistance, inherent antimicrobial resistance, starvation tolerance and surpassing other bacterial competitors, which may explain the occurrence of this feature [11–15].

According to genomic investigations, the pan-genome of *K. pneumoniae* involves a size of about five to six Mbp bearing five to six kilogenes to be encoded. From this number of encodable genes, about seventeen hundred genes are recognized as core genes. The core genome is conserved among bacterial species of *K. pneumoniae*. Typically, the core genes are present in $\geq$95% of the members pertaining to a given species. However, the rest genomic pool includes accessory genes. In other words, the accessory genome is known as dispensable, flexible, adaptive or supplementary genome, which varies among *Klebsiella* spp. The accessory genes are typically present in <95% of the members pertaining to a given species [16–18].

Indeed, progression and development in microbial taxonomic approaches provides easier diagnostic and detective methodologies in association with epidemiological studies, public health surveillance and outbreak investigations. Due to this knowledge, effective approaches such as core genome multilocus sequence typing (cgMLST) can be recruited for new advanced techniques, including dual barcoding approach [19–21].

The *K. pneumoniae* species complex based on genomic phylogenetic structure is categorized into seven major phylogroups comprising Kp1 (*K. pneumoniae subspecies pneumoniae* or *K. pneumoniae sensu stricto*), Kp2 (*K. quasipneumoniae* subsp. *quasipneumoniae*), Kp3 (*K. variicola subsp. variicola*), Kp4 (*K. quasipneumoniae subsp. similipneumoniae*), Kp5 (*K. variicola subsp.* tropica), Kp6 (*K. quasivariicola*) and Kp7 (*K. africana*) [17,19]. In this regard, seven housekeeping genes including *gap*A, *inf*B, *mdh*, *pgi*, *pho*E, *rpo*B and *ton*B are sequenced. Moreover, the K-typing or capsule typing can be achieved through *wzi* gene sequencing or serotyping methods [11].

So, through the MLST typing of the above seven housekeeping genes, several phylogenetic lineages, e.g., clonal groups (CGs) and/or sequence types, exist [22].

As mentioned above, the antimicrobial-resistant and hypervirulent strains of *K. pneumoniae* have raised great concern worldwide. On the other hand, *Klebsiella* spp. are known as significant bacterial agents isolated from patients with ventilator-associated pneumonia (VAP) in ICUs. According to reported results from previous studies, 83% of hospital-acquired pneumonias are associated with VAP [5,23].

Although ß-lactam antimicrobials are known as the first choice for treatment of infections caused by *K. pneumoniae*, the number of ß-lactamase and especially carbapenemase-producing strains considerably increases. Due to this knowledge, the dissemination of ST258 CRKP is a global concern, as ST258 strains are not completely sensitive towards a wide range of antimicrobials comprising aminoglycosides, fluoroquinolones, etc. [24–30].

In accordance with the latest studies, the clonal complex (CC) of CC258 is known as the main CRKP comprising ST11, ST258, ST340, ST437 and ST512. Moreover, there are a wide range of MDR clonal groups (CGs), e.g., CG101, CG490, CG147, CG307, CG152,

CG14/15, CG231, CG43, CG17/20, CG37 and CG29, which are distributed around the world [31–34].

According to recorded reports, about 7.5% of STs (or >115 STs) pertaining to CPKP strains have been recognized in different global geographical regions. In addition, CG258 is thepredominant global CPKP strain with 43 ST members. Among them, ST258, ST11, ST340, ST437 and ST512 are the most predominant members of CG258 worldwide. ST11 ranks first in America (Latin) and Asia, while ST258 are the predominant CRKP strains in America (Latin and North) and some European countries. The ST340 has been reported in Greece and Brazil, and ST512 has been identified in Israel, Italy and Colombia [35].

The latest studies depict ≥1452 STs associated with *K. pneumoniae,* in which 1119 STs are recognized as known strains while the remaining 333 are detected as novel STs. In addition to CG258, CG15 and ST307 carry a huge range of antimicrobial resistance genes that are globally disseminated and are associated with healthcare infectious diseases and nosocomial outbreaks [22].

3. Virulence Factors and Molecular Pathogenesis

In accordance with the latest categorization, *K. pneumoniae* strains are classified into two major pathotypes, including classical *K. pneumoniae* and hypervirulent *K. pneumoniae* (HVKP). Although the classical type is frequent pathogenic agent relating to hospital acquired pneumoniae (HAP), it has limited virulence capability. Furthermore, the classical pathotype easily tends to exchange mobile genetic elements such as plasmids to create MDR strains, while HVKP is recognized as a causative agent of fulminant and invasive diseases and infections in communities. In addition, the HVKP pathotype is capable of bearing plasmids of hypervirulence or carbapenem resistance [36–39]. Hence, the capability of virulence gene acquisition of CRKP is known as a major means of hypervirulent CRKP strains production [40,41]. According to the latest reports, the main portion of HVKP strains is composed of antibiotic-sensitive populations excluding ampicillin; however, in recent years the number of convergent *K. pneumoniae* strains is promoting. The convergent *K. pneumoniae* strains are recognized as MDR HVKP strains bearing aerobactin synthesis locus (*iuc*) and producing ESBL or carbapenemase enzymes. The convergent *K. pneumoniae* strains may originate either from those hypervirulent strains which obtained an MDR plasmid or from MDR strains which acquired a virulence plasmid [42].

It is necessary to mention that the identified CPKP strains may bear different genes such as bla_{IMP}, bla_{KPC} and bla_{NDM}, while the bla_{KPC}-bearing CPKP strains involve the major portion of the isolated cases from clinical samples worldwide [43,44]. As an effective example, bla_{KPC} transmission may occur through a wide range of processes including clonal spread, plasmids and mobile small genetic elements such as transposon (e.g., Tn4401) [35]. Indeed, the Tn4401 is a Tn3-based transposon with a length of 10 Kb which is ended via two genes of Tn3 transpoase (*tnp*A) and Tn3 resolvase (*tnp*R), and two insertion se-quences of ISKpn6 and ISKpn7 [35,45]. The bla_{KPC} is known as a plasmid-borne gene which can be carried by > 40 plasmids. These plasmids originate from different incompatibility (Inc) groups such as A/C, ColE, FIA, I2, IncFII, L/M, N, P, R, U, W and X. The bla_{KPC} carrier plasmids bear a significant number of antimicrobial resistance genes [35,43]. Moreover, K-typing is normally recruited for HVKP categorization. Although the K1 and K2 types are mostly (~70%) belonging to HVKP and may cause invasive infections, some strains of K1 and K2 types do not pertain to HVKP types [5,46–48]. K1, K2, K16, K28, K57 and K63 capsule types are recognized among HVKP strains. The typical phenotypic characteristic of K1 and K2 types is the hypermucoviscous exhibition which can be recognized through a viscous string with a length of more than 5 mm on medium agar [5,49].

Indeed, the integrative conjugal elements and giant plasmids are the effective genetic elements which support the high virulence characteristics in HVKP strains [50–52]. *K. pneumoniae* encompasses four important and effective virulence factors, e.g., adhesive fimbriae (including type 1 type 3 fimbriae), capsule, lipopolysaccharide (LPS) and siderophores [5,23,53–55].

Adhesive fimbriae: *K. pneumoniae* is armed with two important types of fimbriae including type 1 (encoded by fimBEAICDFGH operon) and type 3 (mrkABCDF/mrkABCDEF) fimbriae, which are involved in pathogenesis of the bacteria through attachment to the biotic (human host urothelium) and abiotic (urinary catheter) surfaces to start the process of colonisation, biofilm formation and bacterial invasion (Figure 1) [14,18,56].

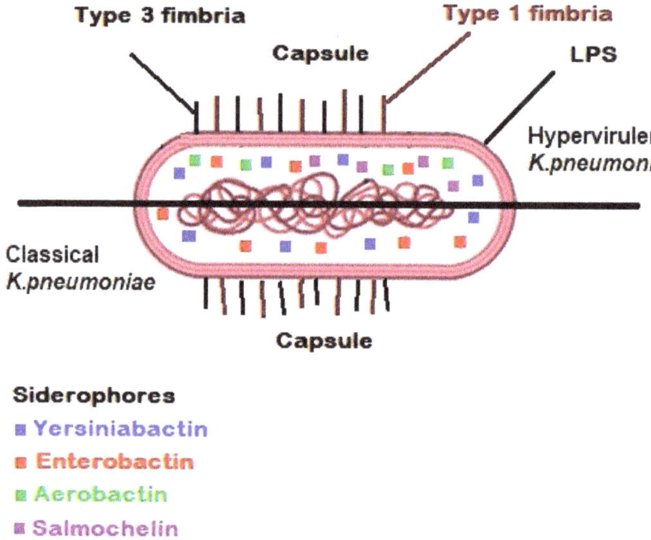

Figure 1. Virulence factors in classical/hypervirulent strains of *K. pneumoniae* [57]. Two types of fimbriae are involved in pathogenesis of the bacteria through attachment to the biotic (human host urothelium) and abiotic (urinary catheter) surfaces to start the process of colonisation, biofilm formation and bacterial invasion. The polysaccharide capsule in *K. pneumoniae* is known as a pivotal virulence factor which acts as the outermost layer in a bacterial cell and interacts with the host. Lipopolysaccharide (LPS) is an effective protective structure against serum complement proteins in parallel with the presence of capsule.

Capsule: The polysaccharide capsule in *K. pneumoniae* is known as a pivotal virulence factor which acts as the outermost layer in a bacterial cell and interacts with the host (Figure 1). All types of this acidic polysaccharide capsule are the product of Wzx/Wzy-dependent polymerization pathway encoding by the *cps* gene cluster. The virulence factor of the capsule covers the *K. pneumoniae* bacterial cells against the host immune system responses such as phagocytosis, complement proteins, opsonophagocytosis, oxidative killing and antimicrobial peptides. In another word, the encapsulated bacterial cells of *K. pneumoniae* are capable of evading the host's immune system through their capsule antigens mimicking the host glycans to survive [27,49,54,55,58,59]. As aforementioned, the K-antigen belonging to *K. pneumoniae* capsule is an effective criterion for classification and serotyping of the pathogenic strain of *K. pneumoniae*. Indeed, sequencing of six genes comprising *galF*, *orf2* (*cpsACP*), *wzi*, *wza*, *wzb* and *wzc* located at the 5′ end of the *cps* gene cluster has shown that these genes are highly conserved, while the mid zone of the *cps* loci encompasses a variable region of nucleotide sequences producing proteins which participate in assembly and polymerization of capsule blocks. Due to this fact, the K-typing method is considered an effective categorization technique. Up to now, >80 serotypes are recognized among pathogenic strains of *K. pneumoniae* according to K-antigen capsule [49,55,58,60,61]. However, up to 70% of *K. pneumoniae* isolated bacterial cells are able to produce a novel capsule or not capable to express any capsule. Hence, this portion of *K. pneumoniae* strains are not typeable through serological methods. Instead, through the contribution of molecular techniques and sequencing technologies, we are able today to investigate the capsule synthesis loci or K-loci belonging to more than 2500 whole genomes of *K. pneumoniae*. The recorded results from previous investigations show 134 distinct K-loci encoding minimally 134 different K-types which can be effective in epidemiological studies in association with *K. pneumoniae* [62]. Capsule is involved in bacterial biofilm formation; the results reported from previous studies depict that unencapsulated strains of *K. pneumoniae* are highly sensitive to host immune responses. Furthermore, the unencapsulated strains of *K. pneumoniae* show reduction in their pathogenicity in mice models [23,54].

The gene clusters encoding capsule are located on chromosome or plasmids. In this regard, the *wzy-K1*, *wzx*, *wzc*, *wza*, *wzb*, *wzi*, *gnd*, *wca*, *cps*A, *cps*B, *cps*G and *gal*F encode exopolysaccharide portion of capsule and are located on chromosome (*wza*, *wzb*, *wzc*, *gnd*, *wca*, *cps*A, *cps*B, *cps*G and *gal*F constitute the cps chromosomal operon gene), while the *rmp*A, *rmp*B and *rmp*A2 genes involved in capsule biosynthesis are locatedon both chromosome and plasmid. Moreover, the genes of *kvr*A, *kvr*B, *rcs*A, *rcs*B, c-*rmp*A and *c-rmp*A2 contribute to capsule biosynthesis and are situated on chromosome. Finally, the genes of *p-rmp*A and *p-rmp*A2, which participate in capsule biosynthesis, are plasmid-borne. *C-rmp*A, *c-rmp*A2, *p-rmp*A and *p-rmp*A2 and *wzy-K1* positively regulate the process of hypercapsulation through affecting the transcripts producing via cps chromosomal operon gene. *Kvr*A, *kvr*B and *rcs*B genes regulate the capsule production through controlling effect on rmpA promoter. Indeed, *rmp*A and *rmp*A2 regulate the mucoidal property in *K. pneumoniae* [54].

Lipopolysaccharide (LPS): LPS is a Gram-negative bacterial endotoxin which is composed of lipid A, O-antigen and an oligosaccharide core. Each constructive part of LPS is respectively encoded by *lpx*, *wbb* and *waa* gene clusters. LPS is an effective protective structure against serum complement proteins in parallel with the presence of capsule (Figure 1). LPS is also a bacterial protector in opposition to the human host humoral immune system. Furthermore, LPS is known as an important inducer biomolecule for toll-like receptor 4 (TLR4), which may activate the expression and secretion of different cytokines and interleukins [23,54,63–67].

Siderophores or iron carriers: The pivotal role of iron related to virulence and pathogenesis of pathogenic microorganisms has been detected. In this regard, there are effective interactions between the iron metabolism and immune cells which affect the pathogenesis of microbial agents (Figure 1) [68–70]. Iron molecules are recognized as competitive resources for pathogenic bacteria, e.g., *K. pneumoniae* survival within their host during a successful infection. Therefore, acquiring and recruiting host iron metals by the pathogenic bacteria is an effective strategy to survive and establish infection within the host in the presence of immune cells, e.g., macrophages (MΦs and neutrophils) and molecules. Indeed, as a first line defensive mechanism in a healthy human host immune system, the iron molecules are normally not free within the plasma. To protect the host from the virulence of pathogenic bacterial cells of *K. pneumoniae*, the iron metals are linked to iron transporters of transferrins and iron-binding immunoglycoproteins of lactoferrins [23,68,70–73].

Iron as an essential element is necessary for both human and microbial pathogens. Iron contributes to different biological features including DNA biosynthesis or replication, transcription, production of energy within mitochondria, central metabolism and enzymatic reactions [73,74]. Hence, the human host body has iron-chelating proteins to bind the iron metals while the pathogens encompass siderophores or iron carriers which bind to iron metal with high affinity. Interestingly, bacterial iron binding proteins are effective competitors to human host iron-chelating proteins. Some bacterial pathogens such as *K. pneumoniae* possess stealth iron carriers. Up to now, several iron scavengers known as siderophores have been recognized among Gram-negative microbial pathogens including enterobactin, aerobactin, yersinobactin, salmochelin, etc., with different levels of affinity for iron molecules. However, *K. pneumoniae* is able to recruit these four iron carriers. According to previous reported results, enterobactin as a highly conserved iron scavenger is the most common siderophore secreted by ~90% of isolated Enterobacterales members. Among the aforementioned iron carriers, enterobactin (encoded by entABCDEF gene cluster upon the chromosome and transported via fepABCDG) has the strongest affinity for iron molecules [54,73,75–78].

4. Mechanisms of Antimicrobial Resistance

K. pneumoniae isolates present resistance to antimicrobial agents through one or more of the following mechanisms:

(a) production of specified enzymes (e.g., β-lactamases or aminoglycoside modifying enzymes) [79,80].
(b) decreased cell permeability through loss of Omps [81].
(c) overexpression of efflux pumps, which are transmembrane proteins, with the antimicrobial agent being usually excreted out of the bacterial cell through an energy-consuming process. For example, an efflux pump called KpnGH contributes to antimicrobial resistance in *K. pneumoniae* [82].
(d) modification of the target of the antimicrobial agent [83].

4.1. B-Lactams—Ambler Classification of β-Lactamases

B-lactam antimicrobials contain a β-lactam ring in their chemical structure. In this group, the following antimicrobials are classified: (a) penicillin and its derivatives (semisynthetic penicillins), (b) cephalosporins and cephamycins, (c) monobactams and (d) carbapenems (imipenem, meropenem, ertapenem and doripenem). B-lactamases are enzymes that hydrolyse the β-lactam ring, inhibiting the action of these antimicrobials [84].

There are two classification schemes of β-lactamases. Initially, according to the initial functional classification system proposed by Bush, β-lactamases are classified in three major groups, based on their substrate and inhibitor profiles. These functional attributes have been associated with molecular structure in a dendrogram for those enzymes with known amino acid sequences [85].

However, the revised molecular classification proposed by Ambler is the most widely used. Based on this classification, only amino acid sequence determination could provide information upon which a molecular phylogeny could be based. According to preliminary data, β-lactamases have a polyphyletic origin. Thus, they are classified in four different classes, designated A, B, C and D [86,87].

Class A β-lactamases are serine-based enzymes. This class includes simple β-lactamases, such as sylfhydryl variable (SHV), temoneira (TEM), cefotaxime hydrolysing capabilities (CTX-M), *Pseudomonas* extended-resistant (PER), Guiana extended-spectrum (GES), Vietnamese extended-spectrum β-lactamase (VEB), integron-borne cephalosporinase (IBC), Serratia fonticola (SFO), Brazil extended-spectrum (BES), Belgium extended-spectrum (BEL) and Tlahuicas Indians (TLA). All these β-lactamases are inhibited both in vivo and in vitro by β-lactamase inhibitors (clavulanate, tazobactam, sulbactam). SHV and TEM can act, due to point mutations, as extended spectrum β-lactamases (ESBLs), while CTX-M is considered the newest ESBL. All the rest could act as ESBLs with milder hydrolytic capacity. ESBLs can potentially be inhibited by clavulanate, but they have an in vivo therapeutic effect only for urinary tract infections (UTIs). Inhibitor-resistant TEMs (IRTs) and inhibitor-resistant SHVs (IRSs), as well as carbapenemases called *K. pneumoniae* carbapenemases (KPCs), are classified in this group [88,89]. KPCs are distinguished in 12 subtypes [90].

Class B β-lactamases include carbapenemases which are called metallo-β-lactamases (MBLs). Their action is based on zinc ions (Zn^{+2}). MBLs hydrolyse all β-lactams except aztreonam, which belongs to monobactams. The most well-known MBLs detected so far are Imipenemase (IMP), Verona integron-encoded MBL (VIM), German imipenemase (GIM), Sao Paulo MBL (SPM), Seoul imipenemase (SIM), Australia imipenemase (AIM), Dutch imipenemase (DIM), New-Delhi MBL (NDM), and the recently detected Tripoli MBL (TMB) and Florence imipenemase (FIM) [91–99]. MBLs are classified further in three subgroups: B1, B2 and B3 [87].

Class C β-lactamases include serine-based enzymes, called cephalosporinases or AmpC β-lactamases. They are distinguished as stable and inducible, and they can be either chromosomally or plasmid-located (AmpC-like). The production of inducible AmpC depends on whether the inducer is weak or strong. They are not inhibited by β-lactamase inhibitors, and they are sensitive to cefepime and carbapenems. *K. pneumoniae* strains mainly transfer AmpC-like enzymes, which are considered to have been transmitted from a bacterial chromosome through plasmid conjugation. AmpC β-lactamases are distinguished in various classes [100].

Class D β-lactamases include serine-based enzymes which are called oxacillinases (OXA). These enzymes are characterized by high heterogeneity regarding their structure and their biochemical characteristics. Therefore, they display a large variety concerning their hydrolytic potential depending on the subtype they belong. They are not inhibited by β-lactamase inhibitors. Some of them act as carbapenemases with a milder hydrolytic capacity compared to carbapenemases belonging to other classes. However, they can provide a high grade of resistance when they co-exist with other resistance mechanisms [101].

4.2. *Decreased Cell Permeability through Loss of Omps*

The contribution of OMP deficiency is considered a secondary mechanism conferring mainly a low level of resistance itself. *OmpA*, *OmpK35*, *OmpK36* and *OmpK37* are the most important OMPs in *K. pneumoniae* strains, with a global concern [102].

OmpA alterations confer resistance to antimicrobial agents, but not to carbapenems [103]. The mutations of *OmpK35* in combination with these of *OmpK36* usually act as a supplementary mechanism of resistance in the emergence of CRKP isolates [104,105]. The downregulation of *OmpK37* has a minor contribution to the appearance of CRKP [106].

4.3. *Transport of Antimicrobial Resistance Genes*

The antimicrobial resistance genes are encompassed in mobile elements such as plasmids, transposons and integrons. These elements are crucially important, as they are involved in the vertical transmission of these genes from *K. pneumoniae* to its descendants, as well as in the horizontal transmission of the genes from a certain *K. pneumoniae* strain to another.

Most plasmids are usually circular double-stranded DNA molecules, but linear plasmids are also detected. The conjugative plasmids are crucial in the transport of antimicrobial resistance genes from a specific *K. pneumoniae* strain to another and they encode all the appropriate factors for this transfer [107]. There is a strong correlation between specific antimicrobial resistance genes and their integration in certain plasmids. Several of them can transfer many copies of these resistance genes, providing even higher grade of resistance [108]. Transposons are small DNA fragments. They are transported from one DNA site to another but do not have the ability of self-replication. The transfer can be conducted either through transposon duplicate and transport of the copy or through cut and transfer of the whole transposon [109].

Integrons are larger genetic elements which can encompass antimicrobial resistance cassettes and are classified in five classes [110]. They can also be incorporated in other mobile genetic elements such as transposons and conjugative plasmids [111].

5. Trends in Molecular Epidemiology

The first MBL detected in a CRKP isolate was IMP-1 in 1996 in Singapore [112]. Since then, CRKP isolates producing IMP have been isolated globally, but mainly in south and southeastern Asia [79,113,114]. VIM MBLs are the most prevalent on a global level. In 2004, an outbreak caused by VIM-1 producing CRKP strains took place in France, after the hospitalisation of a patient in Greece [115]. Since then, several VIM subtypes have been identified, such as VIM-12, VIM-19, VIM-4, VIM-27, VIM-26 and VIM-39, especially in endemic countries for CRKP. These VIM variants are genetically related between them, and they can emerge one from another due to minor genetic events, such as point mutations. The bla_{VIM} genes are commonly integrated in a class 1 integron [116]. The ST147 according to the Institut Pasteur scheme has been the most frequently detected among VIM-producing

K. pneumoniae isolates [117,118]. NDM MBL is a very virulent carbapenemase, as it has a huge capacity to penetrate within the community [119]. A possible explanation could be the presence of a community pool contributing to autochthonous acquisition [120]. It was initially detected in a CRKP isolate in Sweden from the clinical specimen of a patient previously hospitalised in New Delhi, India [121]. Since then, it has disseminated globally and constitutes a threat of major concern [122]. Since its first emergence, ST11 has been the predominant type among NDM producers [123]. All other MBLs are isolated mainly in specific areas and show minor epidemiological concern [91,94–99].

KPC is the most prevalent of all carbapenemases. KPC-1 was initially isolated in the United States of America in 1996 and has expanded rapidly to the east coast. It is considered endemic in many parts of New York [90]. However, the major spread of KPC-producing CRKP began in 2007 after an outbreak of CRKP isolates producing KPC-2 in Crete, Greece. These isolates displayed clonal expansion and they were found to be clonally related with the clone of New York which was previously described [124,125]. Since then, this clone has predominated and was named 'hyperepidemic Greek clone' [126]. According to the Center for Disease Control and Prevention (CDC), around 70% of KPC-2 producing CRKP isolates are assigned to ST258 [127]. ST258 has been associated with multidrug resistance to antimicrobials [128]. However, ST258 KPC-2 producing CRKP isolates are considered low-virulent and are opportunistic pathogens, as only a low proportion of patients colonised with these isolates develop an infection. These CRKP isolates create extended reservoirs with the virulence and mortality rates being relatively low [129]. Patients with co-morbidities and chronic diseases are more vulnerable in suffering from an infection [130]. Recently, KPC-2 CRKP belonging to ST39 have emerged [131].

CRKP isolates harbouring concurrently VIM-1 and KPC-2 are usually assigned in ST147, meaning that they are commonly related with VIM-1 [132]. However, CRKP isolates producing concurrently VIM-1 and KPC-2 have recently been assigned to ST39, implying some kind of relatedness with KPC-producers [133].

Regarding class D carbapenemases, the most prevalent carbapenemase is OXA-48, initially detected in Turkey in 2001. OXA-48 hydrolyses carbapenems in a mild way, conferring a low level of resistance and its action is accompanied with additional resistance mechanisms. Since 2007, OXA-48 producing CRKP isolates have been detected in many countries in Europe and north Africa. However, these isolates are not considered highly virulent. ST11 is the most prevalent among OXA-48 producers [79,134,135]. OXA-162, initially detected in Turkey is a variant of OXA-48, as well as OXA-181 which is the second most prevalent OXA detected worldwide [79,136]. One of the latest OXA subtypes detected is an OXA-48 variant, designated OXA-370, isolated in Brazil in 2014 [137].

The contribution of Omp loss to CRKP emergence is trivial. Omp loss is usually a secondary mechanism which provides low levels of resistance to antimicrobial agents and can act along with carbapenemase action. The most significant Omps in *K. pneumoniae* isolates are *OmpA, OmpK35, OmpK36* and *OmpK37* [102]. Changes in *OmpA* are generally associated with antimicrobial resistance, but not with CRKP appearance [103]. The role of *OmpK35* in carbapenem resistance has been highlighted since 2003, and the contribution of *OmpK36* around 2005 [81,138]. Since then, there has been global concern concerning the role of point mutations in genes encoding *OmpK35* and *OmpK36* as a complementary mechanism in the emergence of CRKP [104,105]. However, in 2012 an outbreak which took place in Greece led to the emergence of clonally related CRKP isolates with resistance to ertapenem exclusively due to down-regulation of *OmpK35* and mutated *OmpK36* [139]. The reduced expression of *OmpK37* has not been associated with the emergence of CRKP [106].

6. Trends in Antimicrobial Treatment

6.1. Colistin

Colistin is an antimicrobial which was discovered in 1949 and belongs to polymyxins (polymyxin E) (Table 1). Its use was abandoned at the beginning of the 1980s due to the high nephrotoxicity observed during its administration [140]. However, due to the spread of antimicrobial resistance and the appearance of CRKP and MDR *K. pneumoniae* (MDRKP), it has revived and is used as first-line treatment for infections caused by these isolates [141].

Colistin has been used widely for the treatment of VAP, bacteremias, abdominal infections and UTIs caused by CRKP and MDRKP [10]. However, the implementation of colistin monotherapy against these infections has been associated with a negative outcome, the emergence of antimicrobial resistance with the emergence of colistin-resistant CRKP [142,143]. Therefore, colistin is usually administered in combined therapeutic protocols along with tigecycline or aminoglycosides, in triple combined schemes along with tigecycline and carbapenems, fosfomycin or aminoglycosides, and in quadruple treatment schemes [1]. In CRKP isolates with a relatively low grade of resistance to carbapenems, therapeutic schemes combining colistin with carbapenems seem to be more effective, while in CRKP isolates with a high grade of carbapenem-resistance, therapeutic regimens including colistin and high dosages of tigecycline, fosfomycin and aminoglycosides present more satisfying results [144]. The advance of knowledge around the dosage of intravenous colistin administration and the progress in the pharmacokinetics of colistin have led to more satisfying therapeutic effects and reduced nephrotoxicity [10,145] As far as the treatment of infections caused by colistin-resistant CRKP isolates is concerned, the combination of colistin with carbapenems of rifampicin has been proven a possible option during the previous years [146,147].

However, it is worth mentioning that despite the fact that colistin is widely used in real-world practice, it is not considered a first-line agent for the treatment of CRKP infections [148,149].

6.2. *Tigecycline*

Tigecycline is a derivative of minocycline and belongs to glycylcyclines. It has a broad spectrum of action and has been used for the treatment of CRKP infections achieving high concentrations in various biological fluids such as lung, skin, soft tissues and bones [150]. When combined with colistin, it presents bactericidal action against CRKP isolates (Table 1) [142,150,151].

6.3. *Fosfomycin*

Fosfomycin is an old antimicrobial agent which has been re-introduced for the treatment of uncomplicated CRKP UTIs (Table 1) [152]. When combined with colistin, its bacterial killing efficacy is greater against CRKP [153].

Over the last five years, several antimicrobials with various activity against MDR Gram-negative bacteria have been launched and approved by the U.S. Food and Drug Administration (FDA) and the European Medical Agency (EMA). These drugs are plazomicin, eravacycline, cefiderocol and temocillin, a β-lactam which has only been approved in Belgium and the United Kingdom. Moreover, ceftolozane–tazobactam, meropenem–vaborbactam, imipenem–cilastatin/relebactam and ceftazidime–avibactam (CAZ-AVI) are antimicrobials that combine β-lactams with β-lactamase inhibitors and are potent alternatives [154].

Table 1. General Characteristics of Antimicrobials [155].

Antimicrobial	PubChem CID	Molecular Formula	Synonyms	Structure	Mode of Action
Colistin	44144393	$C_{52}H_{98}N_{16}O_{13}$	Polymyxin E	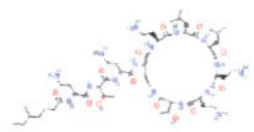	Polycationic peptides which targets bacterial (in particular, Gram-negative bacteria) cell membrane to disrupt it through detergent-like mechanism.
Tigecycline	54686904	$C_{29}H_{39}N_5O_8$	Tygacil	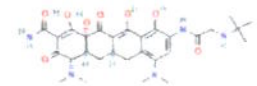	A member of the new class of glycylcyclines. Indeed, glycylcyclines are derived from tetracyclines. Tigecycline targets the ribosomal small subunit of 30S (with higher affinity than tetracyclines) to prevent bacterial protein translation. The attachment of tigecycline to the amino-acyl tRNA molecule inhibits the entrance of the amino-acyl tRNA molecule into the A site of the ribosome to stop the elongation process of the bacterial peptide biosynthesis
Fosfomycin	446987	$C_3H_7O_4P$	Phosphomycin, Phosphono-mycin		Fosfomycin is used against bacterial strains with the property of difficult-to-treat. This antibiotic is the first option against UTIs. Fosfomycin inactivates the UDP-N-acetylglucosamine enolpyruvyl transferase (MurA) enzyme via binding to a cysteine residue of the enzyme's active site. This process results in prevention of peptidoglycan precursor UDP N-acetylmuramic acid (UDP-MurNAc) biosynthesis and in consequence may lead to stopping bacterial cell wall biosynthesis. Therefore, fosfomycin has a bactericidal effect on pathogens.
Plazomicin	42613186	$C_{25}H_{48}N_6O_{10}$	Zemdri	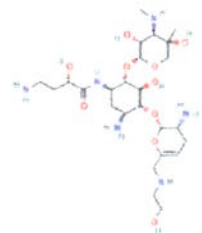	Plazomicin, as a member of aminoglycoside antibiotics, has bactericidal effect through binding to ribosomal small subunit of 30S. This antibiotic changes the spatial structure of the ribosomal A-site (aminoacyl-tRNA site), which may lead to attachment of antibiotic to rRNA molecule. This feature results in mistranslation of mRNA molecules within the process of protein biosynthesis.
Eravacycline	54726192	$C_{27}H_{31}FN_4O_8$	Xerava	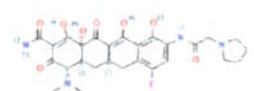	Eravacycline as a fluorocycline antibacterial pertaining to tetracycline class has disruptive effect on bacterial pathogens through targeting their protein biosynthesis processes. This effect is achieved via targeting the ribosomal small subunit of 30S.

Table 1. *Cont.*

Antimicrobial	PubChem CID	Molecular Formula	Synonyms	Structure	Mode of Action
Cefiderocol	77843966	$C_{30}H_{34}ClN_7O_{10}S_2$	SZ34OMG6E8	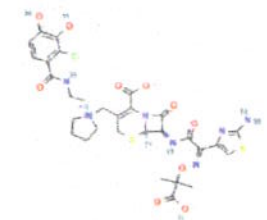	As a cephalosporin drug, has bactericidal effect on aerobic Gram-negative bacteria such as *K.pneumoniae*. Cefiderocol binds to penicillin-binding proteins (PBPs) (in particular with PBP3 and in general with PBP1a, PBP1b, PBP2 and PBP4), inactivating their activities which may lead to inhibition of bacterial cell wall biosynthesis.
Temocillin	171758	$C_{16}H_{18}N_2O_7S_2$	Negaban	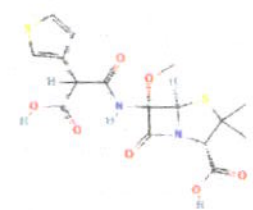	Temocillin acts as inhibitor against bacterial reproduction and growth processes.
Ceftolozan—tazobactam	86291594	$C_{33}H_{42}N_{16}O_{13}S_3$	Zerbaxa	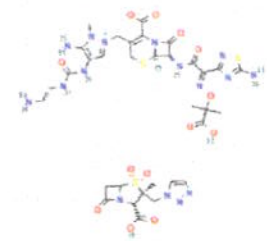	Zerbaxa prevents the growth and reproduction process in bacterial pathogens and has bactericidal effect on UTIs' bacterial causative agents.
Imipenem–cilastatin	17756656	$C_{28}H_{43}N_5O_9S_2$	Thienam	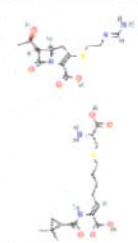	Imipenem–cilastatin prevents the growth and reproduction process in bacterial pathogens. Thienam prevents/antagonizes the process of biosynthesis/actions of the enzymes of bacterial proteases.
Meropenem–vaborbactam	86298703	$C_{29}H_{41}BN_4O_{10}S_2$	Carbavance	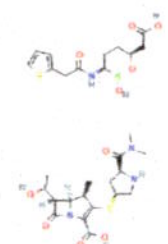	Vaborbactam acts as bacterial serine-ß-lactamases to support the antibacterial effect of penem drugs, e.g., meropenem against CRCKP strains of *K.pneumoniae*, etc.
Ceftazidime–avibactam	90643431	$C_{29}H_{33}N_9O_{13}S_3$	Avycaz	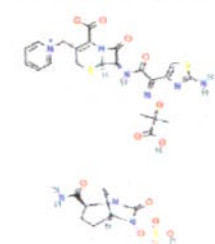	Avycaz prevents the growth and reproduction process in bacterial pathogens. Avibactam prevents/blocks the activities of bacterial ß-lactamases.

6.4. Plazomicin

Plazomicin is a synthetic aminoglycoside which was approved in 2018 for the treatment of complicated UTIs (cUTIs) and pyelonephritis [154]. Plazomicin is effective against CRKP and has been correlated with excellent activity against KPC-producing isolates (92.9%), and OXA-48 isolates (87.0%). However, its action against MBL-producing isolates is limited (40.5%) (Table 1) [156].

6.5. Eravacycline

Eravacycline is a fluorocycline which is two- to fourfold times more active than tigecycline (Table 1) [157]. Eravacycline is active against carbapenem-resistant Gram-negative bacteria (Johnston et al., 2020) and moreover, against a major proportion of NDM- and VIM-producing CRKP (61.3% and 66.7%, respectively) [158]. However, appearance of eravacycline resistance among CRKP due to over-expression of efflux pumps has already occurred [159].

6.6. Cefiderocol

Cefiderocol is a catechol-substituted siderophore (Table 1). It has been approved by the FDA for the treatment of cUTIs in 2019 and for the treatment of ventilator-associated pneumonia (VAP) in 2020 [160]. Cefiderocol inhibits the overwhelming majority of MDR Gram-negative bacteria and is active against CRKP isolates, independently of the existing resistance mechanism [161]. Resistance to cefiderocol has already emerged, especially in MBL-producing CRKP isolates [162]. In addition, co-resistance to cefiderocol and other antimicrobials in KPC-producing CRKP has already been described [163].

6.7. Temocillin

Temocillin is 6-α-methoxy derivative of ticarcillin. Some initial pharmacokinetic properties have displayed some action against KPC-producing CRKP causing UTIs (Table 1) [164]. However, the susceptibility of temocillin against KPC-producing CRKP causing UTIs varies among studies. A study in Poland shows 0% susceptibility, while another one in Greece displays only 8.6% [165,166]. However, a study performed in the UK displays an increased susceptibility of 50.8% [167]. In addition, it is not active against MBL- and OXA-48 producing CRKP isolates [166].

6.8. Ceftolozane–Tazobactam

Ceftolozane–tazobactam is a combination of β-lactam with b-lactamase inhibitor which was approved by the FDA in 2014 for the treatment of cUTIs and intra-abdominal infections (IAI) (Table 1). Furthermore, the approval was extended for VAP in 2019 [154]. However, ceftolozane–tazobactam is mainly active against *K. pneumoniae* isolates producing ESBL, but not against CRKP strains [168].

6.9. Imipenem–Cilastatin/Relebactam

Imipenem–cilastatin/relebactam was approved in 2019 by the FDA for the treatment of cUTIs and IAIs (Table 1). In 2020, it obtained approval for VAP [154]. Relebactam inhibits class A and C β-lactamases. Therefore, it is active against KPC-producing CRKP isolates. However, the addition of relebactam does not restore the activity of imipenem against MBL-producing CRKP [169]. In addition, it does not inhibit adequately OXA-48 producing CRKP [170]. However, resistance to this agent has recently emerged due to genetic rearrangement [171].

6.10. Meropenem–Vaborbactam

Vaborbactam is a cyclic boronate derivative (Table 1). When combined with meropenem, it increases the activity of meropenem against KPC-producing CRKP. It has been approved for the treatment of cUTIs, IAIs and VAP [172]. However, it is ineffective against MBL-producing CRKP, while its action against OXA-48 producers is limited [173]. However, resistance to meropenem–vaborbactam has lately appeared because of mutated OmpK35 and OmpK36 [174].

6.11. Ceftazidime–Avibactam

Avibactam (formerly NXL104, AVE1330A) was patented in 2011 and is a non-β-lactam β-lactamase inhibitor which is active in vitro against Ambler class A and C β-lactamases and displays some activity against some OXA-type β-lactamases, classified in Ambler class D. Avibactam binds covalently to β-lactamases through the creation of a carbamate bond between avibactam's position 7 carbonyl carbon and the same active-site serine that participates in acyl bonding with β-lactam substrates (Table 1) [175].

An initial study conducted in China during 2011–2012 has highlighted the in vitro activity of CAZ-AVI against CRKP and other carbapenem-resistant Gram-negative bacteria producing ESBL, AmpC and KPC. These isolates were clinically the cause for IAIs, UTIs, VAPs and bloodstream infections (BSIs) [176]. Some other studies reach the same conclusions [177,178]. It has also been proven to be active against hypervirulent CRKP isolates [179]. CAZ-AVI is normally not active against MBL-producing CRKP. However, it has been combined with aztreonam in the treatment of some cases of NDM-producing CRKP [180,181]. This combination has been applied recently in a patient with complications of SARS-CoV-2 nosocomial infection [182]. Apart from KPC-producing CRKP, CAZ-AVI has been proved effective and safe in vivo against OXA-48 producers [183].

Several clinical randomised control trials (RCTs) have attempted to investigate the efficacy and safety of CAZ-AVI in treating complicated IAIs and UTIs [184,185]. CAZ-AVI was approved by the FDA in the beginning of 2015 for the treatment of cIAIs (combined with metronidazole) and cUTIs at a dose regimen of 2.5 g every eight hours intravenously [186]. The dosing regimens have been later reviewed again in critically ill patients [187]. In addition, CAZ-AVI has been used in many cases as off-label indication or salvage therapy, with promising clinical and microbiological cure rates [188,189].

In addition, the testing of CAZ-AVI against MDR Gram-negative bacteria causing VAP has showed satisfying results [190]. It has been classified as an emerging drug for the treatment of HAP [191]. Since then, a specific RCT has highlighted the efficacy of CAZ-AVI in the treatment of VAP. Based on the results of this study called pivotal phase III REPROVE trial, the FDA approved the use of CAZ-AVI for treatment of patients with HAP/VAP [192].

In addition, some initial attempts of successful treatment of CRKP BSI have been reported [193]. The in vitro activity of CAZ-AVI against CRKP causing BSIs in cancer patients was later revealed [194]. It has also been effective in vivo against CRKP causing BSIs in hematologic patients [195]. Several studies have highlighted CAZ-AVI with higher clinical cure rates and survival than other drugs in treating CRKP BSIs [196].

However, resistance to CAZ-AVI among KPC-producing CRKP has appeared. Some previous studies underline the inability of avibactam to inhibit several KPC-2 variants [197]. Resistance to KPC-3 variant has been also detected early [198–200]. In addition, resistance has also emerged due to selective pressure during treatment with CAZ-AVI for a KPC-2 CRKP infection [131].

Resistance to CAZ-AVI seems to have a lesser impact on vaborbactam, implying the use of meropenem–vaborbactam previously described as a potent treatment alternative [201]. Notably, meropenem–vaborbactam has been used successfully in combination with aztreonam for the treatment of ceftazidime-resistant CRKP isolates causing BSIs [202]. Moreover, CAZ-AVI resistance could emerge when administered simultaneously with meropenem–vaborbactam for treating CRKP infections [203]. The appearance of ceftazidime-resistant KPC-producing CRKP is very alarming, as it can cause severe outbreaks in SARS-CoV-2 ICUs [204].

6.12. Aztreonam–Avibactam

Aztreonam–avibactam is a combination antimicrobial agent with activity against MBL-producing CRKP. However, it has not been yet approved by FDA [205]. According to recent studies, it is considered an option for the treatment of BSIs caused by colistin-resistant and CAZ-AVI-resistant CRKP isolates [206].

6.13. Guidelines for the Treatment of CRKP Infections

According to the latest guidelines of the European Society of Clinical Microbiology and Infectious Diseases (ESCMID) for patients with severe CRKP infections, meropenem-vaborbactam or ceftazidime–avibactam are recommended, if active in vitro. For patients with CRKP infections due to MBL-producing strains, cefiderocol is conditionally recommended. For non-severe CRKP infections, the use of old antimicrobials is advisable depending on the source of infection, while for cUTIs, aminoglycosides including plazomicin are recommended. If necessary, tigecycline could be used in high doses for the treatment of CRKP pneumonia, but not for BSIs and HAP/VAP [148].

The Infectious Diseases Society of America (IDSA) proposes ciprofloxacin, levofloxacin, trimethoprim–sulfamethoxazole, nitrofurantoin, or a single dose of an aminoglycoside for the treatment of uncomplicated cystitis caused by CRKP, and ciprofloxacin, levofloxacin or trimethoprim–sulfamethoxazole for cUTIs. For patients with severe CRKP infection outside the urinary tract system, ceftazidime–avibactam, meropenem–vaborbactam, and imipenem–cilastatin–relebactam are recommended, while for patients with a diagnosed MBL-producing CRKP infection, ceftazidime–avibactam plus aztreonam, or cefiderocol as monotherapy are proposed. Tigecycline and eravacycline are not recommended as monotherapy for the treatment of CRKP UTIs and BSIs, while according to IDSA, colistin should be avoided for the treatment of CRKP infections due to increased mortality and high nephrotoxicity compared to other antimicrobial options [149].

7. Conclusions

CRKP infections constitute a significant threat for public health. The knowledge of the exact mechanism of CRKP emergence is crucial for the selection of the most appropriate antimicrobial among those most recently launched. Plazomicin, eravacycline, cefiderocol, temocillin, ceftolozane–tazobactam, imipenem–cilastatin/relebactam, meropenem-vaborbactam, ceftazidime–avibactam and aztreonam–avibactam constitute potent alternatives for treating CRKP infections. The evolution of the molecular epidemiology of CRKP strains is dynamic and data and information around it should be continuously updated to diminish the spread of these isolates.

Author Contributions: T.K., K.T. and P.B. have equally contributed to the conception and design of the work and have approved the submitted version of the manuscript. All authors have read and agreed to the published version of the manuscript.

Funding: This research received no external funding.

Institutional Review Board Statement: Not applicable.

Informed Consent Statement: Not applicable.

Data Availability Statement: No new data were created or analyzed in this study. Data sharing is not applicable to this article.

Conflicts of Interest: The authors declare no conflict of interest.

References

1. Katsiari, M.; Panagiota, G.; Likousi, S.; Roussou, Z.; Polemis, M.; Vatopoulos, C.A.; Platsouka, D.E.; Maguina, A. Carbapenem-resistant *Klebsiella pneumoniae* infections in a Greek intensive care unit: Molecular characterisation and treatment challenges. *J. Glob. Antimicrob. Resist.* **2015**, *3*, 123–127. [CrossRef] [PubMed]
2. Zhen, X.; Stalsby Lundborg, C.; Sun, X.; Gu, S.; Dong, H. Clinical and Economic Burden of Carbapenem-Resistant Infection or Colonization Caused by *Klebsiella pneumoniae, Pseudomonas aeruginosa, Acinetobacter baumannii*: A Multicenter Study in China. *Antibiotics* **2020**, *9*, 514. [CrossRef] [PubMed]
3. Ahmadi, Z.; Noormohammadi, Z.; Behzadi, P.; Ranjbar, R. Molecular Detection of gyrA Mutation in Clinical Strains of *Klebsiella pneumoniae. Iran. J. Public Health* **2022**, *51*, 2334–2339. [CrossRef] [PubMed]

4. Sarshar, M.; Behzadi, P.; Ambrosi, C.; Zagaglia, C.; Palamara, A.T.; Scribano, D. FimH and Anti-Adhesive Therapeutics: A Disarming Strategy Against Uropathogens. *Antibiotics* **2020**, *9*, 397. [CrossRef] [PubMed]
5. Martin, R.M.; Bachman, M.A. Colonization, Infection, and the Accessory Genome of *Klebsiella pneumoniae*. *Front. Cell. Infect. Microbiol.* **2018**, *8*, 4. [CrossRef]
6. Cassini, A.; Högberg, L.D.; Plachouras, D.; Quattrocchi, A.; Hoxha, A.; Simonsen, G.S.; Colomb-Cotinat, M.; Kretzschmar, M.E.; Devleesschauwer, B.; Cecchini, M.; et al. Attributable deaths and disability-adjusted life-years caused by infections with antibiotic-resistant bacteria in the EU and the European Economic Area in 2015: A population-level modelling analysis. *Lancet Infect. Dis.* **2019**, *19*, 56–66. [CrossRef]
7. European Centre for Disease Prevention and Control (2019) Surveillance Atlas of Infectious Diseases. Available online: https://atlas.ecdc.europa.eu/public/index.aspx?Dataset=27&HealthTopic=4 (accessed on 30 June 2022).
8. Tesfa, T.; Mitiku, H.; Edae, M.; Assefa, N. Prevalence and incidence of carbapenem-resistant *K. pneumoniae* colonization: Systematic review and meta-analysis. *Syst. Rev.* **2022**, *11*, 240. [CrossRef]
9. Magiorakos, A.P.; Srinivasan, A.; Carey, R.B.; Carmeli, Y.; Falagas, M.E.; Giske, C.G.; Harbarth, S.; Hindler, J.F.; Kahlmeter, G.; Olsson-Liljequist, B.; et al. Multidrug-resistant, extensively drug-resistant and pandrug-resistant bacteria: An international expert proposal for interim standard definitions for acquired resistance. *Clin. Microbiol. Infect.* **2012**, *18*, 268–281. [CrossRef]
10. Petrosillo, N.; Giannella, M.; Lewis, R.; Viale, P. Treatment of carbapenem-resistant *Klebsiella pneumoniae*: The state of the art. *Expert Rev. Anti Infect. Ther.* **2013**, *11*, 159–177. [CrossRef]
11. Gonzalez-Ferrer, S.; Peñaloza, H.F.; Budnick, J.A.; Bain, W.G.; Nordstrom, H.R.; Lee, J.S.; Van Tyne, D. Finding Order in the Chaos: Outstanding Questions in *Klebsiella pneumoniae* Pathogenesis. *Infect. Immun.* **2021**, *89*, e00693-20. [CrossRef]
12. Baker, J.L.; Hendrickson, E.L.; Tang, X.; Lux, R.; He, X.; Edlund, A.; McLean, J.S.; Shi, W. *Klebsiella* and *Providencia* emerge as lone survivors following long-term starvation of oral microbiota. *Proc. Natl. Acad. Sci. USA* **2019**, *116*, 8499–8504. [CrossRef] [PubMed]
13. Hsieh, P.F.; Lu, Y.R.; Lin, T.L.; Lai, L.Y.; Wang, J.T. *Klebsiella pneumoniae* Type VI Secretion System Contributes to Bacterial Competition, Cell Invasion, Type-1 Fimbriae Expression, and In Vivo Colonization. *J. Infect. Dis.* **2019**, *219*, 637–647. [CrossRef] [PubMed]
14. Behzadi, P. Classical chaperone-usher (CU) adhesive fimbriome: Uropathogenic *Escherichia coli* (UPEC) and urinary tract infections (UTIs). *Folia Microbiol.* **2020**, *65*, 45–65. [CrossRef]
15. Dunn, S.J.; Connor, C.; McNally, A. The evolution and transmission of multi-drug resistant *Escherichia coli* and *Klebsiella pneumoniae*: The complexity of clones and plasmids. *Curr. Opin. Microbiol.* **2019**, *51*, 51–56. [CrossRef] [PubMed]
16. Holt, K.E.; Wertheim, H.; Zadoks, R.N.; Baker, S.; Whitehouse, C.A.; Dance, D.; Jenney, A.; Connor, T.R.; Hsu, L.Y.; Severin, J.; et al. Genomic analysis of diversity, population structure, virulence, and antimicrobial resistance in *Klebsiella pneumoniae*, an urgent threat to public health. *Proc. Natl. Acad. Sci. USA* **2015**, *112*, E3574–E3581. [CrossRef]
17. Wyres, K.L.; Lam, M.M.C.; Holt, K.E. Population genomics of *Klebsiella pneumoniae*. *Nat. Rev. Microbiol.* **2020**, *18*, 344–359. [CrossRef]
18. Jahandeh, N.; Ranjbar, R.; Behzadi, P.; Behzadi, E. Uropathogenic *Escherichia coli* virulence genes: Invaluable approaches for designing DNA microarray probes. *Cent. Eur. J. Urol.* **2015**, *68*, 452–458.
19. Hennart, M.; Guglielmini, J.; Bridel, S.; Maiden, M.C.; Jolley, K.A.; Criscuolo, A.; Brisse, S. A Dual Barcoding Approach to Bacterial Strain Nomenclature: Genomic Taxonomy of *Klebsiella pneumoniae* Strains. *Mol. Biol. Evol.* **2022**, *39*, msac135. [CrossRef]
20. Maiden, M.C.J.; Van Rensburg, M.J.J.; Bray, J.; Earle, S.G.; Ford, S.A.; Jolley, K.; McCarthy, N.D. MLST revisited: The gene-by-gene approach to bacterial genomics. *Nat. Rev. Microbiol.* **2013**, *11*, 728–736. [CrossRef]
21. Moura, A.; Criscuolo, A.; Pouseele, H.; Maury, M.M.; Leclercq, A.; Tarr, C.; Björkman, J.T.; Dallman, T.; Reimer, A.; Enouf, V.; et al. Whole genome-based population biology and epidemiological surveillance of Listeria monocytogenes. *Nat. Microbiol.* **2016**, *2*, 16185. [CrossRef]
22. Lam, M.M.C.; Wick, R.R.; Watts, S.C.; Cerdeira, L.T.; Wyres, K.L.; Holt, K.E. A genomic surveillance framework and genotyping tool for *Klebsiella pneumoniae* and its related species complex. *Nat. Commun.* **2021**, *12*, 4188. [CrossRef] [PubMed]
23. Ali, S.; Alam, M.; Hasan, G.M.; Hassan, M.I. Potential therapeutic targets of *Klebsiella pneumoniae*: A multi-omics review perspective. *Brief. Funct. Genom.* **2022**, *21*, 63–77. [CrossRef] [PubMed]
24. Issakhanian, L.; Behzadi, P. Antimicrobial Agents and Urinary Tract Infections. *Curr. Pharm. Des.* **2019**, *25*, 1409–1423. [CrossRef] [PubMed]
25. Behzadi, P.; Garcia-Perdomo, H.A.; Karpinski, T.M.; Issakhanian, L. Metallo-ss-lactamases: A review. *Mol. Biol. Rep.* **2020**, *47*, 6281–6294. [CrossRef]
26. Doorduijn, D.J.; Rooijakkers, S.H.; van Schaik, W.; Bardoel, B.W. Complement resistance mechanisms of *Klebsiella pneumoniae*. *Immunobiology* **2016**, *221*, 1102–1109. [CrossRef]
27. Paczosa, M.K.; Mecsas, J. *Klebsiella pneumoniae*: Going on the Offense with a Strong Defense. *Microbiol. Mol. Biol. Rev. MMBR* **2016**, *80*, 629–661. [CrossRef]
28. Koren, J.; Andrezal, M.; Drahovska, H.; Hubenakova, Z.; Liptakova, A.; Maliar, T. Next-Generation Sequencing of Carbapenem-Resistant *Klebsiella pneumoniae* Strains Isolated from Patients Hospitalized in the University Hospital Facilities. *Antibiotics* **2022**, *11*, 1538. [CrossRef]

29. Foldes, A.; Oprea, M.; Szekely, E.; Usein, C.R.; Dobreanu, M. Characterization of Carbapenemase-Producing *Klebsiella pneumoniae* Isolates from Two Romanian Hospitals Co-Presenting Resistance and Heteroresistance to Colistin. *Antibiotics* **2022**, *11*, 1171. [CrossRef]
30. Conceicao-Neto, O.C.; da Costa, B.S.; Pontes, L.D.S.; Silveira, M.C.; Justo-da-Silva, L.H.; de Oliveira Santos, I.C.; Teixeira, C.B.T.; Oliveira, T.R.T.e; Hermes, F.S.; Galvão, T.C.; et al. Polymyxin Resistance in Clinical Isolates of *K. pneumoniae* in Brazil: Update on Molecular Mechanisms, Clonal Dissemination and Relationship With KPC-Producing Strains. *Front. Cell. Infect. Microbiol.* **2022**, *12*, 898125. [CrossRef]
31. Dong, N.; Yang, X.; Chan, E.W.; Zhang, R.; Chen, S. Klebsiella species: Taxonomy, hypervirulence and multidrug resistance. *EBioMedicine* **2022**, *79*, 103998. [CrossRef]
32. Wyres, K.L.; Holt, K.E. *Klebsiella pneumoniae* Population Genomics and Antimicrobial-Resistant Clones. *Trends Microbiol.* **2016**, *24*, 944–956. [CrossRef] [PubMed]
33. Dong, N.; Zhang, R.; Liu, L.; Li, R.; Lin, D.; Chan, E.W.-C.; Chen, S. Genome analysis of clinical multilocus sequence Type 11 *Klebsiella pneumoniae* from China. *Microb. Genom.* **2018**, *4*, e000149. [CrossRef] [PubMed]
34. Wyres, K.L.; Wick, R.R.; Judd, L.M.; Froumine, R.; Tokolyi, A.; Gorrie, C.L.; Lam, M.M.C.; Duchêne, S.; Jenney, A.; Holt, K.E. Distinct evolutionary dynamics of horizontal gene transfer in drug resistant and virulent clones of *Klebsiella pneumoniae*. *PLoS Genet.* **2019**, *15*, e1008114. [CrossRef] [PubMed]
35. Chen, L.; Mathema, B.; Chavda, K.D.; DeLeo, F.R.; Bonomo, R.A.; Kreiswirth, B.N. Carbapenemase-producing *Klebsiella pneumoniae*: Molecular and genetic decoding. *Trends Microbiol.* **2014**, *22*, 686–696. [CrossRef]
36. Russo, T.A.; Olson, R.; Fang, C.-T.; Stoesser, N.; Miller, M.; MacDonald, U.; Hutson, A.; Barker, J.H.; La Hoz, R.M.; Johnson, J.R.; et al. Identification of Biomarkers for Differentiation of Hypervirulent Klebsiella pneumoniae from Classical *K. pneumoniae*. *J. Clin. Microbiol.* **2018**, *56*, e00776-18. [CrossRef] [PubMed]
37. Tian, D.; Wang, M.; Zhou, Y.; Hu, D.; Ou, H.Y.; Jiang, X. Genetic diversity and evolution of the virulence plasmids encoding aerobactin and salmochelin in *Klebsiella pneumoniae*. *Virulence* **2021**, *12*, 1323–1333. [CrossRef]
38. Tian, D.; Liu, X.; Chen, W.; Zhou, Y.; Hu, D.; Wang, W.; Wu, J.; Mu, Q.; Jiang, X. Prevalence of hypervirulent and carbapenem-resistant *Klebsiella pneumoniae* under divergent evolutionary patterns. *Emerg. Microbes Infect.* **2022**, *11*, 1936–1949. [CrossRef]
39. Lam, M.M.C.; Wyres, K.L.; Duchêne, S.; Wick, R.R.; Judd, L.M.; Gan, Y.-H.; Hoh, C.-H.; Archuleta, S.; Molton, J.S.; Kalimuddin, S.; et al. Population genomics of hypervirulent *Klebsiella pneumoniae* clonal-group 23 reveals early emergence and rapid global dissemination. *Nat. Commun.* **2018**, *9*, 2703. [CrossRef]
40. Lan, P.; Jiang, Y.; Zhou, J.; Yu, Y. A global perspective on the convergence of hypervirulence and carbapenem resistance in *Klebsiella pneumoniae*. *J. Glob. Antimicrob. Resist.* **2021**, *25*, 26–34. [CrossRef]
41. Zhang, N.; Qi, L.; Liu, X.; Jin, M.; Jin, Y.; Yang, X.; Chen, J.; Qin, S.; Liu, F.; Tang, Y.; et al. Clinical and Molecular Characterizations of Carbapenem-Resistant *Klebsiella pneumoniae* Causing Bloodstream Infection in a Chinese Hospital. *Microbiol. Spectr.* **2022**, *10*, e0169022. [CrossRef]
42. Wyres, K.L.; Nguyen, T.N.T.; Lam, M.M.C.; Judd, L.M.; van Vinh Chau, N.; Dance, D.A.B.; Ip, M.; Karkey, A.; Ling, C.L.; Miliya, T.; et al. Genomic surveillance for hypervirulence and multi-drug resistance in invasive *Klebsiella pneumoniae* from South and Southeast Asia. *Genome Med.* **2020**, *12*, 11. [CrossRef] [PubMed]
43. Li, C.; Jiang, X.; Yang, T.; Ju, Y.; Yin, Z.; Yue, L.; Ma, G.; Wang, X.; Jing, Y.; Luo, X.; et al. Genomic epidemiology of carbapenemase-producing *Klebsiella pneumoniae* in china. *Genom. Proteom. Bioinform.* **2022**. [CrossRef] [PubMed]
44. Munoz-Price, L.S.; Poirel, L.; Bonomo, R.A.; Schwaber, M.J.; Daikos, G.L.; Cormican, M.; Cornaglia, G.; Garau, J.; Gniadkowski, M.; Hayden, M.K.; et al. Clinical epidemiology of the global expansion of *Klebsiella pneumoniae* carbapenemases. *Lancet Infect. Dis.* **2013**, *13*, 785–796. [CrossRef] [PubMed]
45. Naas, T.; Cuzon, G.; Villegas, M.V.; Lartigue, M.F.; Quinn, J.P.; Nordmann, P. Genetic structures at the origin of acquisition of the beta-lactamase bla KPC gene. *Antimicrob. Agents Chemother.* **2008**, *52*, 1257–1263. [CrossRef]
46. Lee, I.R.; Molton, J.; Wyres, K.; Gorrie, C.; Wong, J.; Hoh, C.H.; Teo, J.; Kalimuddin, S.; Lye, D.; Archuleta, S.; et al. Differential host susceptibility and bacterial virulence factors driving *Klebsiella* liver abscess in an ethnically diverse population. *Sci. Rep.* **2016**, *6*, 29316. [CrossRef]
47. Marr, C.M.; Russo, T.A. Hypervirulent *Klebsiella pneumoniae*: A new public health threat. *Expert Rev. Anti-Infect. Ther.* **2019**, *17*, 71–73. [CrossRef]
48. Struve, C.; Roe, C.C.; Stegger, M.; Stahlhut, S.G.; Hansen, D.S.; Engelthaler, D.M.; Andersen, P.S.; Driebe, E.M.; Keim, P.; Krogfelt, K.A. Mapping the Evolution of Hypervirulent *Klebsiella pneumoniae*. *mBio* **2015**, *6*, e00630. [CrossRef]
49. Huang, X.; Li, X.; An, H.; Wang, J.; Ding, M.; Wang, L.; Li, L.; Ji, Q.; Qu, F.; Wang, H.; et al. Capsule type defines the capability of *Klebsiella pneumoniae* in evading Kupffer cell capture in the liver. *PLoS Pathog.* **2022**, *18*, e1010693. [CrossRef]
50. Wang, G.; Zhao, G.; Chao, X.; Xie, L.; Wang, H. The Characteristic of Virulence, Biofilm and Antibiotic Resistance of *Klebsiella pneumoniae*. *Int. J. Environ. Res. Public Health* **2020**, *17*, 6278. [CrossRef]
51. Cubero, M.; Grau, I.; Tubau, F.; Pallares, R.; Dominguez, M.A.; Linares, J.; Ardanuy, C. Hypervirulent *Klebsiella pneumoniae* clones causing bacteraemia in adults in a teaching hospital in Barcelona, Spain (2007–2013). *Clin. Microbiol. Infect.* **2016**, *22*, 154–160. [CrossRef]
52. Russo, T.A.; Marr, C.M. Hypervirulent *Klebsiella pneumoniae*. *Clin. Microbiol. Rev.* **2019**, *32*, e00001-19. [CrossRef] [PubMed]

53. Lee, A.H.Y.; Porto, W.F.; Jr, C.d.F.; Dias, S.C.; Alencar, S.A.; Pickard, D.J.; Hancock, R.E.W.; Franco, O.L. Genomic insights into the diversity, virulence and resistance of *Klebsiella pneumoniae* extensively drug resistant clinical isolates. *Microb. Genom.* **2021**, *7*, 000613. [CrossRef]
54. Dai, P.; Hu, D. The making of hypervirulent *Klebsiella pneumoniae*. *J. Clin. Lab. Anal.* **2022**, *36*, e24743. [CrossRef] [PubMed]
55. Li, B.; Zhao, Y.; Liu, C.; Chen, Z.; Zhou, D. Molecular pathogenesis of *Klebsiella pneumoniae*. *Future Microbiol.* **2014**, *9*, 1071–1081. [CrossRef]
56. Ahmadi, M.; Ranjbar, R.; Behzadi, P.; Mohammadian, T. Virulence factors, antibiotic resistance patterns, and molecular types of clinical isolates of *Klebsiella Pneumoniae*. *Expert Rev. Anti-Infect. Ther.* **2022**, *20*, 463–472. [CrossRef] [PubMed]
57. Modified. Available online: Biorender.com (accessed on 20 December 2022).
58. Pan, Y.-J.; Lin, T.-L.; Chen, C.-T.; Chen, Y.-Y.; Hsieh, P.-F.; Hsu, C.-R.; Wu, M.-C.; Wang, J.-T. Genetic analysis of capsular polysaccharide synthesis gene clusters in 79 capsular types of *Klebsiella* spp. *Sci. Rep.* **2015**, *5*, 15573. [CrossRef] [PubMed]
59. Whitfield, C.; Wear, S.S.; Sande, C. Assembly of Bacterial Capsular Polysaccharides and Exopolysaccharides. *Annu. Rev. Microbiol.* **2020**, *74*, 521–543. [CrossRef]
60. Chuang, Y.P.; Fang, C.T.; Lai, S.Y.; Chang, S.C.; Wang, J.T. Genetic determinants of capsular serotype K1 of *Klebsiella pneumoniae* causing primary pyogenic liver abscess. *J. Infect. Dis.* **2006**, *193*, 645–654. [CrossRef]
61. Walker, K.A.; Miller, V.L. The intersection of capsule gene expression, hypermucoviscosity and hypervirulence in *Klebsiella pneumoniae*. *Curr. Opin. Microbiol.* **2020**, *54*, 95–102. [CrossRef]
62. Wyres, K.L.; Wick, R.R.; Gorrie, C.; Jenney, A.; Follador, R.; Thomson, N.R.; Holt, K.E. Identification of *Klebsiella* capsule synthesis loci from whole genome data. *Microb. Genom.* **2016**, *2*, e000102. [CrossRef]
63. Liu, Y.; Bai, J.; Kang, J.; Song, Y.; Yin, D.; Wang, J.; Li, H.; Duan, J. Three Novel Sequence Types Carbapenem-Resistant *Klebsiella pneumoniae* Strains ST5365, ST5587, ST5647 Isolated from Two Tertiary Teaching General Hospitals in Shanxi Province, in North China: Molecular Characteristics, Resistance and Virulence Factors. *Infect. Drug Resist.* **2022**, *15*, 2551–2563. [CrossRef] [PubMed]
64. Behzadi, P.; Garcia-Perdomo, H.A.; Karpinski, T.M. Toll-Like Receptors: General Molecular and Structural Biology. *J. Immunol. Res.* **2021**, *2021*, 9914854. [CrossRef] [PubMed]
65. Behzadi, E.; Behzadi, P. The role of toll-like receptors (TLRs) in urinary tract infections (UTIs). *Cent. Eur. J. Urol.* **2016**, *69*, 404–410.
66. Behzadi, P.; Behzadi, E.; Pawlak-Adamska, E.A. Urinary tract infections (UTIs) or genital tract infections (GTIs)? It's the diagnostics that count. *GMS Hyg. Infect. Control* **2019**, *14*, Doc14.
67. Behzadi, P.; Sameer, A.S.; Nissar, S.; Banday, M.Z.; Gajdács, M.; García-Perdomo, H.A.; Akhtar, K.; Pinheiro, M.; Magnusson, P.; Sarshar, M.; et al. The Interleukin-1 (IL-1) Superfamily Cytokines and Their Single Nucleotide Polymorphisms (SNPs). *J. Immunol. Res.* **2022**, *2022*, 2054431. [CrossRef]
68. Soares, M.P.; Weiss, G. The Iron age of host-microbe interactions. *EMBO Rep.* **2015**, *16*, 1482–1500. [CrossRef]
69. Jaeggi, T.; Kortman, G.A.M.; Moretti, D.; Chassard, C.; Holding, P.; Dostal, A.; Boekhorst, J.; Timmerman, H.M.; Swinkels, D.W.; Tjalsma, H.; et al. Iron fortification adversely affects the gut microbiome, increases pathogen abundance and induces intestinal inflammation in Kenyan infants. *Gut* **2015**, *64*, 731–742. [CrossRef]
70. Aksoyalp, Z.S.; Temel, A.; Erdogan, B.R. Iron in infectious diseases friend or foe?: The role of gut microbiota. *J. Trace Elem. Med. Biol.* **2022**, *75*, 127093. [CrossRef]
71. Shini, V.S.; Udayarajan, C.T.; Nisha, P. A comprehensive review on lactoferrin: A natural multifunctional glycoprotein. *Food Funct.* **2022**, *13*, 11954–11972. [CrossRef]
72. Hu, D.; Li, Y.; Ren, P.; Tian, D.; Chen, W.; Fu, P.; Wang, W.; Li, X.; Jiang, X. Molecular Epidemiology of Hypervirulent Carbapenemase-Producing *Klebsiella pneumoniae*. *Front. Cell. Infect. Microbiol.* **2021**, *11*, 661218. [CrossRef]
73. Murdoch, C.C.; Skaar, E.P. Nutritional immunity: The battle for nutrient metals at the host-pathogen interface. *Nat. Rev. Microbiol.* **2022**, *20*, 657–670. [CrossRef] [PubMed]
74. Lopez, A.; Cacoub, P.; Macdougall, I.C.; Peyrin-Biroulet, L. Iron deficiency anaemia. *Lancet* **2016**, *387*, 907–916. [CrossRef] [PubMed]
75. Hsieh, P.F.; Lin, T.L.; Lee, C.Z.; Tsai, S.F.; Wang, J.T. Serum-induced iron-acquisition systems and TonB contribute to virulence in *Klebsiella pneumoniae* causing primary pyogenic liver abscess. *J. Infect. Dis.* **2008**, *197*, 1717–1727. [CrossRef] [PubMed]
76. Saha, P.; Xiao, X.; Yeoh, B.S.; Chen, Q.; Katkere, B.; Kirimanjeswara, G.S.; Vijay-Kumar, M. The bacterial siderophore enterobactin confers survival advantage to *Salmonella* in macrophages. *Gut Microbes* **2019**, *10*, 412–423. [CrossRef] [PubMed]
77. Raymond, K.N.; Dertz, E.A.; Kim, S.S. Enterobactin: An archetype for microbial iron transport. *Proc. Natl. Acad. Sci. USA* **2003**, *100*, 3584–3588. [CrossRef]
78. Brock, J.H.; Williams, P.H.; Liceaga, J.; Wooldridge, K.G. Relative availability of transferrin-bound iron and cell-derived iron to aerobactin-producing and enterochelin-producing strains of *Escherichia coli* and to other microorganisms. *Infect. Immun.* **1991**, *59*, 3185–3190. [CrossRef]
79. Nordmann, P.; Naas, T.; Poirel, L. Global spread of Carbapenemase-producing Enterobacteriaceae. *Emerg. Infect. Dis.* **2011**, *17*, 1791–1798. [CrossRef]
80. Foudraine, D.E.; Strepis, N.; Klaassen, C.H.W.; Raaphorst, M.N.; Verbon, A.; Luider, T.M.; Goessens, W.H.F.; Dekker, L.J.M. Rapid and Accurate Detection of Aminoglycoside-Modifying Enzymes and 16S rRNA Methyltransferases by Targeted Liquid Chromatography-Tandem Mass Spectrometry. *J. Clin. Microbiol.* **2021**, *59*, e0046421. [CrossRef]

81. Doménech-Sánchez, A.; Martínez-Martínez, L.; Hernández-Allés, S.; Conejo, M.D.C.; Pascual, A.; Tomás, J.M.; Albertí, S.; Benedí, V.J. Role of *Klebsiella pneumoniae* OmpK35 porin in antimicrobial resistance. *Antimicrob. Agents Chemother.* **2003**, *47*, 3332–3335. [CrossRef]
82. Srinivasan, V.B.; Singh, B.B.; Priyadarshi, N.; Chauhan, N.K.; Rajamohan, G. Role of novel multidrug efflux pump involved in drug resistance in *Klebsiella pneumoniae*. *PLoS ONE* **2014**, *9*, e96288. [CrossRef]
83. Nordmann, P.; Dortet, L.; Poirel, L. Carbapenem resistance in Enterobacteriaceae: Here is the storm! *Trends Mol. Med.* **2012**, *18*, 263–272. [CrossRef] [PubMed]
84. Essack, S.Y. The development of beta-lactam antibiotics in response to the evolution of beta-lactamases. *Pharm. Res.* **2001**, *18*, 1391–1399. [CrossRef] [PubMed]
85. Bush, K.; Jacoby, G.A.; Medeiros, A.A. A functional classification scheme for beta-lactamases and its correlation with molecular structure. *Antimicrob. Agents Chemother.* **1995**, *39*, 1211–1233. [CrossRef] [PubMed]
86. Ambler, R.P. The structure of beta-lactamases. *Philos. Trans. R Soc. Lond. B Biol. Sci.* **1980**, *289*, 321–331.
87. Hall, B.G.; Barlow, M. Revised Ambler classification of {beta}-lactamases. *J. Antimicrob. Chemother.* **2005**, *55*, 1050–1051. [CrossRef]
88. Akinci, E.; Vahaboglu, H. Minor extended-spectrum beta-lactamases. *Expert Rev. Anti Infect. Ther.* **2010**, *8*, 1251–1258. [CrossRef]
89. Naas, T.; Poirel, L.; Nordmann, P. Minor extended-spectrum beta-lactamases. *Clin. Microbiol. Infect.* **2008**, *14* (Suppl. 1), 42–52. [CrossRef]
90. Landman, D.; Bratu, S.; Kochar, S.; Panwar, M.; Trehan, M.; Doymaz, M.; Quale, J. Evolution of antimicrobial resistance among *Pseudomonas aeruginosa*, *Acinetobacter baumannii* and *Klebsiella pneumoniae* in Brooklyn, NY. *J. Antimicrob. Chemother.* **2007**, *60*, 78–82. [CrossRef]
91. Toleman, M.A.; Simm, A.M.; Murphy, T.A.; Gales, A.; Biedenbach, D.J.; Jones, R.N.; Walsh, T.R. Molecular characterization of SPM-1, a novel metallo-beta-lactamase isolated in Latin America: Report from the SENTRY antimicrobial surveillance programme. *J. Antimicrob. Chemother.* **2002**, *50*, 673–679. [CrossRef]
92. Senda, K.; Arakawa, Y.; Nakashima, K.; Ito, H.; Ichiyama, S.; Shimokata, K.; Kato, N.; Ohta, M. Multifocal outbreaks of metallo-beta-lactamase-producing *Pseudomonas aeruginosa* resistant to broad-spectrum beta-lactams, including carbapenems. *Antimicrob. Agents Chemother.* **1996**, *40*, 349–353. [CrossRef]
93. Poirel, L.; Naas, T.; Nicolas, D.; Collet, L.; Bellais, S.; Cavallo, J.D.; Nordmann, P. Characterization of VIM-2, a carbapenem-hydrolyzing metallo-beta-lactamase and its plasmid- and integron-borne gene from a *Pseudomonas aeruginosa* clinical isolate in France. *Antimicrob. Agents Chemother.* **2000**, *44*, 891–897. [CrossRef] [PubMed]
94. Castanheira, M.; Toleman, M.A.; Jones, R.N.; Schmidt, F.J.; Walsh, T.R. Molecular characterization of a beta-lactamase gene, blaGIM-1, encoding a new subclass of metallo-beta-lactamase. *Antimicrob. Agents Chemother.* **2004**, *48*, 4654–4661. [CrossRef] [PubMed]
95. Lee, K.; Yum, J.H.; Yong, D.; Lee, H.M.; Kim, H.D.; Docquier, J.D.; Rossolini, G.M.; Chong, Y. Novel acquired metallo-beta-lactamase gene, bla(SIM-1), in a class 1 integron from *Acinetobacter baumannii* clinical isolates from Korea. *Antimicrob. Agents Chemother.* **2005**, *49*, 4485–4491. [CrossRef] [PubMed]
96. Tada, T.; Shimada, K.; Satou, K.; Hirano, T.; Pokhrel, B.M.; Sherchand, J.B.; Kirikae, T. *Pseudomonas aeruginosa* Clinical Isolates in Nepal Coproducing Metallo-beta-Lactamases and 16S rRNA Methyltransferases. *Antimicrob. Agents Chemother.* **2017**, *61*, e00694-17. [CrossRef] [PubMed]
97. Yong, D.; Toleman, M.A.; Bell, J.; Ritchie, B.; Pratt, R.; Ryley, H.; Walsh, T.R. Genetic and biochemical characterization of an acquired subgroup B3 metallo-beta-lactamase gene, blaAIM-1, and its unique genetic context in *Pseudomonas aeruginosa* from Australia. *Antimicrob. Agents Chemother.* **2012**, *56*, 6154–6159. [CrossRef]
98. El Salabi, A.; Borra, P.S.; Toleman, M.A.; Samuelsen, O.; Walsh, T.R. Genetic and biochemical characterization of a novel metallo-beta-lactamase, TMB-1, from an *Achromobacter xylosoxidans* strain isolated in Tripoli, Libya. *Antimicrob. Agents Chemother.* **2012**, *56*, 2241–2245. [CrossRef]
99. Pollini, S.; Maradei, S.; Pecile, P.; Olivo, G.; Luzzaro, F.; Docquier, J.D.; Rossolini, G.M. FIM-1, a new acquired metallo-beta-lactamase from a *Pseudomonas aeruginosa* clinical isolate from Italy. *Antimicrob. Agents Chemother.* **2013**, *57*, 410–416. [CrossRef]
100. Jacoby, G.A. AmpC beta-lactamases. *Clin. Microbiol. Rev.* **2009**, *22*, 161–182, Table of Contents. [CrossRef]
101. Antunes, N.T.; Lamoureaux, T.L.; Toth, M.; Stewart, N.K.; Frase, H.; Vakulenko, S.B. Class D beta-lactamases: Are they all carbapenemases? *Antimicrob. Agents Chemother.* **2014**, *58*, 2119–2125. [CrossRef]
102. Shi, W.; Li, K.; Ji, Y.; Jiang, Q.; Wang, Y.; Shi, M.; Mi, Z. Carbapenem and cefoxitin resistance of *Klebsiella pneumoniae* strains associated with porin OmpK36 loss and DHA-1 beta-lactamase production. *Braz. J. Microbiol.* **2013**, *44*, 435–442. [CrossRef]
103. Llobet, E.; March, C.; Gimenez, P.; Bengoechea, J.A. *Klebsiella pneumoniae* OmpA confers resistance to antimicrobial peptides. *Antimicrob. Agents Chemother.* **2009**, *53*, 298–302. [CrossRef] [PubMed]
104. Partridge, S.R.; Ginn, A.N.; Wiklendt, A.M.; Ellem, J.; Wong, J.S.; Ingram, P.; Guy, S.; Garner, S.; Iredell, J.R. Emergence of blaKPC carbapenemase genes in Australia. *Int. J. Antimicrob. Agents* **2015**, *45*, 130–136. [CrossRef] [PubMed]
105. Tang, H.J.; Ku, Y.H.; Lee, M.F.; Chuang, Y.C.; Yu, W.L. In Vitro Activity of Imipenem and Colistin against a Carbapenem-Resistant *Klebsiella pneumoniae* Isolate Coproducing SHV-31, CMY-2, and DHA-1. *Biomed Res. Int.* **2015**, *2015*, 568079. [CrossRef] [PubMed]
106. Kaczmarek, F.M.; Dib-Hajj, F.; Shang, W.; Gootz, T.D. High-level carbapenem resistance in a *Klebsiella pneumoniae* clinical isolate is due to the combination of bla(ACT-1) beta-lactamase production, porin OmpK35/36 insertional inactivation, and down-regulation of the phosphate transport porin phoe. *Antimicrob. Agents Chemother.* **2006**, *50*, 3396–3406. [CrossRef] [PubMed]

107. Norman, A.; Hansen, L.H.; Sorensen, S.J. Conjugative plasmids: Vessels of the communal gene pool. *Philos. Trans. R Soc. Lond. B Biol. Sci.* **2009**, *364*, 2275–2289. [CrossRef] [PubMed]
108. San Millan, A.; Toll-Riera, M.; Escudero, J.A.; Canton, R.; Coque, T.M.; MacLean, R.C. Sequencing of plasmids pAMBL1 and pAMBL2 from *Pseudomonas aeruginosa* reveals a blaVIM-1 amplification causing high-level carbapenem resistance. *J. Antimicrob. Chemother.* **2015**, *70*, 3000–3003. [CrossRef]
109. Munoz-Price, L.S.; Quinn, J.P. The spread of *Klebsiella pneumoniae* carbapenemases: A tale of strains, plasmids, and transposons. *Clin. Infect. Dis.* **2009**, *49*, 1739–1741. [CrossRef]
110. Labbate, M.; Case, R.J.; Stokes, H.W. The integron/gene cassette system: An active player in bacterial adaptation. *Methods Mol. Biol.* **2009**, *532*, 103–125.
111. Carattoli, A. Importance of integrons in the diffusion of resistance. *Vet. Res.* **2001**, *32*, 243–259. [CrossRef]
112. Koh, T.H.; Babini, G.S.; Woodford, N.; Sng, L.H.; Hall, L.M.; Livermore, D.M. Carbapenem-hydrolysing IMP-1 beta-lactamase in *Klebsiella pneumoniae* from Singapore. *Lancet* **1999**, *353*, 2162. [CrossRef]
113. Li, B.; Xu, X.H.; Zhao, Z.C.; Wang, M.H.; Cao, Y.P. High prevalence of metallo-beta-lactamase among carbapenem-resistant *Klebsiella pneumoniae* in a teaching hospital in China. *Can. J. Microbiol.* **2014**, *60*, 691–695. [CrossRef] [PubMed]
114. Limbago, B.M.; Rasheed, J.K.; Anderson, K.F.; Zhu, W.; Kitchel, B.; Watz, N.; Munro, S.; Gans, H.; Banaei, N.; Kallen, A.J. IMP-producing carbapenem-resistant *Klebsiella pneumoniae* in the United States. *J. Clin. Microbiol.* **2011**, *49*, 4239–4245. [CrossRef] [PubMed]
115. Kassis-Chikhani, N.; Decré, D.; Gautier, V.; Burghoffer, B.; Saliba, F.; Mathieu, D.; Samuel, D.; Castaing, D.; Petit, J.-C.; Dussaix, E.; et al. First outbreak of multidrug-resistant *Klebsiella pneumoniae* carrying blaVIM-1 and blaSHV-5 in a French university hospital. *J. Antimicrob. Chemother.* **2006**, *57*, 142–145. [CrossRef] [PubMed]
116. Karampatakis, T.; Antachopoulos, C.; Iosifidis, E.; Tsakris, A.; Roilides, E. Molecular epidemiology of carbapenem-resistant *Klebsiella pneumoniae* in Greece. *Future Microbiol.* **2016**, *11*, 809–823. [CrossRef]
117. Hasan, C.M.; Turlej-Rogacka, A.; Vatopoulos, A.C.; Giakkoupi, P.; Maatallah, M.; Giske, C.G. Dissemination of blaVIM in Greece at the peak of the epidemic of 2005-2006: Clonal expansion of *Klebsiella pneumoniae* clonal complex 147. *Clin. Microbiol. Infect.* **2014**, *20*, 34–37. [CrossRef] [PubMed]
118. Arca-Suárez, J.; Rodiño-Janeiro, B.K.; Pérez, A.; Guijarro-Sánchez, P.; Vázquez-Ucha, J.C.; Cruz, F.; Gómez-Garrido, J.; Alioto, T.S.; Álvarez-Tejado, M.; Gut, M.; et al. Emergence of 16S rRNA methyltransferases among carbapenemase-producing Enterobacterales in Spain studied by whole-genome sequencing. *Int. J. Antimicrob. Agents* **2022**, *59*, 106456. [CrossRef] [PubMed]
119. Voulgari, E.; Gartzonika, C.; Vrioni, G.; Politi, L.; Priavali, E.; Levidiotou-Stefanou, S.; Tsakris, A. The Balkan region: NDM-1-producing *Klebsiella pneumoniae* ST11 clonal strain causing outbreaks in Greece. *J. Antimicrob. Chemother.* **2014**, *69*, 2091–2097. [CrossRef]
120. Johnson, A.P.; Woodford, N. Global spread of antibiotic resistance: The example of New Delhi metallo-beta-lactamase (NDM)-mediated carbapenem resistance. *J. Med. Microbiol.* **2013**, *62*, 499–513. [CrossRef]
121. Yong, D.; Toleman, M.A.; Giske, C.G.; Cho, H.S.; Sundman, K.; Lee, K.; Walsh, T.R. Characterization of a new metallo-beta-lactamase gene, bla(NDM-1), and a novel erythromycin esterase gene carried on a unique genetic structure in *Klebsiella pneumoniae* sequence type 14 from India. *Antimicrob. Agents Chemother.* **2009**, *53*, 5046–5054. [CrossRef]
122. Moellering, R.C., Jr. NDM-1—A cause for worldwide concern. *N. Engl. J. Med.* **2010**, *363*, 2377–2379. [CrossRef]
123. Izdebski, R.; Biedrzycka, M.; Urbanowicz, P.; Papierowska-Kozdój, W.; Dominiak, M.; Żabicka, D.; Gniadkowski, M. Multiple secondary outbreaks of NDM-producing Enterobacter hormaechei in the context of endemic NDM-producing *Klebsiella pneumoniae*. *J. Antimicrob. Chemother.* **2022**, *77*, 1561–1569. [CrossRef] [PubMed]
124. Maltezou, H.; Giakkoupi, P.; Maragos, A.; Bolikas, M.; Raftopoulos, V.; Papahatzaki, H.; Vrouhos, G.; Liakou, V.; Vatopoulos, A. Outbreak of infections due to KPC-2-producing *Klebsiella pneumoniae* in a hospital in Crete (Greece). *J. Infect.* **2009**, *58*, 213–219. [CrossRef] [PubMed]
125. Pournaras, S.; Protonotariou, E.; Voulgari, E.; Kristo, I.; Dimitroulia, E.; Vitti, D.; Tsalidou, M.; Maniatis, A.N.; Tsakris, A.; Sofianou, D. Clonal spread of KPC-2 carbapenemase-producing *Klebsiella pneumoniae* strains in Greece. *J. Antimicrob. Chemother.* **2009**, *64*, 348–352. [CrossRef] [PubMed]
126. Giakkoupi, P.; Maltezou, H.; Polemis, M.; Pappa, O.; Saroglou, G.; Vatopoulos, A. KPC-2-producing *Klebsiella pneumoniae* infections in Greek hospitals are mainly due to a hyperepidemic clone. *Eurosurveillance* **2009**, *14*, 19218. [CrossRef]
127. Kitchel, B.; Rasheed, J.K.; Patel, J.B.; Srinivasan, A.; Navon-Venezia, S.; Carmeli, Y.; Brolund, A.; Giske, C.G. Molecular epidemiology of KPC-producing *Klebsiella pneumoniae* isolates in the United States: Clonal expansion of multilocus sequence type 258. *Antimicrob. Agents Chemother.* **2009**, *53*, 3365–3370. [CrossRef]
128. Giakkoupi, P.; Papagiannitsis, C.C.; Miriagou, V.; Pappa, O.; Polemis, M.; Tryfinopoulou, K.; Tzouvelekis, L.S.; Vatopoulos, A.C. An update of the evolving epidemic of blaKPC-2-carrying *Klebsiella pneumoniae* in Greece (2009-10). *J. Antimicrob. Chemother.* **2011**, *66*, 1510–1513. [CrossRef]
129. Tzouvelekis, L.S.; Miriagou, V.; Kotsakis, S.D.; Spyridopoulou, K.; Athanasiou, E.; Karagouni, E.; Tzelepi, E.; Daikos, G.L. KPC-producing, multidrug-resistant *Klebsiella pneumoniae* sequence type 258 as a typical opportunistic pathogen. *Antimicrob. Agents Chemother.* **2013**, *57*, 5144–5146. [CrossRef]

130. Borer, A.; Saidel-Odes, L.; Eskira, S.; Nativ, R.; Riesenberg, K.; Livshiz-Riven, I.; Schlaeffer, F.; Sherf, M.; Peled, N. Risk factors for developing clinical infection with carbapenem-resistant *Klebsiella pneumoniae* in hospital patients initially only colonized with carbapenem-resistant *K pneumoniae*. *Am. J. Infect. Control* **2012**, *40*, 421–425. [CrossRef]
131. Räisänen, K.; Koivula, I.; Ilmavirta, H.; Puranen, S.; Kallonen, T.; Lyytikäinen, O.; Jalava, J. Emergence of ceftazidime-avibactam-resistant *Klebsiella pneumoniae* during treatment, Finland, December 2018. *Eurosurveillance* **2019**, *24*, 1900256. [CrossRef]
132. Papagiannitsis, C.C.; Giakkoupi, P.; Kotsakis, S.D.; Tzelepi, E.; Tzouvelekis, L.S.; Vatopoulos, A.C.; Miriagou, V. OmpK35 and OmpK36 porin variants associated with specific sequence types of *Klebsiella pneumoniae*. *J. Chemother.* **2013**, *25*, 250–254. [CrossRef]
133. Karampatakis, T.; Zarras, C.; Pappa, S.; Vagdatli, E.; Iosifidis, E.; Roilides, E.; Papa, A. Emergence of ST39 carbapenem-resistant *Klebsiella pneumoniae* producing VIM-1 and KPC-2. *Microb. Pathog.* **2022**, *162*, 105373. [CrossRef] [PubMed]
134. Poirel, L.; Heritier, C.; Tolun, V.; Nordmann, P. Emergence of oxacillinase-mediated resistance to imipenem in *Klebsiella pneumoniae*. *Antimicrob. Agents Chemother.* **2004**, *48*, 15–22. [CrossRef] [PubMed]
135. Voulgari, E.; Zarkotou, O.; Ranellou, K.; Karageorgopoulos, D.E.; Vrioni, G.; Mamali, V.; Themeli-Digalaki, K.; Tsakris, A. Outbreak of OXA-48 carbapenemase-producing *Klebsiella pneumoniae* in Greece involving an ST11 clone. *J. Antimicrob. Chemother.* **2013**, *68*, 84–88. [CrossRef]
136. Kasap, M.; Torol, S.; Kolayli, F.; Dundar, D.; Vahaboglu, H. OXA-162, a novel variant of OXA-48 displays extended hydrolytic activity towards imipenem, meropenem and doripenem. *J. Enzyme. Inhib. Med. Chem.* **2013**, *28*, 990–996. [CrossRef] [PubMed]
137. Pereira, P.S.; Borghi, M.; de Araújo, C.F.M.; Aires, C.A.M.; Oliveira, J.C.R.; Asensi, M.D.; Carvalho-Assef, A.P.D. Clonal Dissemination of OXA-370-Producing *Klebsiella pneumoniae* in Rio de Janeiro, Brazil. *Antimicrob. Agents Chemother.* **2015**, *59*, 4453–4456. [CrossRef] [PubMed]
138. Loli, A.; Tzouvelekis, L.S.; Tzelepi, E.; Carattoli, A.; Vatopoulos, A.C.; Tassios, P.T.; Miriagou, V. Sources of diversity of carbapenem resistance levels in *Klebsiella pneumoniae* carrying blaVIM-1. *J. Antimicrob. Chemother.* **2006**, *58*, 669–672. [CrossRef]
139. Poulou, A.; Voulgari, E.; Vrioni, G.; Koumaki, V.; Xidopoulos, G.; Chatzipantazi, V.; Markou, F.; Tsakris, A. Outbreak caused by an ertapenem-resistant, CTX-M-15-producing *Klebsiella pneumoniae* sequence type 101 clone carrying an OmpK36 porin variant. *J. Clin. Microbiol.* **2013**, *51*, 3176–3182. [CrossRef]
140. Biswas, S.; Brunel, J.M.; Dubus, J.C.; Reynaud-Gaubert, M.; Rolain, J.M. Colistin: An update on the antibiotic of the 21st century. *Expert Rev. Anti Infect. Ther.* **2012**, *10*, 917–934. [CrossRef]
141. Cassir, N.; Rolain, J.M.; Brouqui, P. A new strategy to fight antimicrobial resistance: The revival of old antibiotics. *Front. Microbiol.* **2014**, *5*, 551. [CrossRef]
142. Pournaras, S.; Vrioni, G.; Neou, E.; Dendrinos, J.; Dimitroulia, E.; Poulou, A.; Tsakris, A. Activity of tigecycline alone and in combination with colistin and meropenem against *Klebsiella pneumoniae* carbapenemase (KPC)-producing Enterobacteriaceae strains by time-kill assay. *Int. J. Antimicrob. Agents* **2011**, *37*, 244–247. [CrossRef]
143. Zarkotou, O.; Pournaras, S.; Tselioti, P.; Dragoumanos, V.; Pitiriga, V.; Ranellou, K.; Prekates, A.; Themeli-Digalaki, K.; Tsakris, A. Predictors of mortality in patients with bloodstream infections caused by KPC-producing *Klebsiella pneumoniae* and impact of appropriate antimicrobial treatment. *Clin. Microbiol. Infect.* **2011**, *17*, 1798–1803. [CrossRef]
144. Rafailidis, P.I.; Falagas, M.E. Options for treating carbapenem-resistant Enterobacteriaceae. *Curr. Opin. Infect. Dis.* **2014**, *27*, 479–483. [CrossRef]
145. Falagas, M.E.; Kasiakou, S.K. Colistin: The revival of polymyxins for the management of multidrug-resistant gram-negative bacterial infections. *Clin. Infect. Dis.* **2005**, *40*, 1333–1341. [CrossRef]
146. Tascini, C.; Tagliaferri, E.; Giani, T.; Leonildi, A.; Flammini, S.; Casini, B.; Lewis, R.; Ferranti, S.; Rossolini, G.M.; Menichetti, F. Synergistic activity of colistin plus rifampin against colistin-resistant KPC-producing *Klebsiella pneumoniae*. *Antimicrob. Agents Chemother.* **2013**, *57*, 3990–3993. [CrossRef]
147. Tuon, F.F.; Santos, T.A.; Almeida, R.; Rocha, J.L.; Cieslinsk, J.; Becker, G.N.; Arend, L.N. Colistin-resistant Enterobacteriaceae bacteraemia: Real-life challenges and options. *Clin. Microbiol. Infect.* **2016**, *22*, e9–e10. [CrossRef]
148. Paul, M.; Carrara, E.; Retamar, P.; Tängdén, T.; Bitterman, R.; Bonomo, R.A.; de Waele, J.; Daikos, G.L.; Akova, M.; Harbarth, S.; et al. European Society of Clinical Microbiology and Infectious Diseases (ESCMID) guidelines for the treatment of infections caused by multidrug-resistant Gram-negative bacilli (endorsed by European society of intensive care medicine). *Clin. Microbiol. Infect.* **2022**, *28*, 521–547. [CrossRef]
149. Tamma, P.D.; Aitken, S.L.; Bonomo, R.A.; Mathers, A.J.; van Duin, D.; Clancy, C.J. Infectious Diseases Society of America 2022 Guidance on the Treatment of Extended-Spectrum beta-lactamase Producing Enterobacterales (ESBL-E), Carbapenem-Resistant Enterobacterales (CRE), and *Pseudomonas aeruginosa* with Difficult-to-Treat Resistance (DTR-*P. aeruginosa*). *Clin. Infect. Dis.* **2022**, *75*, 187–212.
150. Entenza, J.M.; Moreillon, P. Tigecycline in combination with other antimicrobials: A review of in vitro, animal and case report studies. *Int. J. Antimicrob. Agents* **2009**, *34*, 8.e1–8.e9. [CrossRef]
151. Hirsch, E.B.; Tam, V.H. Detection and treatment options for *Klebsiella pneumoniae* carbapenemases (KPCs): An emerging cause of multidrug-resistant infection. *J. Antimicrob. Chemother.* **2010**, *65*, 1119–1125. [CrossRef]
152. Mezzatesta, M.L.; La Rosa, G.; Maugeri, G.; Zingali, T.; Caio, C.; Novelli, A.; Stefani, S. In vitro activity of fosfomycin trometamol and other oral antibiotics against multidrug-resistant uropathogens. *Int. J. Antimicrob. Agents* **2017**, *49*, 763–766. [CrossRef]

153. Sharma, R.; Garcia, E.; Diep, J.K.; Lee, V.H.; Minhaj, F.; Jermain, B.; Ellis-Grosse, E.J.; Abboud, C.S.; Rao, G.G. Pharmacodynamic and immunomodulatory effects of polymyxin B in combination with fosfomycin against KPC-2-producing *Klebsiella pneumoniae*. *Int. J. Antimicrob. Agents* **2022**, *59*, 106566. [CrossRef]
154. Yusuf, E.; Bax, H.I.; Verkaik, N.J.; van Westreenen, M. An Update on Eight "New" Antibiotics against Multidrug-Resistant Gram-Negative Bacteria. *J. Clin. Med.* **2021**, *10*, 1068. [CrossRef]
155. PubChem Compounds. Available online: https://pubchem.ncbi.nlm.nih.gov/docs/compounds (accessed on 16 December 2022).
156. Castanheira, M.; Sader, H.S.; Mendes, R.E.; Jones, R.N. Activity of Plazomicin Tested against Enterobacterales Isolates Collected from U.S. Hospitals in 2016–2017: Effect of Different Breakpoint Criteria on Susceptibility Rates among Aminoglycosides. *Antimicrob. Agents Chemother.* **2020**, *64*, e02418-19. [CrossRef]
157. Livermore, D.M.; Mushtaq, S.; Warner, M.; Woodford, N. In Vitro Activity of Eravacycline against Carbapenem-Resistant Enterobacteriaceae and *Acinetobacter baumannii*. *Antimicrob. Agents Chemother.* **2016**, *60*, 3840–3844. [CrossRef]
158. Maraki, S.; Mavromanolaki, V.E.; Magkafouraki, E.; Moraitis, P.; Stafylaki, D.; Kasimati, A.; Scoulica, E. Epidemiology and in vitro activity of ceftazidime-avibactam, meropenem-vaborbactam, imipenem-relebactam, eravacycline, plazomicin, and comparators against Greek carbapenemase-producing *Klebsiella pneumoniae* isolates. *Infection* **2022**, *50*, 467–474. [CrossRef]
159. Zheng, J.X.; Lin, Z.W.; Sun, X.; Lin, W.H.; Chen, Z.; Wu, Y.; Qi, G.B.; Deng, Q.W.; Qu, D.; Yu, Z.J. Overexpression of OqxAB and MacAB efflux pumps contributes to eravacycline resistance and heteroresistance in clinical isolates of *Klebsiella pneumoniae*. *Emerg. Microbes Infect.* **2018**, *7*, 139. [CrossRef]
160. Zhanel, G.G.; Golden, A.R.; Zelenitsky, S.; Wiebe, K.; Lawrence, C.K.; Adam, H.J.; Idowu, T.; Domalaon, R.; Schweizer, F.; Zhanel, M.A.; et al. Cefiderocol: A Siderophore Cephalosporin with Activity Against Carbapenem-Resistant and Multidrug-Resistant Gram-Negative Bacilli. *Drugs* **2019**, *79*, 271–289. [CrossRef]
161. Falcone, M.; Tiseo, G.; Nicastro, M.; Leonildi, A.; Vecchione, A.; Casella, C.; Forfori, F.; Malacarne, P.; Guarracino, F.; Barnini, S.; et al. Cefiderocol as Rescue Therapy for *Acinetobacter baumannii* and Other Carbapenem-resistant Gram-negative Infections in Intensive Care Unit Patients. *Clin. Infect. Dis.* **2021**, *72*, 2021–2024. [CrossRef]
162. Yao, J.; Wang, J.; Chen, M.; Cai, Y. Cefiderocol: An Overview of Its in-vitro and in-vivo Activity and Underlying Resistant Mechanisms. *Front. Med.* **2021**, *8*, 741940. [CrossRef]
163. Poirel, L.; Sadek, M.; Kusaksizoglu, A.; Nordmann, P. Co-resistance to ceftazidime-avibactam and cefiderocol in clinical isolates producing KPC variants. *Eur. J. Clin. Microbiol. Infect. Dis.* **2022**, *41*, 677–680. [CrossRef]
164. Adams-Haduch, J.M.; Potoski, B.A.; Sidjabat, H.E.; Paterson, D.L.; Doi, Y. Activity of temocillin against KPC-producing *Klebsiella pneumoniae* and Escherichia coli. *Antimicrob. Agents Chemother.* **2009**, *53*, 2700–2701. [CrossRef] [PubMed]
165. Kuch, A.; Zieniuk, B.; Żabicka, D.; Van de Velde, S.; Literacka, E.; Skoczyńska, A.; Hryniewicz, W. Activity of temocillin against ESBL-, AmpC-, and/or KPC-producing Enterobacterales isolated in Poland. *Eur. J. Clin. Microbiol. Infect. Dis.* **2020**, *39*, 1185–1191. [CrossRef]
166. Tsakris, A.; Koumaki, V.; Politi, L.; Balakrishnan, I. Activity of temocillin against KPC-producing Enterobacteriaceae clinical isolates. *Int. J. Antimicrob. Agents* **2020**, *55*, 105843. [CrossRef] [PubMed]
167. Woodford, N.; Pike, R.; Meunier, D.; Loy, R.; Hill, R.; Hopkins, K.L. In vitro activity of temocillin against multidrug-resistant clinical isolates of *Escherichia coli, Klebsiella spp.* and *Enterobacter spp.*, and evaluation of high-level temocillin resistance as a diagnostic marker for OXA-48 carbapenemase. *J. Antimicrob. Chemother.* **2014**, *69*, 564–567. [CrossRef]
168. Karlowsky, J.A.; Lob, S.H.; Khan, A.; Chen, W.-T.; Woo, P.C.Y.; Seto, W.H.; Ip, M.; Leung, S.; Wong, Q.W.-L.; Chau, R.W.Y.; et al. Activity of ceftolozane/tazobactam against Gram-negative isolates among different infections in Hong Kong: SMART 2017–2019. *J. Med. Microbiol.* **2022**, *71*, 001487. [CrossRef] [PubMed]
169. Mansour, H.; Ouweini, A.E.L.; Chahine, E.B.; Karaoui, L.R. Imipenem/cilastatin/relebactam: A new carbapenem beta-lactamase inhibitor combination. *Am. J. Health Syst. Pharm.* **2021**, *78*, 674–683. [CrossRef]
170. Haidar, G.; Clancy, C.J.; Chen, L.; Samanta, P.; Shields, R.K.; Kreiswirth, B.N.; Nguyen, M.H. Identifying Spectra of Activity and Therapeutic Niches for Ceftazidime-Avibactam and Imipenem-Relebactam against Carbapenem-Resistant Enterobacteriaceae. *Antimicrob. Agents Chemother.* **2017**, *61*, e00642-17. [CrossRef] [PubMed]
171. Gaibani, P.; Bovo, F.; Bussini, L.; Lazzarotto, T.; Amadesi, S.; Bartoletti, M.; Viale, P.; Ambretti, S. Dynamic evolution of imipenem/relebactam resistance in a KPC-producing *Klebsiella pneumoniae* from a single patient during ceftazidime/avibactam-based treatments. *J. Antimicrob. Chemother.* **2022**, *77*, 1570–1577. [CrossRef]
172. Mouktaroudi, M.; Kotsaki, A.; Giamarellos-Bourboulis, E.J. Meropenem-vaborbactam: A critical positioning for the management of infections by Carbapenem-resistant Enterobacteriaceae. *Expert Rev. Anti Infect. Ther.* **2022**, *20*, 809–818. [CrossRef]
173. Castanheira, M.; Doyle, T.B.; Kantro, V.; Mendes, R.E.; Shortridge, D. Meropenem-Vaborbactam Activity against Carbapenem-Resistant Enterobacterales Isolates Collected in U.S. Hospitals during 2016 to 2018. *Antimicrob. Agents Chemother.* **2020**, *64*, e01951-19. [CrossRef]
174. Gaibani, P.; Lombardo, D.; Bussini, L.; Bovo, F.; Munari, B.; Giannella, M.; Bartoletti, M.; Viale, P.; Lazzarotto, T.; Ambretti, S. Epidemiology of Meropenem/Vaborbactam Resistance in KPC-Producing *Klebsiella pneumoniae* Causing Bloodstream Infections in Northern Italy, 2018. *Antibiotics* **2021**, *10*, 536. [CrossRef] [PubMed]
175. Zhanel, G.G.; Lawson, C.D.; Adam, H.; Schweizer, F.; Zelenitsky, S.; Lagacé-Wiens, P.R.; Denisuik, A.; Rubinstein, E.; Gin, A.S.; Hoban, D.J.; et al. Ceftazidime-avibactam: A novel cephalosporin/beta-lactamase inhibitor combination. *Drugs* **2013**, *73*, 159–177. [CrossRef] [PubMed]

176. Wang, X.; Zhang, F.; Zhao, C.; Wang, Z.; Nichols, W.W.; Testa, R.; Li, H.; Chen, H.; He, W.; Wang, Q.; et al. In vitro activities of ceftazidime-avibactam and aztreonam-avibactam against 372 Gram-negative bacilli collected in 2011 and 2012 from 11 teaching hospitals in China. *Antimicrob. Agents Chemother.* **2014**, *58*, 1774–1778. [CrossRef] [PubMed]
177. Sader, H.S.; Castanheira, M.; Flamm, R.K.; Farrell, D.J.; Jones, R.N. Antimicrobial activity of ceftazidime-avibactam against Gram-negative organisms collected from U.S. medical centers in 2012. *Antimicrob. Agents Chemother.* **2014**, *58*, 1684–1692. [CrossRef]
178. Flamm, R.K.; Farrell, D.J.; Sader, H.S.; Jones, R.N. Ceftazidime/avibactam activity tested against Gram-negative bacteria isolated from bloodstream, pneumonia, intra-abdominal and urinary tract infections in US medical centres (2012). *J. Antimicrob. Chemother.* **2014**, *69*, 1589–1598. [CrossRef]
179. Yu, F.; Lv, J.; Niu, S.; Du, H.; Tang, Y.-W.; Bonomo, R.A.; Kreiswirth, B.N.; Chen, L. In Vitro Activity of Ceftazidime-Avibactam against Carbapenem-Resistant and Hypervirulent *Klebsiella pneumoniae* Isolates. *Antimicrob. Agents Chemother.* **2018**, *62*, e01031-18. [CrossRef]
180. Benchetrit, L.; Mathy, V.; Armand-Lefevre, L.; Bouadma, L.; Timsit, J.F. Successful treatment of septic shock due to NDM-1-producing *Klebsiella pneumoniae* using ceftazidime/avibactam combined with aztreonam in solid organ transplant recipients: Report of two cases. *Int. J. Antimicrob. Agents* **2020**, *55*, 105842. [CrossRef]
181. Bocanegra-Ibarias, P.; Camacho-Ortiz, A.; Garza-Gonzalez, E.; Flores-Trevino, S.; Kim, H.; Perez-Alba, E. Aztreonam plus ceftazidime-avibactam as treatment of NDM-1-producing *Klebsiella pneumoniae* bacteraemia in a neutropenic patient: Last resort therapy? *J. Glob. Antimicrob. Resis.t* **2020**, *23*, 417–419. [CrossRef]
182. Perrotta, F.; Perrini, M.P. Successful Treatment of *Klebsiella pneumoniae* NDM Sepsis and Intestinal Decolonization with Ceftazidime/Avibactam Plus Aztreonam Combination in a Patient with TTP Complicated by SARSCoV-2 Nosocomial Infection. *Medicina* **2021**, *57*, 424. [CrossRef]
183. De la Calle, C.; Rodríguez, O.; Morata, L.; Marco, F.; Cardozo, C.; García-Vidal, C.; Del Río, A.; Feher, C.; Pellicé, M.; Puerta-Alcalde, P.; et al. Clinical characteristics and prognosis of infections caused by OXA-48 carbapenemase-producing Enterobacteriaceae in patients treated with ceftazidime-avibactam. *Int. J. Antimicrob. Agents* **2019**, *53*, 520–524. [CrossRef]
184. Mawal, Y.; Critchley, I.A.; Riccobene, T.A.; Talley, A.K. Ceftazidime-avibactam for the treatment of complicated urinary tract infections and complicated intra-abdominal infections. *Expert Rev. Clin. Pharmacol.* **2015**, *8*, 691–707. [CrossRef] [PubMed]
185. Sousa, A.; Pérez-Rodríguez, M.T.; Soto, A.; Rodríguez, L.; Perez-Landeiro, A.; Martínez-Lamas, L.; Nodar, A.; Crespo, M. Effectiveness of ceftazidime/avibactam as salvage therapy for treatment of infections due to OXA-48 carbapenemase-producing Enterobacteriaceae. *J. Antimicrob. Chemother.* **2018**, *73*, 3170–3175. [CrossRef] [PubMed]
186. Thompson, C.A. Ceftazidime with beta-lactamase inhibitor approved for complicated infections. *Am. J. Health Syst. Pharm.* **2015**, *72*, 511. [CrossRef] [PubMed]
187. Stein, G.E.; Smith, C.L.; Scharmen, A.; Kidd, J.M.; Cooper, C.; Kuti, J.; Mitra, S.; Nicolau, D.P.; Havlichek, D.H. Pharmacokinetic and Pharmacodynamic Analysis of Ceftazidime/Avibactam in Critically Ill Patients. *Surg. Infect.* **2019**, *20*, 55–61. [CrossRef]
188. Dietl, B.; Martinez, L.M.; Calbo, E.; Garau, J. Update on the role of ceftazidime-avibactam in the management of carbapenemase-producing Enterobacterales. *Future Microbiol.* **2020**, *15*, 473–484. [CrossRef]
189. Chen, F.; Zhong, H.; Yang, T.; Shen, C.; Deng, Y.; Han, L.; Chen, X.; Zhang, H.; Qian, Y. Ceftazidime-Avibactam as Salvage Treatment for Infections Due to Carbapenem-Resistant *Klebsiella pneumoniae* in Liver Transplantation Recipients. *Infect. Drug Resist.* **2021**, *14*, 5603–5612. [CrossRef] [PubMed]
190. Sader, H.S.; Castanheira, M.; Flamm, R.K.; Mendes, R.E.; Farrell, D.J.; Jones, R.N. Ceftazidime/avibactam tested against Gram-negative bacteria from intensive care unit (ICU) and non-ICU patients, including those with ventilator-associated pneumonia. *Int. J. Antimicrob. Agents* **2015**, *46*, 53–59. [CrossRef]
191. Liapikou, A.; Torres, A. Emerging drugs for nosocomial pneumonia. *Expert Opin. Emerg. Drugs* **2016**, *21*, 331–341. [CrossRef]
192. Torres, A.; Rank, D.; Melnick, D.; Rekeda, L.; Chen, X.; Riccobene, T.; Critchley, I.A.; Lakkis, H.D.; Taylor, D.; Talley, A.K. Randomized Trial of Ceftazidime-Avibactam vs Meropenem for Treatment of Hospital-Acquired and Ventilator-Associated Bacterial Pneumonia (REPROVE): Analyses per US FDA-Specified End Points. *Open Forum. Infect. Dis.* **2019**, *6*, ofz149. [CrossRef]
193. Wu, G.; Abraham, T.; Lee, S. Ceftazidime-Avibactam for Treatment of Carbapenem-Resistant Enterobacteriaceae Bacteremia. *Clin. Infect. Dis.* **2016**, *63*, 1147–1148. [CrossRef]
194. Hachem, R.; Reitzel, R.; Rolston, K.; Chaftari, A.M.; Raad, I. Antimicrobial Activities of Ceftazidime-Avibactam and Comparator Agents against Clinical Bacteria Isolated from Patients with Cancer. *Antimicrob. Agents Chemother.* **2017**, *61*, e02106-16. [CrossRef] [PubMed]
195. Castón, J.J.; Lacort-Peralta, I.; Martín-Dávila, P.; Loeches, B.; Tabares, S.; Temkin, L.; Torre-Cisneros, J.; Paño-Pardo, J.R. Clinical efficacy of ceftazidime/avibactam versus other active agents for the treatment of bacteremia due to carbapenemase-producing Enterobacteriaceae in hematologic patients. *Int. J. Infect. Dis.* **2017**, *59*, 118–123. [CrossRef] [PubMed]
196. Shields, R.K.; Nguyen, M.H.; Chen, L.; Press, E.G.; Potoski, B.A.; Marini, R.V.; Doi, Y.; Kreiswirth, B.N.; Clancy, C.J. Ceftazidime-Avibactam Is Superior to Other Treatment Regimens against Carbapenem-Resistant *Klebsiella pneumoniae* Bacteremia. *Antimicrob. Agents Chemother.* **2017**, *61*, e00883-17. [CrossRef]
197. Papp-Wallace, K.M.; Winkler, M.L.; Taracila, M.A.; Bonomo, R.A. Variants of beta-lactamase KPC-2 that are resistant to inhibition by avibactam. *Antimicrob. Agents Chemother.* **2015**, *59*, 3710–3717. [CrossRef] [PubMed]

198. Humphries, R.M.; Yang, S.; Hemarajata, P.; Ward, K.W.; Hindler, J.A.; Miller, S.A.; Gregson, A. First Report of Ceftazidime-Avibactam Resistance in a KPC-3-Expressing *Klebsiella pneumoniae* Isolate. *Antimicrob. Agents Chemother.* **2015**, *59*, 6605–6607. [CrossRef]
199. Livermore, D.M.; Warner, M.; Jamrozy, D.; Mushtaq, S.; Nichols, W.W.; Mustafa, N.; Woodford, N. In vitro selection of ceftazidime-avibactam resistance in Enterobacteriaceae with KPC-3 carbapenemase. *Antimicrob. Agents Chemother.* **2015**, *59*, 5324–5330. [CrossRef]
200. Shields, R.K.; Chen, L.; Cheng, S.; Chavda, K.D.; Press, E.G.; Snyder, A.; Pandey, R.; Doi, Y.; Kreiswirth, B.N.; Nguyen, M.H.; et al. Emergence of Ceftazidime-Avibactam Resistance Due to Plasmid-Borne blaKPC-3 Mutations during Treatment of Carbapenem-Resistant *Klebsiella pneumoniae* Infections. *Antimicrob. Agents Chemother.* **2017**, *61*, e02097-16. [CrossRef]
201. Tsivkovski, R.; Lomovskaya, O. Potency of Vaborbactam Is Less Affected than That of Avibactam in Strains Producing KPC-2 Mutations That Confer Resistance to Ceftazidime-Avibactam. *Antimicrob. Agents Chemother.* **2020**, *64*, e01936-19. [CrossRef]
202. Belati, A.; Bavaro, D.F.; Diella, L.; De Gennaro, N.; Di Gennaro, F.; Saracino, A. Meropenem/Vaborbactam Plus Aztreonam as a Possible Treatment Strategy for Bloodstream Infections Caused by Ceftazidime/Avibactam-Resistant *Klebsiella pneumoniae*: A Retrospective Case Series and Literature Review. *Antibiotics* **2022**, *11*, 373. [CrossRef]
203. Dulyayangkul, P.; Douglas, E.J.A.; Lastovka, F.; Avison, M.B. Resistance to Ceftazidime/Avibactam plus Meropenem/Vaborbactam When Both Are Used Together Is Achieved in Four Steps in Metallo-beta-Lactamase-Negative *Klebsiella pneumoniae*. *Antimicrob. Agents Chemother.* **2020**, *64*, e00409-20. [CrossRef]
204. Bianco, G.; Boattini, M.; Bondi, A.; Comini, S.; Zaccaria, T.; Cavallo, R.; Costa, C. Outbreak of ceftazidime-avibactam resistant *Klebsiella pneumoniae* carbapenemase (KPC)-producing Klebsiella pneumoniae in a COVID-19 intensive care unit, Italy: Urgent need for updated diagnostic protocols of surveillance cultures. *J. Hosp. Infect.* **2022**, *122*, 217–219. [CrossRef] [PubMed]
205. Ransom, E.; Bhatnagar, A.; Patel, J.B.; Machado, M.-J.; Boyd, S.; Reese, N.; Lutgring, J.D.; Lonsway, D.; Anderson, K.; Brown, A.C.; et al. Validation of Aztreonam-Avibactam Susceptibility Testing Using Digitally Dispensed Custom Panels. *J. Clin. Microbiol.* **2020**, *58*, e01944-19. [CrossRef] [PubMed]
206. Yu, W.; Luo, Q.; Shen, P.; Chen, Y.; Xu, H.; Xiao, Y.; Qiu, Y. New options for bloodstream infections caused by colistin- or ceftazidime/avibactam-resistant *Klebsiella pneumoniae*. *Int. J. Antimicrob. Agents* **2021**, *58*, 106458. [CrossRef] [PubMed]

antibiotics

Article

Assessment of Colistin Heteroresistance among Multidrug-Resistant *Klebsiella pneumoniae* Isolated from Intensive Care Patients in Europe

Anouk J. M. M. Braspenning †, Sahaya Glingston Rajakani †, Adwoa Sey, Mariem El Bounja, Christine Lammens, Youri Glupczynski and Surbhi Malhotra-Kumar *

Laboratory of Medical Microbiology, Vaccine & Infectious Disease Institute, Universiteit Antwerpen, 2610 Antwerp, Belgium
* Correspondence: surbhi.malhotra@uantwerpen.be; Tel.: +32-3-265-27-52
† These authors contributed equally to this work.

Abstract: Heteroresistance (HR) to colistin is especially concerning in settings where multi-drug-resistant (MDR) *K. pneumoniae* are prevalent and empiric use of colistin might lead to treatment failures. This study aimed to assess the frequency of occurrence of colistin HR (CHR) among (MDR) *K. pneumoniae* (n = 676) isolated from patients hospitalized in 13 intensive care units (ICUs) in six European countries in a clinical trial assessing the impact of decolonization strategies. All isolates were whole-genome-sequenced and studied for in vitro colistin susceptibility. The majority were colistin-susceptible (CS) (n = 597, MIC $\leq$ 2 µg/mL), and 79 were fully colistin-resistant (CR) (MIC > 2 µg/mL). A total of 288 CS isolates were randomly selected for population analysis profiling (PAP) to assess CHR prevalence. CHR was detected in 108/288 CS *K. pneumoniae*. No significant association was found between the occurrence of CHR and country, MIC-value, K-antigen type, and O-antigen type. Overall, 92% (617/671) of the *K. pneumoniae* were MDR with high prevalence among CS (91%, 539/592) and CR (98.7%, 78/79) isolates. In contrast, the proportion of carbapenemase-producing *K. pneumoniae* (CP-Kpn) was higher among CR (72.2%, 57/79) than CS isolates (29.3%, 174/594). The proportions of MDR and CP-Kpn were similar among CHR (MDR: 85%, 91/107; CP-Kpn: 29.9%, 32/107) and selected CS isolates (MDR: 84.7%, 244/288; CP-Kpn: 28.1%, 80/285). WGS analysis of PAP isolates showed diverse insertion elements in *mgrB* or even among technical replicates underscoring the stochasticity of the CHR phenotype. CHR isolates showed high sequence type (ST) diversity (Simpson's diversity index, SDI: 0.97, in 52 of the 85 STs tested). CR (SDI: 0.85) isolates were highly associated with specific STs (ST101, ST147, ST258/ST512, $p \leq 0.003$). The widespread nature of CHR among MDR *K. pneumoniae* in our study urge the development of rapid HR detection methods to inform on the need for combination regimens.

Keywords: colistin resistance; mechanisms; mgrB; population analysis profiling; whole-genome sequencing

Citation: Braspenning, A.J.M.M.; Rajakani, S.G.; Sey, A.; El Bounja, M.; Lammens, C.; Glupczynski, Y.; Malhotra-Kumar, S. Assessment of Colistin Heteroresistance among Multidrug-Resistant *Klebsiella pneumoniae* Isolated from Intensive Care Patients in Europe. *Antibiotics* **2024**, *13*, 281. https://doi.org/10.3390/antibiotics13030281

Academic Editor: Theodoros Karampatakis

Received: 5 March 2024
Revised: 15 March 2024
Accepted: 18 March 2024
Published: 20 March 2024

1. Introduction

Klebsiella pneumoniae frequently cause community-acquired and nosocomial infections such as pneumonia, urinary tract infection, liver abscesses and bloodstream infections [1]. In recent years, antimicrobial resistance in *K. pneumoniae* has become problematic [2,3]. In particular, resistance to carbapenems is frequently associated with resistance to multiple classes of other antibiotics which leads to limited possibilities for clinical management as alternative treatment options are limited and lead to higher rates of treatment failures [4]. In such cases, last-resort antibiotics such as colistin [5,6] may be used in association with other antibiotics. However, the high worldwide prevalence of carbapenem-resistant, multi-drug resistant (MDR) *K. pneumoniae* has fueled and increased the use of colistin over the last years, accelerating the emergence of isolates resistant to this compound [3]. Besides full colistin resistance (CR), colistin heteroresistance (CHR) has also been increasingly reported over the last years [7,8]. Heteroresistance (HR) is defined as a phenotype in which a bacterial isolate contains subpopulations of cells that show a substantial reduction in antibiotic susceptibility compared with the main population (minimum inhibitory concentration (MIC) increase of at least eight-fold), allowing for these cells to grow in the presence of the antibiotic [9]. Detection of these subpopulations can be challenging and raise concern as their frequency may rise during antibiotic exposure and possibly lead to treatment failure [9–11]. However, the relevance of CHR in causing reduced clinical effectiveness and negatively affecting the treatment outcome remains controversial [7]. Recent studies have highlighted the high prevalence of CHR, but these were often limited by small sample size (mostly relying on local data) and limited investigation of the characteristics of the CHR *K. pneumoniae* strains. In this study, we took advantage of the unique availability of a large collection of clinical isolates from various intensive care units (ICUs) across Europe to gain more insight into the frequency of occurrence of CHR among colistin-sensitive (CS) *K. pneumoniae* and to investigate its possible association with epidemiological and clinical characteristics such as country of origin, the different therapeutic intervention strategies to which the patients were exposed during their stay in the ICU, and sequence type (ST). We also investigated CHR *K. pneumoniae* strains for their possible association with capsular polysaccharide antigen types (K-antigen type), lipopolysaccharide antigen types (O-antigen type), and colistin MIC value.

2. Materials and Methods

2.1. Sample and Isolate Collection

Klebsiella pneumoniae (n = 676) collected during a clinical trial as part of the R-GNOSIS project (Resistance in Gram-Negative Organisms: Studying Intervention Strategies) (NCT02208154) were utilized. Isolates were collected from patients hospitalized in ICUs from 13 sites in six European countries: Belgium (n = 239), Spain (n = 201), Portugal (n = 61), Italy (n = 143), Slovenia (n = 9), and United Kingdom (n = 23) between 1 December 2013 and 31 May 2017. The reason for the lower number of isolates in Slovenia and United Kingdom were two-fold. Firstly, only a single ICU in each country participated in the trial in contrast to other included countries. Secondly, the baseline colonization rates with *Enterobacteriaceae* resistant to third-generation cephalosporins in Slovenia and the United Kingdom were rather low [12]. ICU patients with an expected duration of invasive mechanical ventilation of minimally 24 h were included while those who were not intubated nor mechanically ventilated and those that stayed in the unit for less than 24 h were not included in the study. In the primary R-GNOSIS clinical trial, participants were assigned to three different groups of intervention, namely chlorhexidine mouthwash (CHX), selective oropharyngeal decontamination (SOD), and selective digestive tract decontamination (SDD), aimed at reducing the risk of bloodstream infections due to MDR-Gram-negative bacteria among ventilated patients in ICUs with moderate to high prevalence of antibiotic resistance [12]. Both SDD and SOD topical decontamination treatments consist of an oropharyngeal paste and enteral suspension containing antimicrobials that includes colistin (as well as tobramycin and nystatin) (Supplementary Table S1). After a baseline period (6–14 months), each intervention was implemented for periods of six months in a random order for each ICU and was separated by a one-month wash-out/in period. Samples and subsequently culture-isolated microorganisms were categorized into three categories: surveillance, point prevalence survey (PPS), or clinical. Surveillance samples were from patients directly undergoing interventions (i.e., who had received the decontamination treatment) and were taken twice weekly. PPS samples were collected monthly from all patients present in the ward at that time, including those not undergoing the interventions. Clinical samples were those obtained when needed for the clinical management of the patients [12]. Isolates originated from various body sites: respiratory (more specifically aspirate, throat swab, sputum, bronchoalveolar lavage or non-directed bronchoalveolar lavage), blood, groin, or rectum.

2.2. MIC Determination

All 676 first patient isolates were subjected to colistin susceptibility testing using either an E-test (bioMérieux, Marcy l'Etoile, France) or automatic testing method (BD Phoenix™ (BD Diagnostics, Le Pont de Claix, France), Sensititre™ (Thermo Fisher Scientific, Waltham, MA, USA), Vitek® (bioMérieux, Marcy l'Etoile, France), MicroScan (Beckman Coulter, San Diego, CA, USA)) [12]. Of the 676 isolates, 79 (11.7%) were determined to be fully CR. It must be noted, however, that these methods are not recommended for colistin susceptibility testing and it has been shown that they may underestimate CR rates [13], with the exception of Sensititre™ which has been found to perform well when compared to the reference method [14]. Therefore, to confirm the susceptibility of the selected isolates (see population analysis profiling assay), the colistin MIC was also determined by broth microdilution using the MICRONAUT MIC-Strip Colistin (MERLIN Diagnostika GmbH, Berlin, Germany). In the case of one skipped well, the result was determined disregarding this well (i.e., the skipped well was not seen as the lowest concentration showing no growth). In the case of multiple skipped wells, the test was repeated. An isolate was classified as susceptible (MIC $\leq$ 2 mg/L) or resistant (MIC > 2 mg/L) based on the epidemiological cut-off (ECOFF) values provided by EUCAST [15]. Two CS strains, *Escherichia coli* ATCC 25922 (MIC: 0.25–2 mg/L) and *P. aeruginosa* ATCC 27853 (MIC: 0.5–4 mg/L), and two CR strains, *K. pneumoniae* 08400 (MIC: 64 mg/L) and *E. coli* NCTC 13846 *mcr-1* positive (MIC: 4 mg/L), were used as controls [16].

2.3. Population Analysis Profiling (PAP) Assay

For HR screening, we randomly selected 288 isolates of the 597 CS (MIC $\leq$ 2 mg/L) isolates for population analysis profiling (PAP) based on the two following criteria: (1) Number of isolates selected per country had to reflect the proportion of isolates contributed by each country and also (2) to match with the distribution of isolates found during the baseline and the three intervention strategies (Supplementary Table S1).

For the PAP assay, 0.5 MacFarland (McF) bacterial suspensions were prepared using the BD PhoenixSpec™ nephelometer (BD Diagnostics, Le Pont de Claix, France) starting from an overnight culture on Columbia blood agar (Oxoid Ltd., Basingstoke, UK) with 5% defibrinated horse blood (International Medical Products, Oudergem, Belgium). A 100 µL aliquot of this suspension was spirally plated using the Eddy Jet (IUL instruments S.A., Barcelona, Spain) on a series of cation adjusted Mueller Hinton Agar (CAMHA) (BD Diagnostics, Le Pont de Claix, France) plates containing colistin (Sigma-Aldrich, St. Louis, MO, USA) in increasing concentrations (0 mg/L, colistin free; 1 mg/L; 2 mg/L; 4 mg/L, 8 mg/L; 16 mg/L). The number of colonies were counted after 24 h of aerobic incubation of the plates at 37 °C, and a graph of the $\log_{10}$ CFU/mL was plotted against the increasing colistin concentrations. One CS strain, *E. coli* ATCC 25922 (MIC: 0.25–2 mg/L), one CR strain, *K. pneumoniae* 08400 (MIC: 64 mg/L), and two CHR strains, *K. pneumoniae* ATCC 13883 (MIC: 1 mg/L) and *K. pneumoniae* IT0244CP (MIC: 0.5 mg/L), were used as quality controls [16].

Since there is no consensus on the definition of CHR, we used two previously defined schemes (Figure 1). Classification 1 (C1) was based on the classification as used by Band et al. [17] while Classification 2 (C2) was based on Andersson et al. [9] with the additional requirement that growth at a frequency of minimally 1×10^{-7} must be observed at least at 4 mg/L even if the eight-fold MIC of the isolate was below 4 mg/L. More specifically, this means that for a MIC of $\leq$0.5 mg/L, growth at a frequency of 1×10^{-7} must be observed on plates containing 1, 2 and 4 mg/L of colistin, for a MIC of 1 mg/L the former must be observed additionally on the plates containing 8 mg/L of colistin and for a MIC of 2 mg/L also on plates containing 16 mg/L. This additional requirement was put in place to account for the possibility of false positives occurring due to the inoculum effect. The frequency was determined for each concentration of colistin using the following calculation:

$$Frequency = \frac{\text{CFU/mL } colistin\ plate}{\text{CFU/mL } colistin\ free\ plate}$$

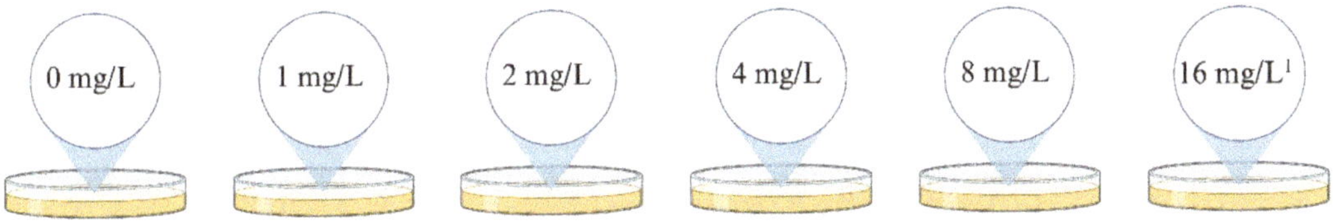

Classification 1					
Growth	Growth	Growth	$\geq 1\times10^{-6}$ [2]	$\geq 1\times10^{-6}$ [2]	NA
Classification 2: for MIC ≤0.5 mg/L					
Growth	$\geq 1\times10^{-7}$	$\geq 1\times10^{-7}$	$\geq 1\times10^{-7}$	NR	NA
Classification 2: for MIC 1 mg/L					
Growth	$\geq 1\times10^{-7}$	$\geq 1\times10^{-7}$	$\geq 1\times10^{-7}$	$\geq 1\times10^{-7}$	NA
Classification 2: for MIC 2 mg/L					
Growth	$\geq 1\times10^{-7}$	$\geq 1\times10^{-7}$	$\geq 1\times10^{-7}$	$\geq 1\times10^{-7}$	$\geq 1\times10^{-7}$

Figure 1. Classification schemes for colistin heteroresistance (CHR). Figure describes the growth requirements, more specifically the frequency of growth required, for an isolate to be determined CHR. In the case there is no frequency requirement at a specific concentration of colistin but only the requirement that there is visible growth on the plate, this is indicated as "Growth". For Classification 2, the requirements to be fulfilled depend on the MIC for colistin of the isolates. NR = no requirement, NA = not applicable, MIC = minimum inhibitory concentration. [1] Plates containing 16 mg/L of colistin were only included for isolates with a MIC of 2 mg/L. [2] Frequency of $\geq 1 \times 10^{-6}$ only required for either 4 mg/L or 8 mg/L, not for both though it is allowed. Created with BioRender.com.

2.4. Whole-Genome Sequencing

Whole-genome sequencing (WGS) was employed to determine the ST, O-antigen type, and K-antigen type of the 676 *K. pneumoniae* isolates as well as to look for mutations in known (hetero)resistance genes for colistin. Strains were cultured on CAMHA and incubated for 16–20 h at 35–37 °C. After incubation, one colony of the CAMHA plate was inoculated in a polypropylene tube containing 4 mL of cation adjusted Mueller Hinton Broth (CAMHB) (BD Diagnostics, Le Pont de Claix, France) and incubated again for 16–20 h at 35–37 °C. A negative growth control was prepared containing only 4 mL of CAMHB. DNA extraction was performed using the MasterPure™ Complete DNA and RNA Purification Kit (Epicentre Biotechnologies, Madison, WI, USA) following manufacturer's instructions. DNA was further purified using the DNA Clean & ConcentratorTM-10 kit (Zymo Research, Irvine, CA, USA) following instructions as provided by the manufacturer. Library preparation was performed using the Nextera® XT DNA Sample Preparation Kit and the Nextera® XT Index Kit v2 Set A (Illumina, San Diego, CA, USA) in conjunction with the Zephyr® G3 NGS liquid handler (PerkinElmer, Waltham, MA, USA), containing heating and shaking modules controlled by the Inheco Multi TEC Controller (INHECO GmbH, Martinsried, Germany). Sequencing was performed with the MiSeq sequencer (Illumina, San Diego, CA, USA). Data were analyzed using BacPipe v6.0 [18] and CLC Genomics Workbench software (Qiagen, Hilden, Germany).

2.5. Statistical Analysis

To determine whether there was an association between CHR/CR and country, MIC-value, intervention strategy, ST, O-antigen, or K-antigen type and between country, ST, and MIC-value, a Pearson Chi-square test or Fisher Exact test were used. When conditions for a Chi-square test or Fisher Exact test were not met (i.e., no cells with expected values < 1, and no more than 20% of cells with values < 5), a Monte Carlo simulation was used. Since this is not an exact method, in contrast to the regular Fisher Exact test, the p-value was given with the 99% confidence interval (99% CI). p-values less than 0.05 were considered statistically significant. In the case of a statistically significant association, a pairwise z-test with Bonferroni correction was used to assess which groups had an association. All analyses were performed using the IBM® SPSS® Statistics software version 28.0.1.1 (IBM Corp., Armonk, NY, USA).

3. Results

3.1. Population Analysis Profiling (PAP)

The 288 selected isolates were tested by PAP assay in 11 distinct runs. Results of the PAP assay for the control strains in the 11 full runs are shown in Figure 2. One of the two HR control strains, IT0244CP, grew on all colistin-containing plates in all 11 runs, though the frequency varied from run to run. The frequency threshold of 1×10^{-6} was only reached twice (C1); still, IT0244CP always fulfilled C2. The second CHR control strain *K. pneumoniae* ATCC 13883 also showed some variability in the frequency of the resistant subpopulation but fulfilled C1 in all 11 runs.

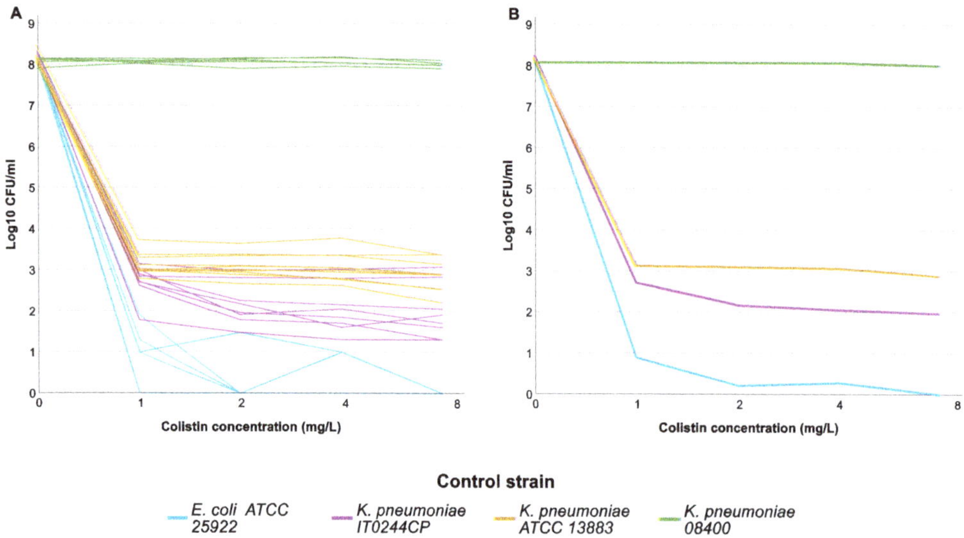

Figure 2. Population analysis profiling (PAP) assay results of the control strains. Graph represents the $\log_{10}$ colony forming units (CFU)/mL per concentration of colistin used in the agarplates of the PAP assay. (**A**) Individual results for each strain for each run, graph illustrates the intra-run variation for the different control strains. (**B**) Average result for each control strain, graph illustrates the overall result of the control strains.

Overall, out of the 288 isolates tested, 25 were classified as being CHR based on the more stringent criteria of C1 whilst 108 isolates were classified as being CHR based on the less stringent C2 criteria. All isolates that fulfilled C1 also fulfilled C2 (Table 1 and Supplementary Table S2).

Table 1. Table contains a detailed breakdown of the PAP assay results (n = 288) including the reasons why isolates did not fulfil the definition of colistin heteroresistance (CHR) and the number of isolates per observed result. Of note, for some isolates that did not fulfil the definition of CHR but did show growth > 2 mg/L, there were multiple reasons why they were not classified as CHR. Sub-reasons are listed in order of importance. Each isolate was only included once, and if it fulfilled multiple sub-reasons, it was only included in the sub-reason considered most important. MIC = minimum inhibitory concentration, C1 = Classification 1, C2 = Classification 2.

	Observed Results	No. of Isolates
Fulfilling C1	Growth on the plates containing 4 and/or 8 mg/L of colistin with a frequency of at least 1×10^{-6}	25
Fulfilling only C2	Growth on at least all the plates containing colistin at a concentration up to and including eight-fold the MIC of the isolate at a frequency of minimally 1×10^{-7}, minimum concentration at which there should be growth was 4 mg/L	83
Not fulfilling either classification but growth > 2 mg/L	Growth on 4 and/or 8 mg/L plate but frequency < 1×10^{-7}	45
	For MIC 0.0625–0.5 mg/L: growth with frequency $\geq 1 \times 10^{-7}$ on 8 mg/L plate but <1×10^{-7} on 4 mg/L plate	8
	For MIC 1–2 mg/L: growth with frequency $\geq 1 \times 10^{-7}$ on 4 and/or 8 mg/L plate but <1×10^{-7} on plates $\geq$ eight-fold the MIC	30
	Growth with frequency $\geq 1 \times 10^{-7}$ on plates $\geq$ eight-fold the MIC but frequency of 1×10^{-7} not reached on all plates < eight-fold the MIC	19
No growth at 4 and 8 mg/L		71
No growth at 1, 2, 4 and 8 mg/L		7

3.2. *High Prevalence of Multi-Drug-Resistant and Carbapenemase-Producing Isolates*

Overall, 617 out of 671 (92%) of the *K. pneumoniae* isolates, for which the information was available, were classified as MDR based on Magiorakos et al. [19]. MDR prevalence was high among both CS (91%, 539/592) and CR (98.7%, 78/79) isolates. In contrast, the proportion of carbapenemase-producing *K. pneumoniae* (CP-Kpn) was higher among CR isolates (72.2%, 57/79) than among CS isolates (29.3%, 174/594). On the other hand, similar proportions of MDR and of CP-Kpn were observed among CHR (MDR: 85%, 91/107; CP-Kpn: 29.9%, 32/107) and selected CS isolates (MDR: 84.7%, 244/288; CP-Kpn: 28.1%, 80/285) (Supplementary Tables S3 and S4).

3.3. *Analysis of Association between Colistin-Resistant K. pneumoniae, ST, O-Antigen Type and K-Antigen Type*

To investigate the distribution of CR within various STs, we studied those STs that were represented at least 10 times in our collection. The 79 CR *K. pneumoniae* isolates in our study were distributed among 14 different STs with three STs (ST101, ST147, ST258/512) accounting for 62% (n = 49) of the total CR isolates. CR was significantly associated with specific STs (p at least 0.003); for instance, ST101 (CR: 12.7%, 10/79, CS: 2.7%, 16/597), ST147 (CR: 30.4%, 24/79, CS: 6.9%, 41/597) and ST258/ST512 (CR: 19%, 15/79, CS: 4.9%, 29/597) (p for all <0.05) showed higher proportions among CR strains than among CS strains. Conversely, ST15 (CR: 3.8%, 3/79, CS: 11.4%, 68/597), ST307 (CR: 3.8%, 3/79, CS: 13.6%, 81/597) and ST405 (CR: 5.1%, 4/79, CS: 10.1%, 60/597) (p for all <0.05) had higher proportions of CS isolates compared to CR. We could not document exclusive association either with CS or with CR of any ST that was present in the collection at least 10 times (Figure 3A).

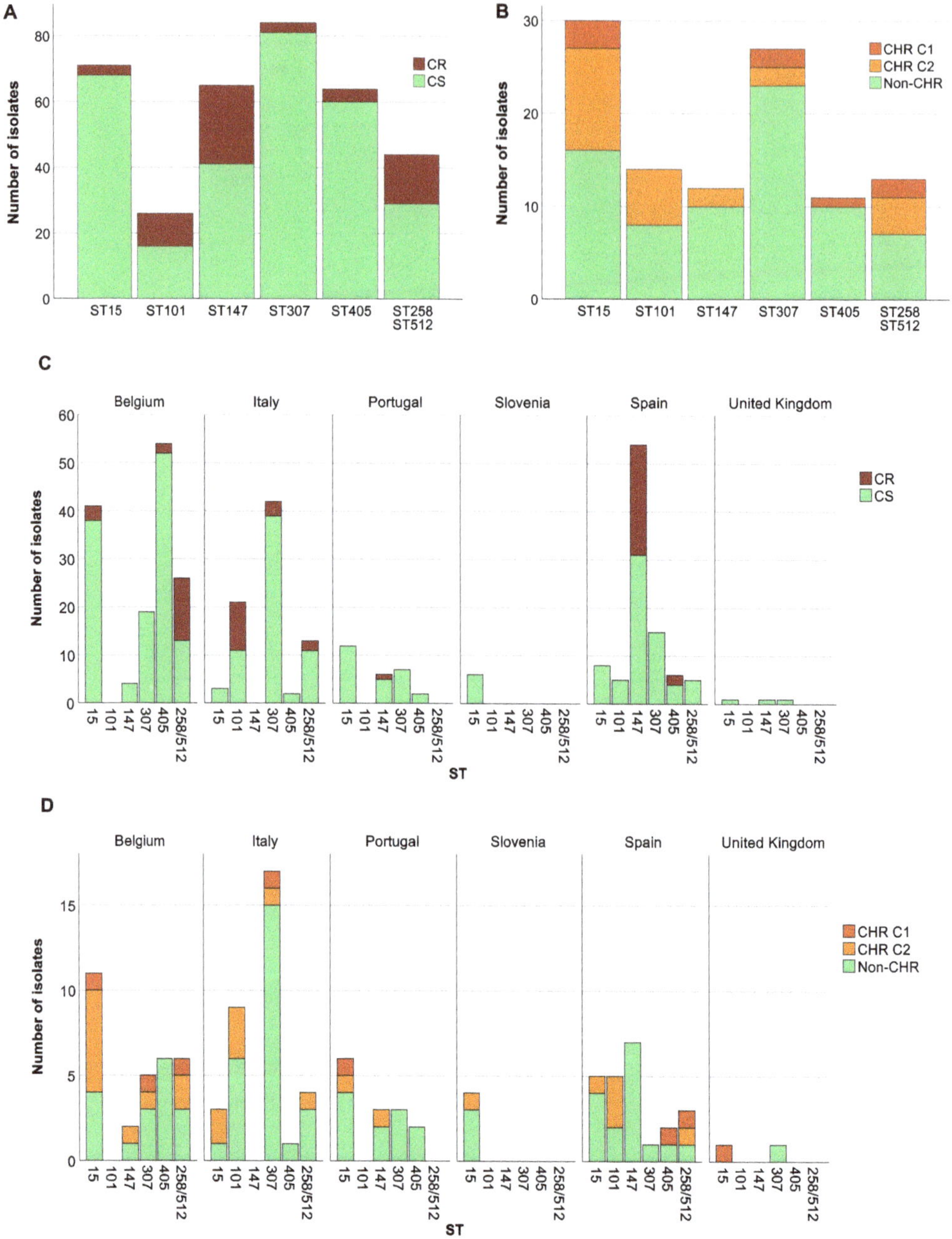

Figure 3. Distribution of isolates across sequence types (STs). Graphs show the number of isolates per ST as well as the number of colistin-resistant (CR)/colistin-heteroresistant (CHR) and colistin-susceptible (CS)/non-CHR isolates. Only STs with a statistically significant association with CR and/or CHR are shown. Of note, Classification 1 (C1) + Classification 2 (C2) represents the total amount of isolates fulfilling C2 whilst C2 alone represents isolates only fulfilling C2. (**A**) CR per ST, (**B**) CHR per ST, (**C**) CR per ST per country, Slovenia (n = 9) and United Kingdom (n = 23), were not taken into further consideration due to the low number of isolates, (**D**) CHR per ST per country.

CS isolates ($n = 597$) were distributed over 87 STs (Simpson's diversity index, SDI: 0.94) while the 288 CS isolates tested on the PAP assay belonged to 85 STs (SDI: 0.97). Of these, CHR was found among 52 different STs and showed a remarkably higher genetic diversity (SDI: 0.97) compared to ST distribution among CR isolates (SDI: 0.85). Ten STs (ST11, ST15, ST45, ST101, ST147, ST258/512, ST307, ST405, ST409 and ST437) were common to both CR and CHR isolates.

Besides an association between CR and STs, country and STs also showed a strong association (p at least <0.001). *K. pneumoniae* belonging to ST15 were spread across all six countries; however, the proportion of ST15 was higher in Belgium (17.2%, 41/239) and Portugal (19.7%, 12/61) than in Italy (2.1%, 3/143) and Spain (4%, 8/201) ($p < 0.05$). ST307 was present in five of the six countries. In this case, the proportion of ST307 isolates in Italy (29.4%, 42/143) was significantly different compared to Belgium (7.9%, 19/239) and Spain (7.5%, 15/201) ($p < 0.05$) but not when compared to Portugal (11.5%, 7/61) ($p > 0.05$). ST147 was not found in Italy and most of the isolates were from Spain (26.9%, 54/201) for which the proportion also significantly differed from both Belgium (1.7%, 4/239) and Portugal (9.8%, 6/61) ($p < 0.05$). In contrast to the previous STs which were found in almost all countries, ST101 and ST258/ST512 were only found in two and three countries, respectively. ST101 showed a statistically significant difference between the two countries with a higher proportion in Italy (14.7%, 21/143) compared to Spain (2.5%, 5/201) ($p < 0.05$). ST258/ST512 showed a statistically significant difference between Belgium (10.5%, 25/239) and Spain (2.5%, 5/201) ($p < 0.05$), but not between Italy (9.1%, 13/143) and the other two countries ($p > 0.05$). Among ST101, all CR isolates were isolated in Italy, for ST147 all but one were isolated in Spain (one CR isolate from Portugal), and for ST258/ST512 all but two from Belgium (two CR isolates from Italy) (Figure 3C).

For the K-antigen and O-antigen types, a tight association is known to exist with specific STs and therefore the significant differences found in the proportion of CR and CS within a ST were likewise reflected in similar differences in the proportion of K and O antigen types (p for both ≤ 0.001).

3.4. A Higher Proportion of CR among Specific STs Was Not Reflected in CHR Proportions

For CHR and ST, an association was found (p at least 0.021). In contrast, CHR isolates did not show any clear association with K-antigen and O-antigen types ($p \geq 0.169$). ST307 and ST405 were found to have a higher proportion of non-CHR isolates (ST307: CHR: 3.7%, 4/108, non-CHR: 12.8%, 23/180, $p < 0.05$ and ST405: CHR: 0.9%, 1/108, non-CHR: 5.6%, 10/180, $p < 0.05$) (Figure 3B). Selected isolates from both STs were spread across different countries (Figure 3D). There was no association between STs found to have a higher proportion of CR isolates and CHR. Of the 39 isolates tested for CHR with STs ST101, ST147, and ST258/ST512, only 2 were determined to be CHR based on C1 and 14 exclusively based on C2.

3.5. Analysis of Association between Colistin (Hetero)Resistant K. pneumoniae, Country, Intervention Strategy, and MIC-Value

The highest proportion of CR *K. pneumoniae* was found in Spain (44.3%, 35/79, $p < 0.05$) (Figure 4A). There was also a difference in the distribution of CR across the baseline, CHX, SDD, and SOD ($p = 0.001$) with a higher proportion of CR isolates in patients in all three intervention strategy groups compared to the baseline period ($p < 0.05$) but no significant difference in distribution of CR was identified between the three intervention groups (Supplementary Figure S1). This could mainly be explained by the increase in the proportion of CR ST147 isolates in Spain during the intervention periods (for all three groups) compared to the baseline (8.8–15.6% versus 3.8%). In contrast, no significant association was found between increase in CHR with any country ($p > 0.723$) (Figure 4B) or between CHR and any intervention strategy ($p > 0.668$). CHR isolates were present in the highest proportions in ST15 followed by ST101 (Figure 3B). Of note, CHR isolates that matched the stricter definition (i.e., C1) were found in higher proportion in ST15 followed by ST307, but they were not found at all among the ST101 and ST147 isolates (Figure 3B).

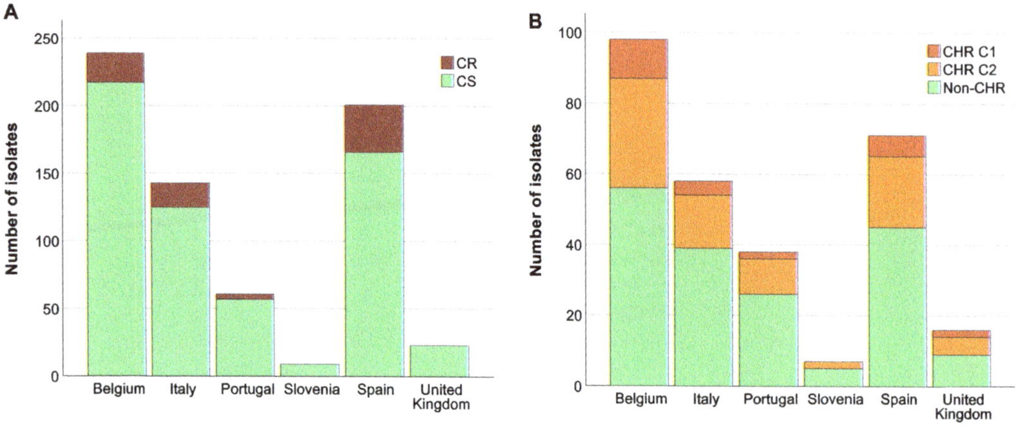

Figure 4. Distribution of isolates across countries. Graphs show the number of isolates per country as well as the number of colistin-resistant (CR)/colistin-heteroresistant (CHR) and colistin-susceptible (CS)/non-CHR isolates. Of note, Classification 1 (C1) + Classification 2 (C2) represents the total amount of isolates fulfilling C2 whilst C2 alone represents isolates only fulfilling C2. (**A**) CR per country, (**B**) CHR per country.

We could not find any association between the colistin MIC value of the 288 selected isolates and countries (p at least 0.599) (Figure 5B), nor was an association present between ST and MIC values (p at least 0.08) (Figure 5C). Remarkably, however, there was a trend towards higher colistin MIC values for isolates classified as CHR based on C1 compared to CHR isolates fulfilling only C2 (Figure 5A).

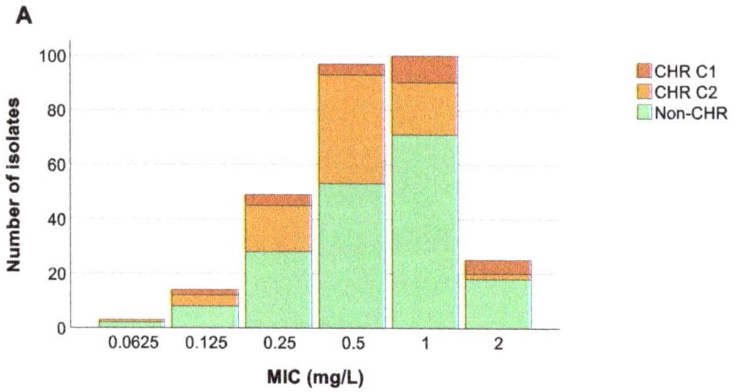

Figure 5. *Cont.*

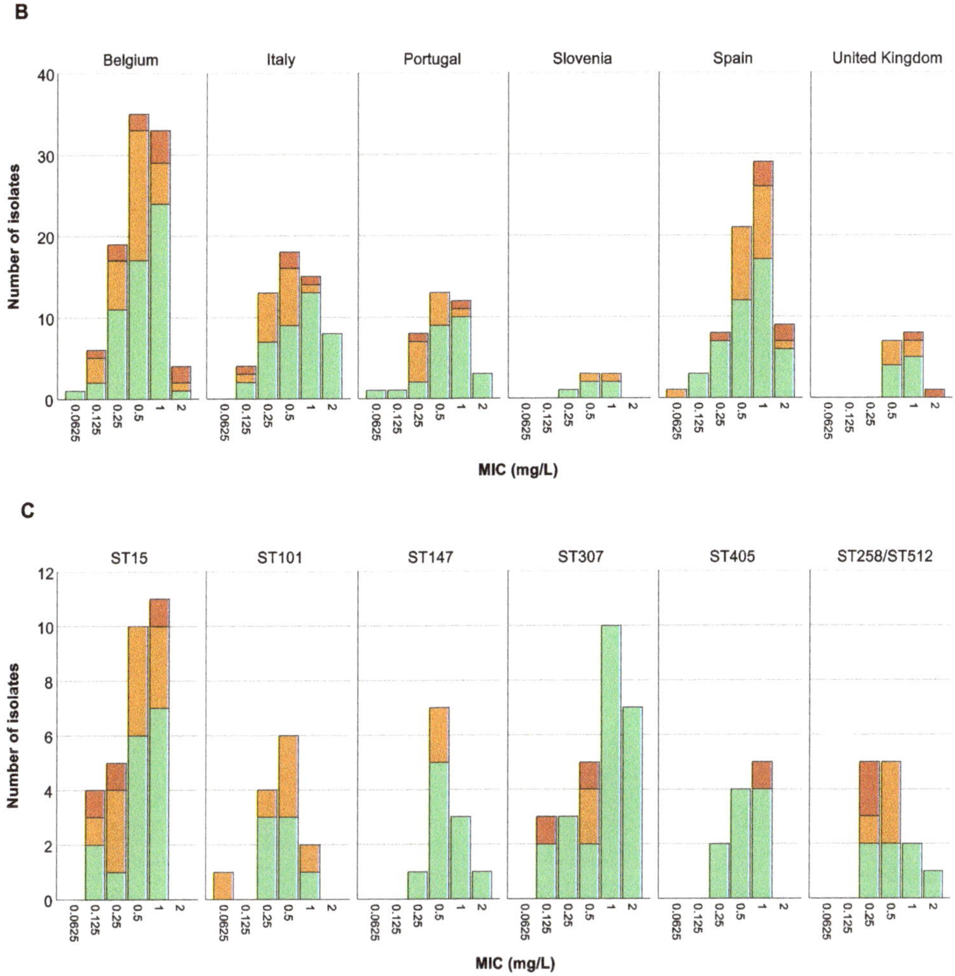

Figure 5. Distribution of isolates across minimum inhibitory concentration (MIC) values. Graphs show the number of isolates per MIC as well as the number of colistin-heteroresistant (CHR) and non-CHR isolates. Of note, Classification 1 (C1) + Classification 2 (C2) represents the total amount of isolates fulfilling C2 whilst C2 alone represents isolates only fulfilling C2. (**A**) CHR per MIC value, (**B**) CHR per MIC value per country, (**C**) CHR per MIC value per sequence type (ST). Only STs with a statistically significant association with CR and/or CHR are shown.

3.6. Colistin Resistance Mechanisms in Resistant Subpopulations

Isolates AN1505CP2 and IT0244CP were previously already determined to be CHR [16]. In this study, these findings were confirmed with both isolates fulfilling C1, and colonies from each plate were sequenced (data not published). Mutations found in colistin resistance-associated genes (*mgrB*, *phoP*, *phoQ*, *pmrA*, *pmrB*, *pmrD*, *arnA*, *kpnEF*, *kpnF*, *crrB* and *acrB*) are summarized in Table 2. The resistant subpopulations of AN1505CP2 and IT0244CP both showed disruptions of the *mgrB* gene by various insertion sequences (IS1R, IS1X2, ISKpn34, IS903B).

Table 2. Summary of mutations found in CR and CHR isolates. The table contains information on the mutations found in the resistant subpopulation of two confirmed CHR isolates (AN1505CP2 and IT0244CP) at different concentrations of the PAP assay plates on three separate assays. Additionally, the table contains information on three CR isolates with the same ST as IT0244CP (ST409). For AN1505CP2, there were no CR isolates with the same ST (ST323). CR = colistin-resistant, CHR = colistin-heteroresistant, PAP = population analysis profiling, MIC = minimum inhibitory concentration, ST = sequence type, IS = insertion sequence.

Isolate ID	MIC (mg/L)	ST	PAP Assay Plate Conc. (mg/L)	Mutations in *mgrB*	Other Mutations
IT0307CP (CR)	128	ST409		IS1R of IS1 family interruption at nt 107	
IT0636C (CR)	128	ST409		ISKpn34 of IS3 family interruption at nt 46	
IT0915C (CR)	64	ST409		IS903B of IS5 family interruption at nt 34	
IT0244CP (CHR 1st PAP)	0.5	ST409	2	ISKpn34 of IS3 family interruption at promoter	
			8	IS903B of IS5 family interruption at nt 117	
			16	IS1S of IS1 family interruption at promoter	
IT0244CP (CHR 3rd PAP)	0.5	ST409	2	IS1X2 of IS1 family interruption at nt 123	
AN1505CP2 (CHR 1st PAP)	1	ST323	4	Deleted	
			8	IS903B of IS5 family interruption at nt 70	
AN1505CP2 (CHR 2nd PAP)	1	ST323	8		*pmrB:* T157P
AN1505CP2 (CHR 3rd PAP)	1	ST323	8	Q30X	
			16	Deleted	

For IT0244CP, the disruption of *mgrB* was found across different concentrations of colistin in the same PAP assay (but these were linked with different insertion sequences and interruption at different nucleotide positions) (Table 2). In contrast, the resistant subpopulation of AN1505CP2 did show differences in the type of mutations found in *mgrB* between different concentrations of colistin in the same PAP assay run. To assess whether the mechanism of resistance was ST-specific or stochastic, we looked into CR isolates with the same ST as IT0244CP (ST409). For AN1505CP2 (ST323), in our collection, we could not find any CR isolates that belong to the same ST (Table 2). For ST409, both the CR and CHR isolates showed disruptions of *mgrB* by insertion sequences. However, we did not find reproducibility in the mutations selected even for the same (CHR) isolate between technical replicates of the PAP assay, highlighting the stochasticity of the selection process.

4. Discussion

In general, research around HR has been proven to be challenging, especially due to a lack of standardization of identification methods as well as of precise classification criteria. In the past, it has therefore also been difficult to acquire reliable estimates on the prevalence of HR as the sample sizes of the performed studies have been generally small and, due to a lack of standardization, both in definitions of HR but also in assays, it has been difficult to compile and compare different studies [1,9]. There are various ways through which HR can be screened, including methods that are used routinely for example for MIC determination, but they have been deemed to be not performant enough [9,20,21].

A strength of this study was its large sample size compared to previous studies on CHR using the PAP assay [22–25]. Within this study, two different classification schemes for CHR were used. C1 was considered to be more stringent due to the higher frequency requirement, and the number of isolates fulfilling this more stringent classification CHR was thus limited (n = 25). The majority of isolates categorized as CHR fulfilled the second less stringent classification (n = 108) which was considered to be less stringent due to a lower frequency requirement though this classification included an additional requirement with regard to the concentration at which growth should occur. It has to be acknowledged that there were also numerous isolates that could not be classified as CHR by any of these two classifications but did show growth on PAP assay plates > 2 mg/L colistin. Most often, this was growth at frequency $< 1 \times 10^{-7}$ (often only one colony). We cannot rule out that these colonies were spontaneous mutants and were thus not caused by the isolate being CHR or were due to an inoculum effect. They were therefore not classified as CHR.

This study was also unique in the investigated associations, which, to our knowledge, have not been studied previously in *K. pneumoniae*. A study on CHR in *A. baumannii* reported the absence of association between CHR and MIC value and clonal complexes [24]; however, these results are not necessarily applicable to CHR in *K. pneumoniae*. In this study, we could not find any association between CHR and country, intervention strategy, K-antigen type, or O-antigen type or colistin MIC value. For the ST, an association was found, however, only when C2 was used. In contrast, an association was found between CR and country, ST, O-antigen type, and K-antigen type. It is interesting to note that, based on an annual report from the European Centre for Disease Prevention and Control (ECDC) on antimicrobial consumption, Spain, which was found to have a higher number of CR isolates, also had a relatively higher consumption of polymyxins in hospitals compared to other European countries at least in the last period of the trial in which these isolates were collected (2016–2017) [26]. Also during 2017, colistin consumption in Spain remained relatively much higher compared to other countries included here [26]. Unfortunately, no data were available for a larger part of the duration of the trial (2013–2016).

For the ST, O-antigen, and K-antigen type, we often saw a relationship between the different molecular indicators and the STs which were found to have a statistically significant difference in the amount of CS and CR isolates, e.g., a specific K-antigen type was only present in combination with a specific ST and they were both found to have a higher proportion of CR isolates. For CHR, interestingly, no such pattern was found. CHR STs found to have an association with CR in this study are especially interesting since more than 90% of them are known to be associated with multidrug resistance [27–29]. In this study, MDR rates were high in CS, CHR, and CR isolates (>75%) and around one-third of the isolates determined to be CHR were CP-Kpn. CHR in MDR CP-Kpn is especially important to report as in most sites with a historically high proportion of CP-Kpn, colistin is empirically prescribed, and knowledge on the predilection towards development of CHR might facilitate the use of tailored antibiotic combinations instead.

In contrast to CS and CHR, the proportion of CR CP-Kpn isolates was relatively high (±30% vs. ±70%). Since CP-Kpn are more likely to be exposed to colistin, it is to be expected that CR rates will increase over time in this subset of isolates which is also reflected in this study with the far majority of CP-Kpn being CR. Interestingly, rates of CHR among CP-Kpn were similar to those of CS among CP-Kpn, and CHR rates also did not differ significantly between the baseline, CHX, SOD, and SDD. This also suggests that colistin exposure through SOD and SDD had no association with selection of isolates with a CHR phenotype.

For CR, we did find an association with CHX, SOD, and SDD. The association found between CR and CHX, SOD and SDD could, however, be linked to the relatively higher proportion of ST147 isolates, which was similarly found to be associated with CR compared to the baseline. Additionally, most ST147 isolates were originating from Spain. It was beyond the scope of this study to investigate the exact influence of each parameter on the increased proportion of CR isolates.

Finally, a closer look at the mechanism of colistin resistance for the resistant subpopulation of two CHR isolates showed that mutations in colistin resistance-conferring genes assessed in this study could not always be identified for each isolate. A recent study on CHR in wild-type *K. pneumoniae* isolates also reported that for 28% of mutants sequenced, no genetic modification was found in the panel of genes assessed [30]. Disruption of *mgrB* by insertion sequences was the most commonly found genetic modification in the resistant subpopulations. Additionally, a nonsense mutation and complete deletion of *mgrB* was found once and twice, respectively. A mutation outside of *mgrB* was only found once. The same study in wild-type *K. pneumoniae* isolates also reported a high number of *mgrB* genetic modifications (54%) [30]. A complete deletion was only found in 4% of the mutants while disruption by insertion sequences and other amino acid alterations were found in 28% and 22% of mutants, respectively [30]. Luo et al. also reported in their study that there was a high rate of *mgrB* insertional mutations and no mutations in *pmrAB* or *phoPQ* and stated that this was consistent with previous findings which showed that those genes had a significantly lower mutation rate compared to *mgrB*. However, they also stated that the high amount of *mgrB* disruptions may be related to the high prevalence of ST11 since this ST showed a significantly higher rate of *mgrB* disruptions compared to other STs in their study [31].

Though this study helps in expanding the knowledge on CHR in a clinical setting, there are also some limitations. Firstly, only limited sequencing data were available for CHR isolates. Future studies are needed to further assess the diversity of mechanisms of CHR and whether these mechanisms are ST-specific. Secondly, we cannot exclude confounding factors such as the usage of more colistin in some local settings, outbreaks with specific STs, and the prevalence of CP-Kpn which may vary between sites. Thirdly, we only studied the CS population for CHR. However, CHR can also exist as a (sub)proportion of CR isolates. Additionally, we only focused on CHR as a phenomenon in which there is a minor subpopulation with a MIC above the breakpoint in a major population with a MIC below the breakpoint. However, HR may also occur in entirely susceptible populations [32].

Given the large number of isolates screened, this study is a step forward in elucidating the prevalence and burden of CHR in common ST lineages of *K. pneumoniae*. Our data prompt for the development of more robust and simple diagnostics to enable implementation of HR detection on a larger scale, and for more structured studies to quantify the actual impact of CHR on treatment failures in patients receiving colistin.

Supplementary Materials: The following supporting information can be downloaded at: https://www.mdpi.com/article/10.3390/antibiotics13030281/s1, Table S1: Summary of interventions including amount of isolates within each intervention per sample category. Table provides an overview of the contents of the different intervention strategies as well as the number of isolates sorted per sample category for both the complete collection (n = 676) and the selected subset (n = 288) of isolates. Surv = surveillance, PPS = point prevalence survey, Clin = clinical, CS = colistin sulphate, TBS = tobramycin sulphate, NYS = nystatin. [1] 2% mouthwash replaced by 1% oral gel after reports of oral mucosal adverse events. [2] Normally, regiment includes four days of IV cephalosporin, not included because of settings of moderate/high resistance. [3] Either sampling during 1 month wash-out/in period between intervention strategies or interruption of the intervention. Table S2: Overview of selected isolates. Table offers an overview of the selected isolates (n = 288) and their characteristics. BEL = Belgium, ESP = Spain, UK = United Kingdom, SVN = Slovenia, ITA = Italy, PRT = Portugal, Cat. = category, Surv = surveillance, PPS = point prevalence survey, Clin = clinical, CHX = chlorhexidine digluconate, SOD = selective oropharyngeal decontamination, SDD = selective digestive tract decontamination, ST = sequence type, K-antigen type = capsular polysaccharide antigen type, O-antigen type = lipopolysaccharide antigen type, Col MIC = colistin minimum inhibitory concentration, CHR = colistin-heteroresistant, MDR = multidrug-resistant, CP = carbapenemase producing, Undef. = undefined, ND = no data. Table S3: Number of MDR and CP-Kpn isolates. Tables gives summary of the number of MDR and CP-Kpn CS and CR isolates. For CS, MDR, and CP-Kpn, classification was not possible for five and three isolates, respectively. CS = colistin-susceptible, CR = colistin-resistant, MDR = multidrug-resistant, CP-Kpn = carbapenemase-producing *K. pneumoniae.* Table S4: Number of MDR and CP-Kpn isolates. Tables gives summary of the number of MDR and CP-Kpn selected CS and CHR isolates. For CS, MDR, and CP-Kpn, classification was not possible for four and three isolates, respectively, of which one isolate was also CHR following C2. For CHR, C1 + C2 represents the total amount of isolates fulfilling C2 whilst C2 alone represents isolates only fulfilling C2. CS = colistin-susceptible, CHR = colistin-heteroresistant, C1 = Classification 1, C2 = Classification 2, MDR = multidrug-resistant, CP-Kpn = carbapenemase-producing *K. pneumoniae.* Figure S1: Distribution of isolates across baseline and intervention strategies. Graphs show the number of isolates per strategy as well as the number of CR/CHR and CS/non-CHR isolates. Of note, C1 + C2 represents the total amount of isolates fulfilling C2 whilst C2 alone represents isolates only fulfilling C2. (A) CR per intervention strategy, (B) CHR per intervention strategy.

Author Contributions: Conceptualization and study design: S.M.-K. and Y.G.; Experimental execution: A.J.M.M.B., S.G.R., A.S., M.E.B. and C.L.; Bioinformatics: S.G.R.; Manuscript writing: S.M.-K., Y.G. and A.J.M.M.B.; Manuscript review and editing: all authors. All authors have read and agreed to the published version of the manuscript.

Funding: S.G.R. and part of this study are funded by University of Antwerp BOF Doctoral Project Funds (BOF-DOCPRO 2019–project ID 40179). S.M.-K. gratefully acknowledges University of Antwerp Methusalem funding (Vaccine & Infectious Diseases Excellence in Antwerp: Infectious disease prevention, control, and management in a One Health policy context (VAX-IDEA)).

Institutional Review Board Statement: Not applicable.

Informed Consent Statement: Informed consent was obtained from all subjects involved in the main clinical trial (NCT02208154). Bacterial isolates obtained from patient samples were utilized in the current study.

Data Availability Statement: All sequenced data generated and analyzed in this study were deposited at NCBI under Bioproject ID PRJNA948355.

Acknowledgments: We would like to thank Sabine Chapelle, Gert Leten, Liesbeth Bryssinck and Joyce Jacobs for excellent technical and administrative support.

Conflicts of Interest: The authors declare no conflicts of interest.

References

1. Roch, M.; Sierra, R.; Andrey, D.O. Antibiotic heteroresistance in ESKAPE pathogens, from bench to bedside. *Clin. Microbiol. Infect.* **2022**, *29*, 320–325. [CrossRef] [PubMed]
2. Della Rocca, M.T.; Foglia, F.; Crudele, V.; Greco, G.; De Filippis, A.; Franci, G.; Finamore, E.; Galdiero, M. Antimicrobial resistance changing trends of Klebsiella pneumoniae isolated over the last 5 years. *New Microbiol.* **2022**, *45*, 338–343. [PubMed]
3. Mohd Asri, N.A.; Ahmad, S.; Mohamud, R.; Mohd Hanafi, N.; Mohd Zaidi, N.F.; Irekeola, A.A.; Shueb, R.H.; Yee, L.C.; Mohd Noor, N.; Mustafa, F.H.; et al. Global Prevalence of Nosocomial Multidrug-Resistant Klebsiella pneumoniae: A Systematic Review and Meta-Analysis. *Antibiotics* **2021**, *10*, 1508. [CrossRef] [PubMed]
4. Wang, M.G.; Earley, M.; Chen, L.; Hanson, B.M.; Yu, Y.S.; Liu, Z.Y.; Salcedo, S.; Cober, E.; Li, L.J.; Kanj, S.S.; et al. Clinical outcomes and bacterial characteristics of carbapenem-resistant complex among patients from different global regions (CRACKLE-2): A prospective, multicentre, cohort study. *Lancet Infect. Dis.* **2022**, *22*, 401–412. [CrossRef] [PubMed]
5. Ah, Y.M.; Kim, A.J.; Lee, J.Y. Colistin resistance in Klebsiella pneumoniae. *Int. J. Antimicrob. Agents* **2014**, *44*, 8–15. [CrossRef]
6. Ahmed, M.A.G.E.S.; Zhong, L.L.; Shen, C.; Yang, Y.Q.; Doi, Y.; Tian, G.B. Colistin and its role in the Era of antibiotic resistance: An extended review (2000–2019). *Emerg. Microbes Infect.* **2020**, *9*, 868–885. [CrossRef]
7. Band, V.I.; Satola, S.W.; Burd, E.M.; Farley, M.M.; Jacob, J.T.; Weiss, D.S. Carbapenem-Resistant *Klebsiella pneumoniae* Exhibiting Clinically Undetected Colistin Heteroresistance Leads to Treatment Failure in a Murine Model of Infection. *mBio* **2018**, *9*, e02448-17. [CrossRef]
8. Moosavian, M.; Shoja, S.; Nashibi, R.; Ebrahimi, N.; Tabatabaiefar, M.A.; Rostami, S.; Peymani, A. Post Neurosurgical Meningitis due to Colistin Heteroresistant Acinetobacter baumannii. *Jundishapur J. Microbiol.* **2014**, *7*, e12287. [CrossRef]
9. Andersson, D.I.; Nicoloff, H.; Hjort, K. Mechanisms and clinical relevance of bacterial heteroresistance. *Nat. Rev. Microbiol.* **2019**, *17*, 479–496. [CrossRef]
10. Stojowska-Swedrzynska, K.; Lupkowska, A.; Kuczynska-Wisnik, D.; Laskowska, E. Antibiotic Heteroresistance in Klebsiella pneumoniae. *Int. J. Mol. Sci.* **2021**, *23*, 449. [CrossRef] [PubMed]
11. Dewachter, L.; Fauvart, M.; Michiels, J. Bacterial Heterogeneity and Antibiotic Survival: Understanding and Combatting Persistence and Heteroresistance. *Mol. Cell* **2019**, *76*, 255–267. [CrossRef] [PubMed]

12. Wittekamp, B.H.; Plantinga, N.L.; Cooper, B.; Lopez-Contreras, J.; Coll, P.; Mancebo, J.; Wise, M.P.; Morgan, M.P.G.; Depuydt, P.; Boelens, J.; et al. Decontamination Strategies and Bloodstream Infections With Antibiotic-Resistant Microorganisms in Ventilated Patients A Randomized Clinical Trial. *J. Am. Med. Assoc.* **2018**, *320*, 2087–2098. [CrossRef] [PubMed]
13. Anantharajah, A.; Glupczynski, Y.; Hoebeke, M.; Bogaerts, P.; Declercq, P.; Denis, O.; Descy, J.; Flore, K.; Magerman, K.; Rodriguez-Villalobos, H.; et al. Multicenter study of automated systems for colistin susceptibility testing. *Eur. J. Clin. Microbiol. Infect. Dis.* **2021**, *40*, 575–579. [CrossRef] [PubMed]
14. Richter, S.S.; Karichu, J.; Otiso, J.; Van Heule, H.; Keller, G.; Cober, E.; Rojas, L.J.; Hujer, A.M.; Hujer, K.M.; Marshall, S.; et al. Evaluation of Sensititre Broth Microdilution Plate for determining the susceptibility of carbapenem-resistant Klebsiella pneumoniae to polymyxins. *Diagn. Microbiol. Infect. Dis.* **2018**, *91*, 89–92. [CrossRef] [PubMed]
15. European Committee on Antimicrobial Susceptibility Testing. Breakpoint Tables for Interpretation of MICs and Zone Diameters; 2023. Available online: https://www.eucast.org/ (accessed on 15 December 2023).
16. Rajakani, S.G.; Xavier, B.B.; Sey, A.; Mariem, E.; Lammens, C.; Goossens, H.; Glupczynski, Y.; Malhotra-Kumar, S. Insight into Antibiotic Synergy Combinations for Eliminating Colistin Heteroresistant. *Genes* **2023**, *14*, 1426. [CrossRef] [PubMed]
17. Band, V.I.; Hufnagel, D.A.; Jaggavarapu, S.; Sherman, E.X.; Wozniak, J.E.; Satola, S.W.; Farley, M.M.; Jacob, J.T.; Burd, E.M.; Weiss, D.S. Antibiotic combinations that exploit heteroresistance to multiple drugs effectively control infection. *Nat. Microbiol.* **2019**, *4*, 1627–1635. [CrossRef] [PubMed]
18. Xavier, B.B.; Mysara, M.; Bolzan, M.; Ribeiro-Goncalves, B.; Alako, B.T.F.; Harrison, P.; Lammens, C.; Kumar-Singh, S.; Goossens, H.; Carrico, J.A.; et al. BacPipe: A Rapid, User-Friendly Whole-Genome Sequencing Pipeline for Clinical Diagnostic Bacteriology. *iScience* **2020**, *23*, 100769. [CrossRef] [PubMed]
19. Magiorakos, A.P.; Srinivasan, A.; Carey, R.B.; Carmeli, Y.; Falagas, M.E.; Giske, C.G.; Harbarth, S.; Hindler, J.F.; Kahlmeter, G.; Olsson-Liljequist, B.; et al. Multidrug-resistant, extensively drug-resistant and pandrug-resistant bacteria: An international expert proposal for interim standard definitions for acquired resistance. *Clin. Microbiol. Infect.* **2012**, *18*, 268–281. [CrossRef]
20. Juhasz, E.; Ivan, M.; Pinter, E.; Pongracz, J.; Kristof, K. Colistin resistance among blood culture isolates at a tertiary care centre in Hungary. *J. Glob. Antimicrob. Resist.* **2017**, *11*, 167–170. [CrossRef]
21. Sherman, E.X.; Wozniak, J.E.; Weiss, D.S. Methods to Evaluate Colistin Heteroresistance in Acinetobacter baumannii. *Methods Mol. Biol.* **2019**, *1946*, 39–50. [CrossRef]
22. Morales-Leon, F.; Lima, C.A.; Gonzalez-Rocha, G.; Opazo-Capurro, A.; Bello-Toledo, H. Colistin Heteroresistance among Extended Spectrum beta-lactamases-Producing Klebsiella pneumoniae. *Microorganisms* **2020**, *8*, 1279. [CrossRef]
23. Band, V.I.; Satola, S.W.; Smith, R.D.; Hufnagel, D.A.; Bower, C.; Conley, A.B.; Rishishwar, L.; Dale, S.E.; Hardy, D.J.; Vargas, R.L.; et al. Colistin Heteroresistance Is Largely Undetected among Carbapenem-Resistant Enterobacterales in the United States. *mBio* **2021**, *12*, e02881-20. [CrossRef]
24. Kon, H.; Hameir, A.; Nutman, A.; Temkin, E.; Keren Paz, A.; Lellouche, J.; Schwartz, D.; Weiss, D.S.; Kaye, K.S.; Daikos, G.L.; et al. Prevalence and Clinical Consequences of Colistin Heteroresistance and Evolution into Full Resistance in Carbapenem-Resistant Acinetobacter baumannii. *Microbiol. Spectr.* **2023**, *11*, e05093-22. [CrossRef]
25. Howard-Anderson, J.; Davis, M.; Page, A.M.; Bower, C.W.; Smith, G.; Jacob, J.T.; Andersson, D.I.; Weiss, D.S.; Satola, S.W. Prevalence of colistin heteroresistance in carbapenem-resistant Pseudomonas aeruginosa and association with clinical outcomes in patients: An observational study. *J. Antimicrob. Chemother.* **2022**, *77*, 793–798. [CrossRef]
26. European Centre for Disease Prevention Control. Antimicrobial Consumption in the EU/EEA (ESAC-Net)—Annual Epidemiological Report 2021. 2022. Available online: https://www.ecdc.europa.eu/ (accessed on 12 January 2024).
27. Bialek-Davenet, S.; Criscuolo, A.; Ailloud, F.; Passet, V.; Jones, L.; Delannoy-Vieillard, A.S.; Garin, B.; Le Hello, S.; Arlet, G.; Nicolas-Chanoine, M.H.; et al. Genomic definition of hypervirulent and multidrug-resistant Klebsiella pneumoniae clonal groups. *Emerg. Infect. Dis.* **2014**, *20*, 1812–1820. [CrossRef] [PubMed]
28. Peirano, G.; Chen, L.; Kreiswirth, B.N.; Pitout, J.D.D. Emerging Antimicrobial-Resistant High-Risk Klebsiella pneumoniae Clones ST307 and ST147. *Antimicrob. Agents Chemother.* **2020**, *64*, e01148-20. [CrossRef] [PubMed]
29. Rodrigues, C.; Desai, S.; Passet, V.; Gajjar, D.; Brisse, S. Genomic evolution of the globally disseminated multidrug-resistant Klebsiella pneumoniae clonal group 147. *Microb. Genom.* **2022**, *8*, 000737. [CrossRef] [PubMed]
30. Sanchez-Leon, I.; Perez-Nadales, E.; Marin-Sanz, J.A.; Garcia-Martinez, T.; Martinez-Martinez, L. Heteroresistance to colistin in wild-type Klebsiella pneumoniae isolates from clinical origin. *Microbiol. Spectr.* **2023**, *11*, e0223823. [CrossRef] [PubMed]
31. Luo, Q.; Xu, L.; Wang, Y.; Fu, H.; Xiao, T.; Yu, W.; Zhou, W.; Zhang, K.; Shen, J.; Ji, J.; et al. Clinical relevance, mechanisms and evolution of polymyxin B heteroresistance carbapenem-resistant Klebsiella pneumoniae: A genomic, retrospective cohort study. *Clin. Microbiol. Infect.* **2024**. [CrossRef]
32. El-Halfawy, O.M.; Valvano, M.A. Antimicrobial heteroresistance: An emerging field in need of clarity. *Clin. Microbiol. Rev.* **2015**, *28*, 191–207. [CrossRef] [PubMed]

antibiotics

Article

Virulence of *Pseudomonas aeruginosa* in Cystic Fibrosis: Relationships between Normoxia and Anoxia Lifestyle

Rosanna Papa [1,†], Esther Imperlini [2,†], Marika Trecca [1], Irene Paris [1], Gianluca Vrenna [3], Marco Artini [1,*] and Laura Selan [1]

1 Department of Public Health and Infectious Diseases, Sapienza University, p. le Aldo Moro 5, 00185 Rome, Italy; rosanna.papa@uniroma1.it (R.P.); marika.trecca@uniroma1.it (M.T.); irene.paris@uniroma1.it (I.P.); laura.selan@uniroma1.it (L.S.)
2 Department for Innovation in Biological, Agro-Food and Forest Systems, University of Tuscia, 01100 Viterbo, Italy; imperlini@unitus.it
3 Research Unit of Diagnostical and Management Innovations, Children's Hospital and Institute Research Bambino Gesù, 00165 Rome, Italy; gianluca.vrenna@opbg.net
* Correspondence: marco.artini@uniroma1.it; Tel.: +39-0649694298
† These authors contributed equally to this work.

Abstract: The airways of cystic fibrosis (CF) patients are colonized by many pathogens and the most common is *Pseudomonas aeruginosa*, an environmental pathogen that is able to infect immunocompromised patients thanks to its ability to develop resistance to conventional antibiotics. Over 12% of all patients colonized by *P. aeruginosa* harbour multi-drug resistant species. During airway infection in CF, *P. aeruginosa* adopts various mechanisms to survive in a hostile ecological niche characterized by low oxygen concentration, nutrient limitation and high osmotic pressure. To this end, *P. aeruginosa* uses a variety of virulence factors including pigment production, biofilm formation, motility and the secretion of toxins and proteases. This study represents the first report that systematically analyzes the differences in virulence features, in normoxia and anoxia, of clinical *P. aeruginosa* isolated from CF patients, characterized by multi- or pan-drug antibiotic resistance compared to antibiotic sensitive strains. The virulence features, such as biofilm formation, protease secretion and motility, are highly diversified in anaerobiosis, which reflects the condition of chronic CF infection. These findings may contribute to the understanding of the real-world lifestyle of pathogens isolated during disease progression in each particular patient and to assist in the design of therapeutic protocols for personalized medicine.

Keywords: oxygen concentration; antimicrobial resistance; biofilm; pyocyanin; pyoverdine; motility; proteases

Citation: Papa, R.; Imperlini, E.; Trecca, M.; Paris, I.; Vrenna, G.; Artini, M.; Selan, L. Virulence of *Pseudomonas aeruginosa* in Cystic Fibrosis: Relationships between Normoxia and Anoxia Lifestyle. *Antibiotics* **2024**, *13*, 1. https://doi.org/10.3390/antibiotics13010001

Academic Editor: Theodoros Karampatakis

Received: 27 November 2023
Revised: 14 December 2023
Accepted: 17 December 2023
Published: 19 December 2023

1. Introduction

Cystic fibrosis (CF) is a life-threatening autosomal recessive multiorgan disorder, caused by mutations in the gene encoding the CF transmembrane conductance regulator (CFTR) that mediates chloride transport through the mucus-producing cells [1]. Functional failure of CFTR results in the production of thick and sticky mucus whose retention leads to serious bacterial infections [2]. The most affected organ is the lung whose recurrent chronic infections and local airway inflammation are the main cause of morbidity and mortality in CF patients, who have a life expectancy of 38 years [3,4].

The airways of CF patients are colonized by many pathogens, where the most common is the opportunistic Gram-negative bacterium *Pseudomonas aeruginosa*; in particular, it is the most predominant bacterium in children affected by CF and its prevalence increases with the age of these patients [5]. In general, *P. aeruginosa* is an environmental pathogen able to infect immunocompromised patients, thus accounting for 10% of nosocomial infections [6]. These infections are caused by *P. aeruginosa* developing resistance to conventional antibiotics and represent a major healthcare concern: notably the 12.3% of all patients colonized by *P. aeruginosa* who harbour multi-drug resistant (MDR) species (Cystic Fibrosis Foundation Patient Registry, Annual Data Report 2021). *P. aeruginosa* is highly pathogenic compared to other Gram-negative bacteria, due to the production of several virulence factors [6,7]. These factors, combined with its adaptability, enable *P. aeruginosa* to invade eukaryotic cells and, hence, to colonize CF airways over other pathogens [8]. Therefore, it is not surprising that *P. aeruginosa* survives in the hostile CF lung environment. Although such environments are characterized by viscous and stagnant mucus, providing niches with low levels of oxygen and fluctuating pH, and by the presence of other competitive bacteria and antibiotic molecules, *P. aeruginosa* is able to evade the host immune system [7,8]. To this end, *P. aeruginosa* uses a variety of virulence factors including biofilm formation, motility and the secretion of toxins and proteases [7,9]. These virulence factors allow *P. aeruginosa* persistence in hostile niches of the host where bacterial communities form a mature biofilm surrounded by a robust protective matrix: *P. aeruginosa* growing as biofilm is difficult to eradicate [10]. Many lines of experimental evidence suggest that *P. aeruginosa* growing as biofilm in the lung of CF patients, is subjected to hypoxic or even anaerobic conditions as the chronic infection progresses [11–14]. This suggests that oxygen concentration may play a key role in the biofilm formation, and in the stability of MDR-*P. aeruginosa* species and their persistence in CF airway mucus.

By contrast, pyocyanin is an organic blue-green pigment produced by *P. aeruginosa* in an oxygen-dependent biosynthetic pathway [15]. It is characterized by a redox-activity-generating reactive oxygen species (ROS), in particular a superoxide, and by inducing cellular oxidative stress. Furthermore, it is able to alter the host immune response with different mechanisms and promote the evasion of the immune response, establishing chronicity of infections [16].

In the present study, we investigated the modulation of *P. aeruginosa* virulence factors in clinical strains isolated from CF patients characterized by multi- (MDR) and pan-drug (PDR) antibiotic resistance profiles compared to antibiotic-sensitive strains (WT), all grown in planktonic and biofilm form under normoxic and anoxic conditions. These findings may contribute to the further understanding of the complex landscape of *P. aeruginosa* phenotype modulation, with particular attention to virulence factors in clinical isolates from CF patients, in relation to their antibiotic resistance profile and to anoxia.

2. Results

2.1. Pyocyanin Production in P. aeruginosa during CF Infection

Pyocyanin production was analyzed at 24 h and 48 h in normoxia (Figure S1, Supplementary Materials), since its biosynthesis is strongly associated with the presence of oxygen. As expected, its production was strongly strain-dependent but generally it is higher at 24 h rather than 48 h. Moreover, the higher pyocyanin producers were WT (wild type, sensitive strains) and MDR strains (except for PDR4), as reported in Figure 1. The linear decrease in pyocyanin production in the three groups is particularly pronounced at 48 h (Figure 1). The evidenced differences between pyocyanin production for WT and MDR strains at 24 and 48 h were statistically significative (p value < 0.01). Differences in PDR strains were not significative.

2.2. Pyoverdine Production in P. aeruginosa during CF Infection

In conditions of poor oxygenation, a lower production of pyoverdine was previously observed [17]. Based on these considerations, the pyoverdine production was assessed only in aerobic conditions at 24, 48 and 72 h. As reported in Figure S2 in Supplementary Materials, pyoverdine production was observed mainly after 48 h and 72 h of growth (exceptions are represented by some WT and MDR strains). PDR strains appeared to be weak pyoverdine producers both at 48 h and 72 h. At 72 h there was a notable increase in production by the MDR class, which brings the observed values considerably outside the scale in which the values fall for the remaining part of the bacterial collection. As can be seen in Figure 2, pyoverdine is produced at a very low level from PDR strains, suggesting that it is not a key factor for them. Statistical analyses performed to compare pyoverdine production at different times within the same bacterial class revealed that the data obtained were significative (p value < 0.0001 for all conditions, except for MDRs at 24 h compared with 72 h with p value < 0.05).

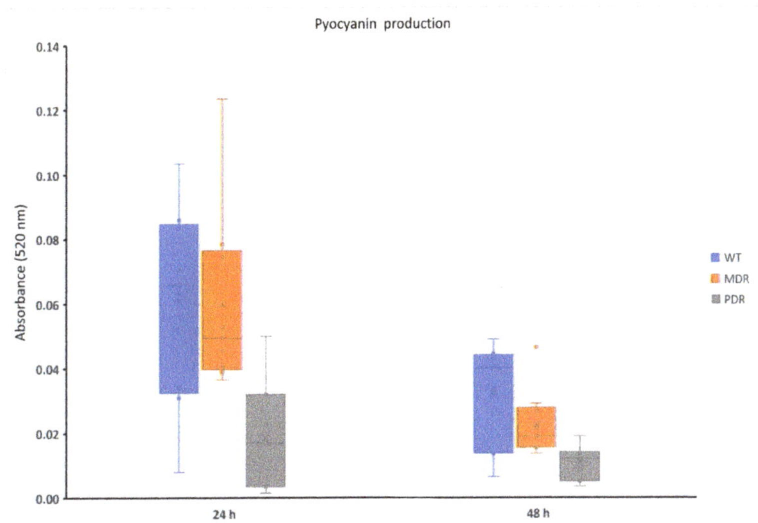

Figure 1. Pyocyanin production of WT, MDR and PDR strains at 24 h and 48 h in normoxia. WT: wild type, sensitive strains; MDR: multi-drug resistant strains; PDR: pan-drug resistant strains.

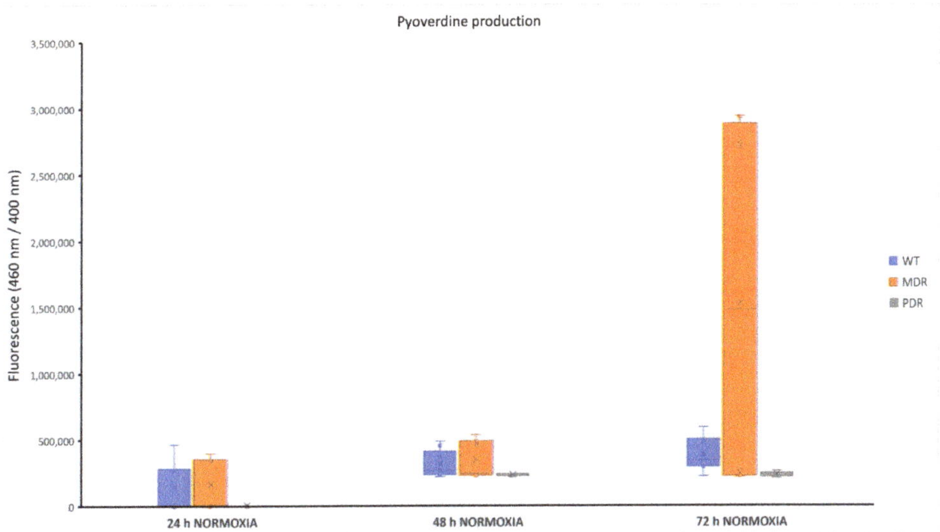

Figure 2. Pyoverdine production of WT, MDR and PDR strains at 24 h, 48 h and 72 h. WT: wild type, sensitive strains; MDR: multi-drug resistant strains; PDR: pan-drug resistant strains.

2.3. *Biofilm Formation in P. aeruginosa during CF Infection*

Biofilm formation of bacterial strains derived from CF patients characterized by different antimicrobial profiles was assessed in anoxic and normoxic conditions after 18 h of incubation at 37 °C. All bacterial strains were able to produce biofilm with different capabilities (Figure S3, Supplementary Materials). Biofilm content is independent from the antimicrobial profile of bacterial strains, while significative differences were observed for two tested conditions (anoxia and normoxia). Figure 3 reports the trend of the biofilm amount produced by WT, MDR and PDR. WT-grouped strains were isolated in the early stages of infections and sensitive to antibiotics, while the MDRs and PDRs included isolates in later stages of disease. In normoxic conditions, no large variations were apparent between the three groups, while substantial differences can be observed in anaerobic conditions, where WTs produced a small quantity of biofilm compared with PDRs, and especially compared with MDRs. Differences in biofilm content for each bacterial class in normoxia and anoxia were statistically significative only for WTs (p value < 0.05). These data could be indicative for the ability of *P. aeruginosa* clinical strains to adapt to the environment during the progression of infection. In fact, MDR and PDR strains seemed to better adapt to the hostile niches of a fibrotic lung, characterized by reduced oxygen diffusion.

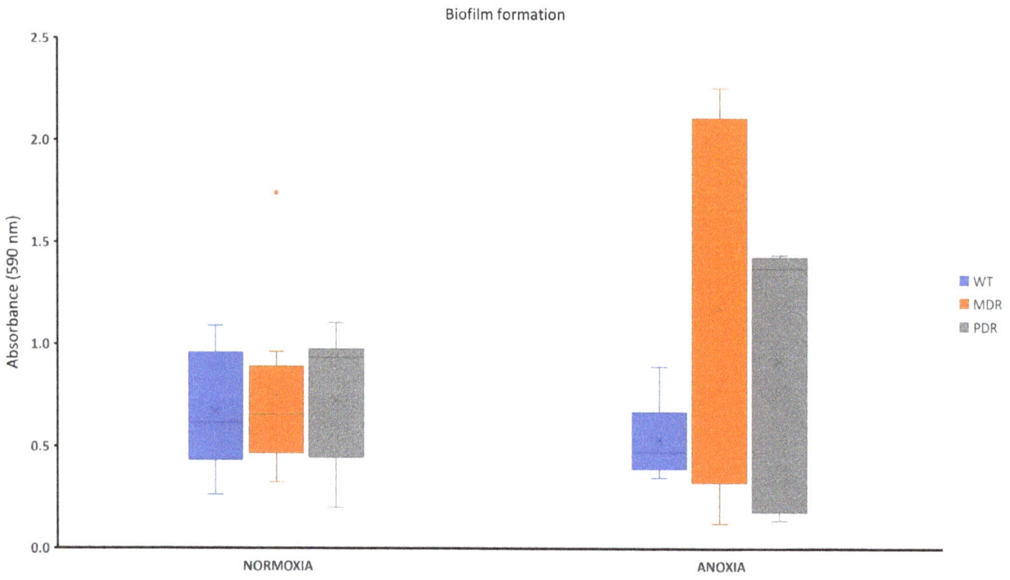

Figure 3. Biofilm formation of WT, MDR and PDR strains in normoxia and anoxia. WT: wild type, sensitive strains; MDR: multi-drug resistant strains; PDR: pan-drug resistant strains.

2.4. *Proteolytic Content in P. aeruginosa during CF Infection*

The extracellular proteases secreted by *P. aeruginosa* clinical strains were quantitatively and qualitatively evaluated. Quantitative analysis was performed by azocasein assay. Using azocasein as a substrate it was possible to spectrophotometrically measure the absorbance of the azocompound released by the proteolytic action of the proteases. The concentration of this azocompound is proportional to the concentration of the proteases. As reported in Figure S4 in Supplementary Materials, the analysis was performed by analyzing the proteolytic content released into the supernatant after 24 h and 48 h of incubation at 37 °C, in the presence or absence of oxygen. The obtained data from the analysis of bacterial cultures in the presence of oxygen (normoxia) showed a comparable proteolytic activity either at 24 h and 48 h of incubation at 37 °C, for almost all the analyzed clinical strains, with the exception of the strains WT8, PDR6 and PDR8A. In normoxia, we found a content of proteases in the supernatant of WT strains comparable to that of MDR strains, while PDR strains proved to be weak producers of proteases. These data confirm the lower expression of virulence factors in clinical strains with higher antimicrobial resistance. In anoxic conditions, by contrast, the proteolytic content of WT strains showed a notable increase after 24 h of growth. No differences were evidenced for MDR and PDR strains (Figure 4). The differences between proteolytic content in WTs in normoxia and Figure 4 anoxia were statistically significative only at 24 h (p value < 0.05). Within the MRD group, the differences evidenced in normoxia and anoxia were statistically significative both at 24 and 48 h (p value < 0.0001 at 24 h and p value < 0.05 at 48 h, respectively). Conversely, for the PDR group the obtained data were not significative. Overall, these data suggest a higher virulence in anoxic conditions in the first 24 h of infection.

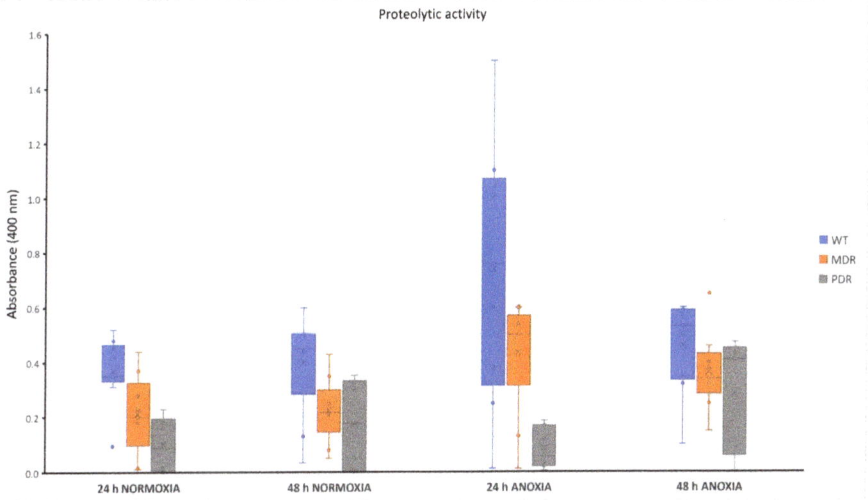

Figure 4. Protease production of WT, MDR and PDR strains at 24 h and 48 h in normoxia and anoxia. WT: wild type, sensitive strains; MDR: multi-drug resistant strains; PDR: pan-drug resistant strains.

The extracellular proteases secreted by clinical *P. aeruginosa* strains were qualitatively analyzed by using gelatin-zymography. This assay was performed on the culture supernatants of three selected *P. aeruginosa* clinical strains for each bacterial group (high, low and medium protease production for each group): WT2, WT4 and WT5 for WT strains; MDR1, MDR5 and MDR7 for MDR strains; PDR2, PDR5 and PDR7 for PDR strains. Therefore, zymography analysis was performed on the culture supernatants of all these strains at 24 h and 48 h in normoxia or anoxia, and after protein separation on SDS-PAGE gels containing a synthetic substrate (gelatin). After run, SDS-PAGE gels were incubated in a proper activation buffer and stained, thus allowing the detection of the different clear bands corresponding to the proteases/gelatinases active in the bacterial cultures of all the tested clinical strains (Figure 5). Regardless of growth conditions, a gelatin-degrading proteolytic band of approximately 46 kDa was clearly detectable in all clinical strains (Figure 5). However, this gelatinolytic band was considerable in all WT tested strains, in normoxia at 24 h and even more so at 48 h, as well as in anoxia at both time points. The same 46 kDa proteolytic band was also observed in the MDR1, MDR5 and PDR5 strains (Figure 5). Interestingly, the same gelatinolytic band at 46 kDa was just visible in the PRD2 strain in normoxia at 24 h and 48 h, but more visible and active in anoxia (Figure 5). By contrast, 46 kDa gelatinase was more active in MDR7 in normoxia than in anoxia at both time points; whereas in PDR7 it was only active at 48 h both in normoxia and anoxia. At

24 h, this PRD strain, in fact, showed slightly active or inactive protease in normoxia and anoxia, respectively (Figure 5).

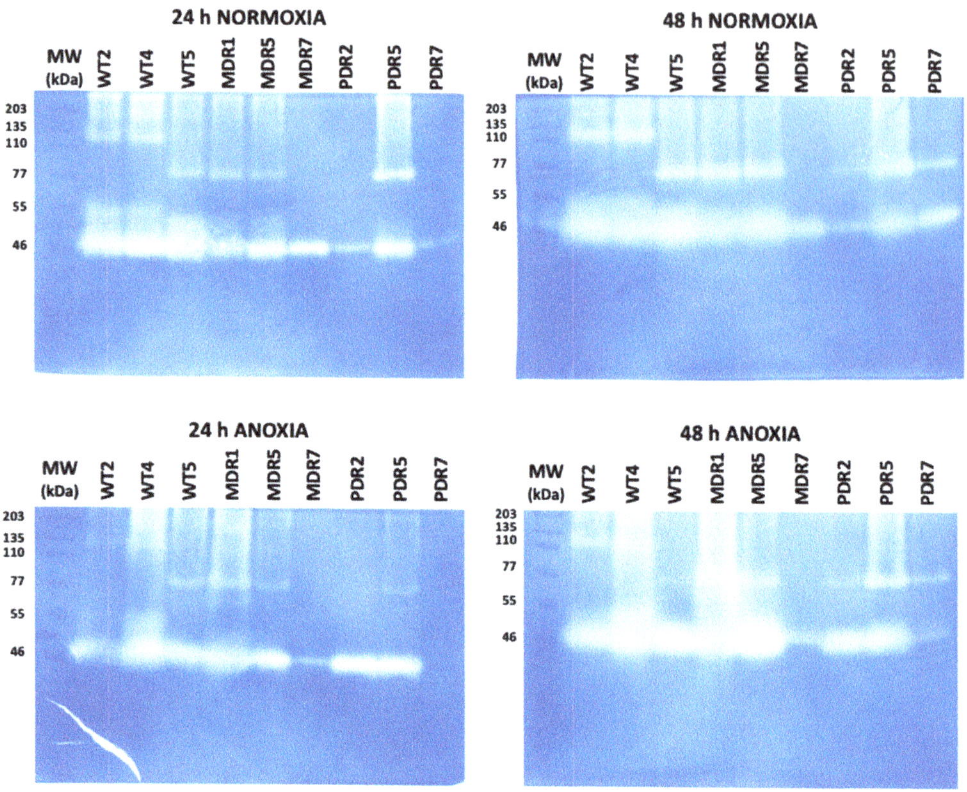

Figure 5. Gelatin-zymography to assay proteases secreted by *P. aeruginosa* clinical strains at 24 h and 48 h in normoxia and anoxia. Protein samples from unconcentrated culture supernatants were separated by using 10% SDS-PAGE gel with 0.2% gelatin. All gels were incubated in a development buffer and destained with 0.5% Coomassie blue R-250, thus obtaining clear bands corresponding to active proteases on a blue background. MW, molecular weight in kDa of protein markers. WT: wild type, sensitive strains; MDR: multi-drug resistant strains; PDR: pan-drug resistant strains.

On the other hand, the gelatinolytic activities of clinical strains differed for molecular weight > 60 kDa, and it is evident that there is no similarity in their electrophoretic patterns due to the presence or absence of 110 kDa and 77 kDa bands (Figure 5). In conclusion, these data indicate that clinical isolates differentially secrete proteases dependent on their growth phase and condition.

2.5. Motility in *P. aeruginosa* during CF Infection

In the present study we analyzed swimming and swarming motility at 24 h, 48 h and 72 h, in both normoxia and anoxia.

Swimming motility is mediated by a polar flagellum that allows *P. aeruginosa* to move in a liquid medium. In normoxia, all strains except WT8, MDR7, MDR8 and PDR3 showed a swimming motility with different capabilities (Figure 6, left panels). In anoxic conditions, a reduction of swimming motility was apparent for almost all bacterial strains regardless their antimicrobial profiles (Figure 6, right panels).

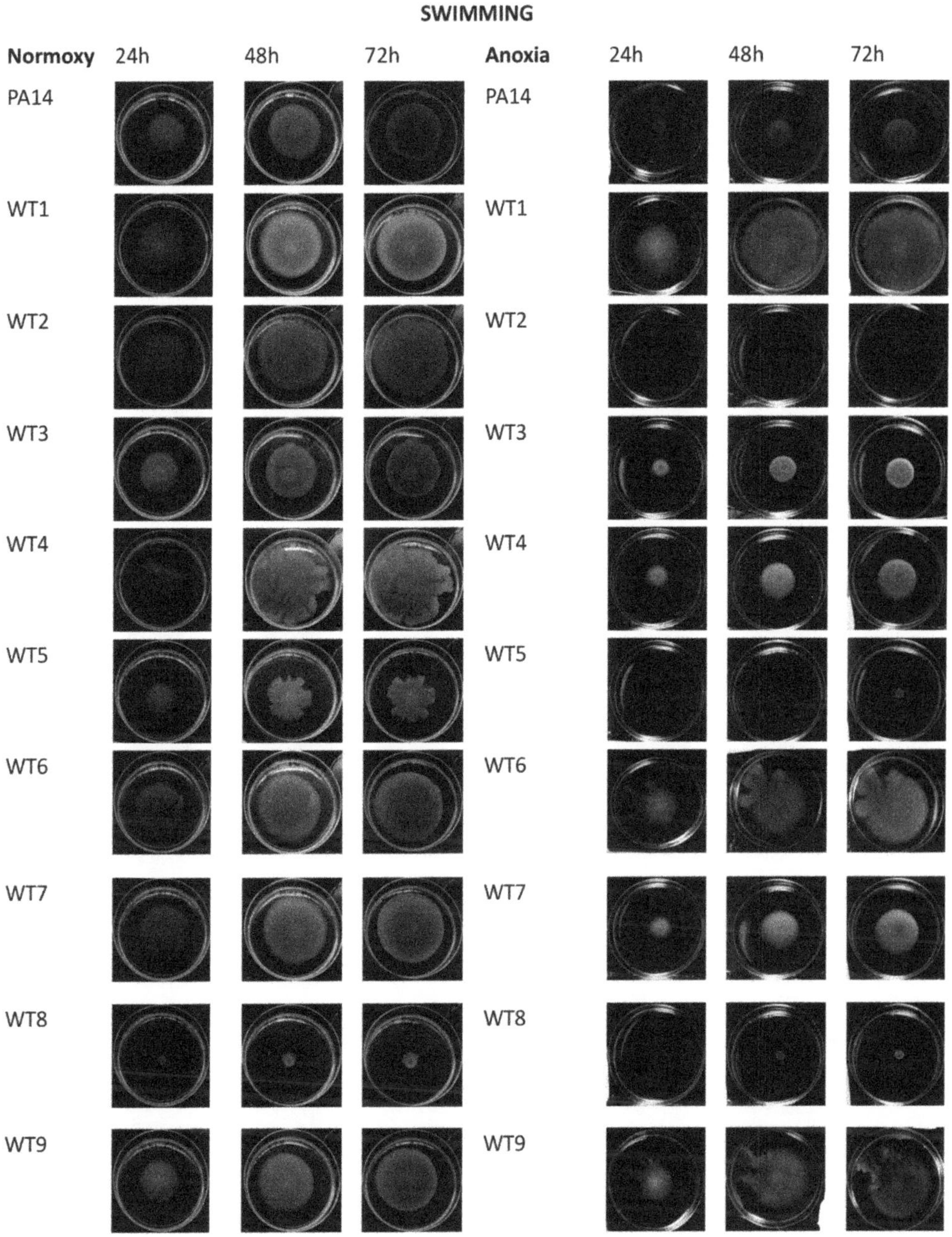

Figure 6. *Cont.*

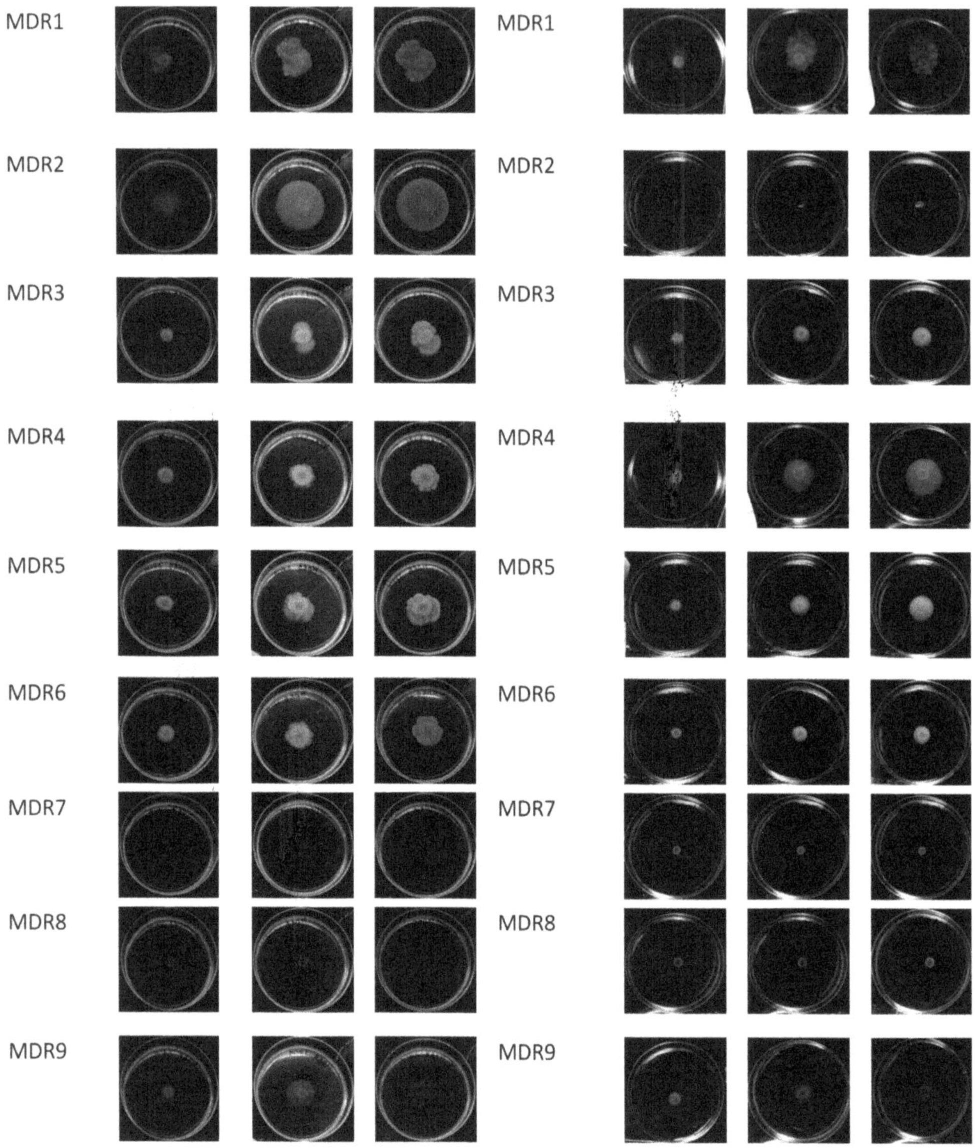

Figure 6. *Cont.*

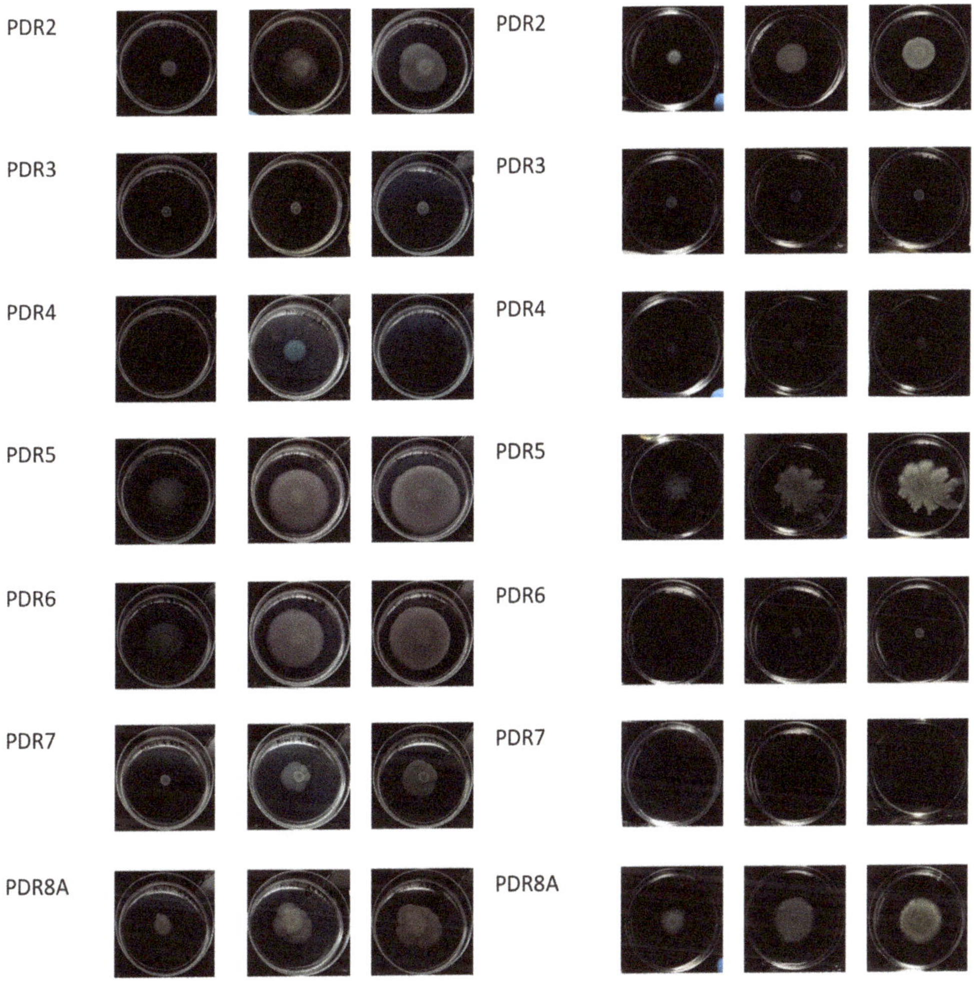

Figure 6. Motility assay: swimming assay of *P. aeruginosa* bacterial strains in normoxia (left panel) and anoxia (right panel) measured at 24 h, 48 h and 72 h of bacterial growth. WT: wild type, sensitive strains; MDR: multi-drug resistant strains; PDR: pan-drug resistant strains.

By optical comparison of the plates grown in the presence or absence of oxygen, phenotypical variations in the colonies were observed. In particular, as expected, in anoxia a reduction of pigments was observed (particularly in strains PA14, WT2, WT3, WT6, WT7, MDR7, MDR8, MDR9, PDR4, PDR5, PDR6).

Swarming motility is defined by the rapid and coordinated translocation of a bacterial population across a semisolid surface. All WT strains, except WT8, showed a swarming motility in the presence of oxygen (Figure 7, left panel). Conversely MDR strains showed a reduced swarming motility except MDR1, MDR2 and MDR9 strains; in PRD strains only PDR5, PDR6 and partially PDR8A showed swarming activity (Figure 7, left panel). This kind of motility was completely inhibited in anoxic conditions, where only WT3 showed a very slight swarming motility. Diameters of swimming and swarming halos were reported in Tables S1 and S2 in the Supplementary Materials section.

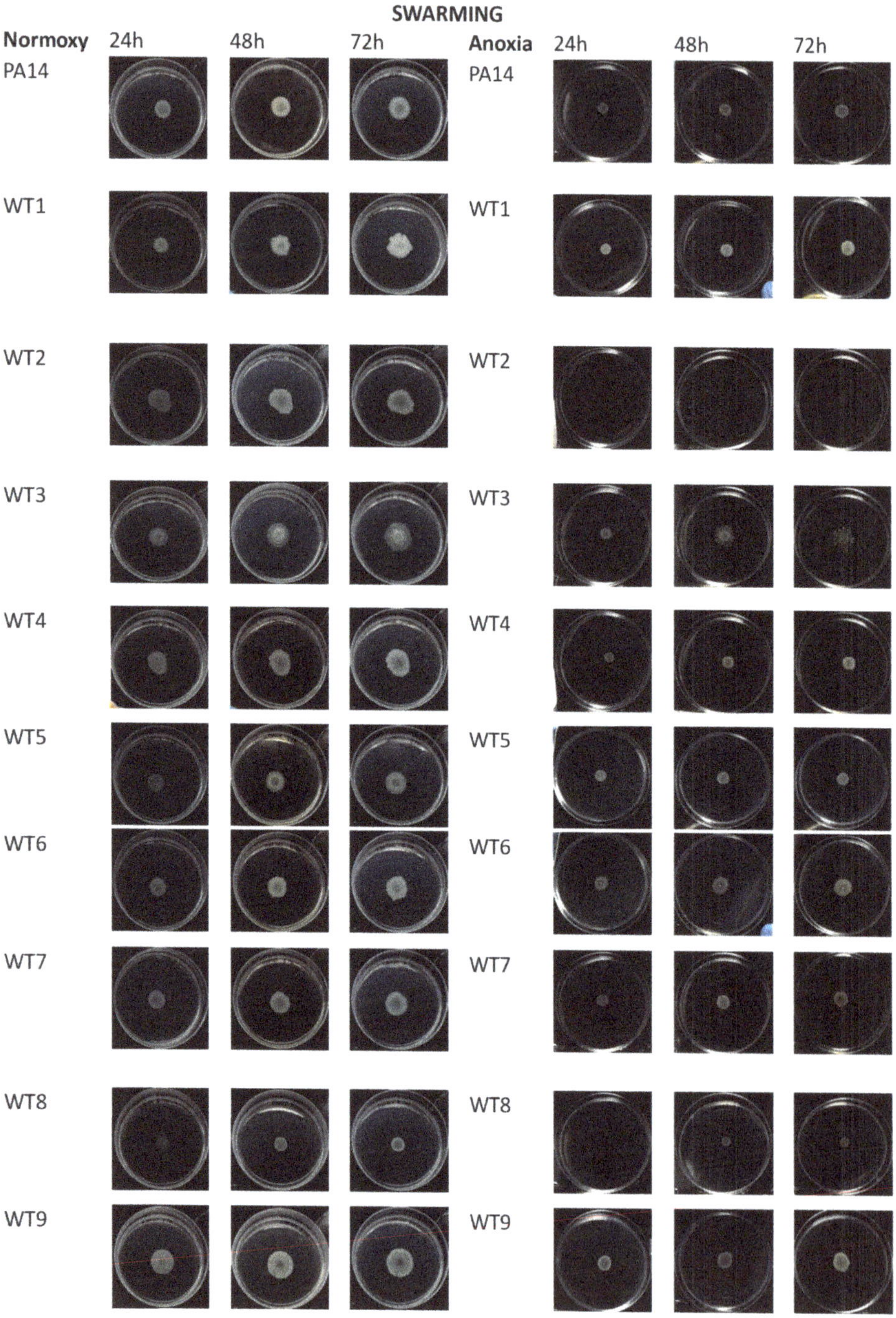

Figure 7. *Cont.*

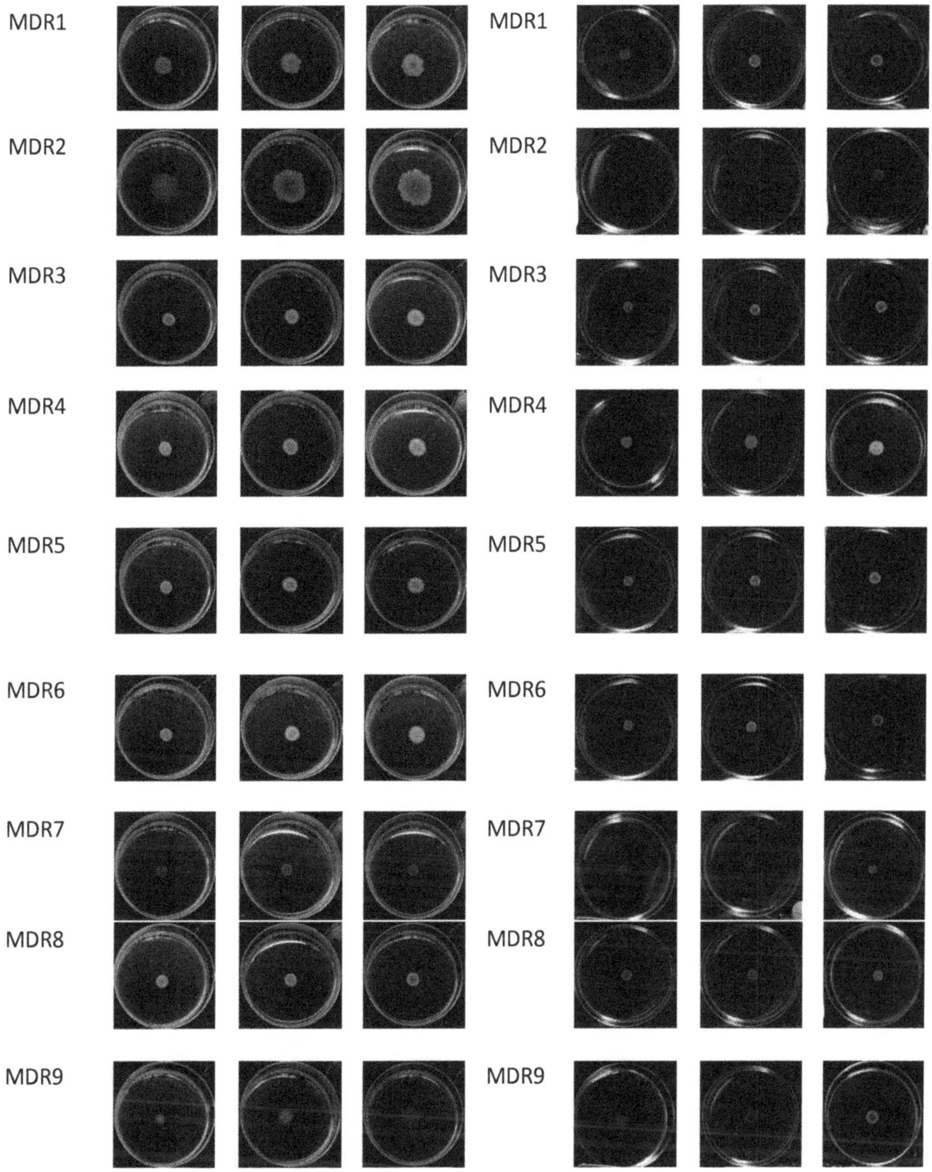

Figure 7. *Cont.*

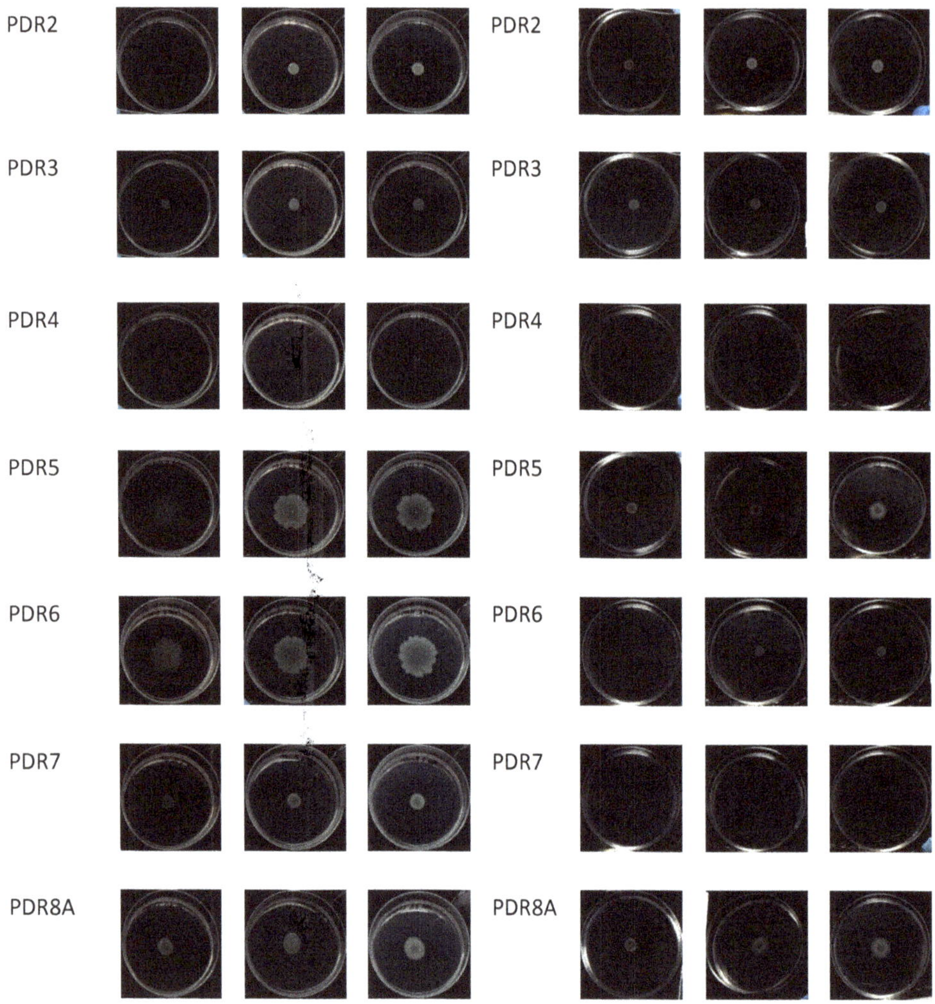

Figure 7. Swarming motility of *P. aeruginosa* bacterial strains in normoxia (left panel) and anoxia (right panel) measured at 24 h, 48 h and 72 h of bacterial growth. WT: wild type, sensitive strains; MDR: multi-drug resistant strains; PDR: pan-drug resistant strains.

3. Discussion

During chronic infection of CF airways, *P. aeruginosa* adapts its phenotype within the specific ecological niches of the altered lung parenchyma. Thanks to its high plasticity, it displays several mechanisms that enable it to persist in an ecological niche characterized by low oxygen concentration, nutrient limitation, high osmotic pressure and oxidative stress, and in competition with other microorganisms.

These modifications can occur early or late during the progression of the colonization [18]. For example, flagella and pili allow bacterial cells to reach and establish themselves in the pulmonary niche in the early phase of infection, but their expression is afterwards downregulated in order to activate a more complex machinery able to guarantee its survival in the inhospitable environment of CF lung [19].

Changes in other virulence factors such as an increase in antimicrobial resistance and/or modifications in the biofilm lifestyle, protease secretion, pyocyanin and pyoverdine production, are also involved in the transition to chronic pathophenotype. This mechanism of adaptation is also reflected on surface protein pattern, where substantial differences appeared in chronic antibiotic resistance of *P. aeruginosa* isolates compared to early antibiotic sensitive *P. aeruginosa* strains [20].

CF progression leads also to a considerably reduced oxygen concentration in the lungs; as a consequence, some bacteria switch to anaerobic metabolism while infection shifts towards chronicity, especially during late-stage disease [12]. Furthermore, a hypoxic microenvironment increases bacterial multidrug resistance by elevating the expression of multidrug efflux pumps, as recently reported [21]. In particular, antibiotic efflux pumps are modulated by two-component systems, such as the RstA/RstB system. Regulator RstA positively regulates the enzymes involved in the anaerobic nitrate respiratory chain. On the other hand, overexpression of efflux pumps leads to enhanced consumption of oxygen, causing microenvironmental hypoxia, which in turn promotes anaerobic nitrate respiration in *P. aeruginosa* [21].

A previous study evaluated how the transition from a condition of normoxia to anoxia can influence the coexistence of *S. aureus* and *P. aeruginosa* in the lungs of CF patients, and concluded that this switch drives the two pathogens to colonize different regions of lung [22].

In this study, we systematically analyzed the differences in virulence features of clinical *P. aeruginosa* strains characterized by multi- and pan-drug antibiotic resistance profiles compared to antibiotic sensitive strains. These virulence features, such as biofilm, protease secretion and motility, are highly diversified in anaerobiosis, which reflects the condition of CF chronic infection.

Firstly, we characterized pigment production along the clinical bacterial strains. Pyocyanin is a secondary metabolite secreted by *P. aeruginosa* in aerobic conditions, which can undergo a redox cycle resulting in the generation of reactive oxygen species (ROS) harmful to the host cell [23]. For this reason, it was only assessed in the presence of oxygen. As expected, pyocyanin production decreased in PDR and MDR strains compared to WT ones at 24 h and mainly at 48 h of growth (Figure 2). This latter finding is strongly supported by previous reports in the literature, where pyocyanin synthesis appears to be fundamental in the early stages of infection until the establishment of chronic infections [24]. Once bacteria have adapted to the host, in fact, they reduce the expression of virulence factors to escape the attention of the immune system.

No significative differences were observed in pyoverdine production among WT, MDR and PDR groups, except for four MDR strains whose production at 72 h of growth was at least one order of magnitude higher than that observed in the other analyzed strains (Figure 3). Pyoverdine is a fluorescent siderophore produced by *Pseudomonas* species, with the function of intracellular iron acquisition [25]. Furthermore, it induces an iron accumulation that is toxic for eukaryotic cells. In fact, the elevated reactivity of the soluble Fe^{2+} ion for hydrogen peroxide and for oxygen, in general, leads to the production of highly reactive oxygen species (ROS), which culminates in the damage of essential cellular components causing oxidative stress and cellular death [26]. Pyoverdine is usually implicated in acute illness, but also in the production of mature biofilms, so its role in the progression of the disease is still unclear [27].

Regarding biofilm formation, WT, MDR and PDR strains did not show significative differences when they were grown in the presence of oxygen; conversely in anoxia, while WT strains reduced their ability to form biofilm, MDR and PDR strains showed a significative increase in the amount of biofilm (Figure 3). This observation could be explained with the adaptation of multi-resistant strains to persist and colonize the CF lung in the absence of oxygen, due to the thick mucus, which occurs after the onset of bacterial infection. Conversely, this mechanism of adaptation seems to be adopted less in WT strains isolated from early infections, possibly because bacterial cells still colonize microenvironments where anoxia is not extended.

Protease expression and activity are known to depend on bacterial growth phases and/or conditions, thus contributing to specific environmental adaptations of bacteria that lead to differences in antimicrobial profiles [28,29]. As expected, the secretion of proteases was higher in the WT group derived from early infection. In this condition, bacteria still express virulence factors to counteract the host defenses. This is mainly evidenced in anoxia, a typical condition of a chronic infection where bacteria are usually multi-resistant rather than antibiotic-sensitive strains. However, protease quantitative data, although unexpected, were in line with those obtained from gelatin-zymography. The main proteolytic band at 46 KDa, visible in all tested clinical strains, was more active in the WT group in normoxia, as well as in anoxia. In addition to the similarities in the proteolytic patterns among the three groups of bacteria, the qualitative analysis of clinical isolates also revealed differences within the same group or depending on the growth phase and condition (Figure 5).

With regard to bacterial motility, however, it is well known that bacteria lose the ability to express the flagellum when infection progresses from acute to chronic form. This trend was mainly highlighted when bacteria were grown in anoxia (especially for the MDR and PDR strains). However, there are exceptions, represented for example, by the PDR5 and PDR8A strains, that are still able to move. Swarming motility, on the other hand, was completely absent in anoxia for almost all bacterial strains.

In conclusion, our data suggest that it is fundamental to analyze the virulence factors produced by clinical *P. aeruginosa* during CF progression rather than in conventional laboratory conditions characterized by an optimal oxygen tension. By contrast, we think it is mandatory to analyze them in a condition as close as possible to the real condition of a CF lung: a highly hostile environment characterized by largely anoxic regions, where bacteria adapt their virulence arsenal to persist and escape the immune system response.

4. Materials and Methods

4.1. Ethics Approval and Informed Consent

This study was approved by the Ethical Committee of Pediatric Hospital and Institute of Research Bambino Gesù (OPBG) in Rome, Italy (No. 1437_OPBG_2017 of July 2017). The study was conducted in respect of the Declaration of Helsinki as statement of ethical principles for medical research involving human subjects. All participants, or the legal guardians of those included in the study, signed an informed consent form.

4.2. Bacterial Strains and Growth Conditions

Clinical strains of *P. aeruginosa* were isolated from airways of CF patients in follow-up to OPBG. Bacterial strains were classified as follows: (i) nine antibiotic-sensitive strains conventionally classified as wild type (WT) were isolated from CF patients with recent infections (<12 months); (ii) nine multi-drug resistant non-mucoid strains (MDR) were isolated from CF patients with chronic colonization (4–15 years); (iii) seven pan-drug resistant non-mucoid strains (PDR) were isolated from CF patients with chronic colonization (4–15 years). These strains were previously classified for their antimicrobial profiles in Montemari et al. [20]. Reference strain *P. aeruginosa* PA14 was used. According to The European Committee on Antimicrobial Susceptibility Testing (http://www.eucast.org, accessed on 20 October 2023), WT strains are defined as sensitive to all antimicrobials, while MDR strains are resistant to at least one agent in three or more antimicrobial categories and PDR strains are resistant to all antibiotics in all classes. Antimicrobial profiles for all tested strains are summarized in Table 1. Bacteria were grown in Brain Heart Infusion broth (BHI, Oxoid, Basingstoke, UK). Planktonic condition was performed at 37 °C under orbital shaking (180 rpm), while biofilm formation was performed at 37 °C in static conditions. For anaerobiosis Ruskinn Concept 400 Workstation was used (LabTech, Heathfield, UK).

Table 1. Antimicrobial profiles of clinical and PA14 strains [20].

Strain	Fluoroquinilones		Penicillins	Monobactams	Cephalosporins				Aminoglycosides		Carbapenems	
	CIP 5 µg	LEV 5 µg	TZP 30–6 µg	ATM 30 µg	CAZ 10 µg	FEP 30 µg	CZA 10–4 µg	C/T 30–10 µg	AK 30 µg	TOB 10 µg	IM 10 µg	MRP 10 µg
PA14	S	S	S	S	S	S	-	-	S	S	S	S
WT1	I	R	I	I	I	I	-	-	S	S	I	S
WT2	I	I	I	I	I	I	-	-	S	S	I	S
WT3	I	I	I	-	I	I	S	S	S	S	I	S
WT4	I	I	I	-	I	I	S	S	S	S	I	S
WT5	I	I	S	-	I	I	-	-	S	I	-	S
WT6	I	I	I	-	I	I	S	S	S	S	I	S
WT7	I	I	I	I	I	I	-	-	S	S	I	S
WT8	I	I	I	-	I	I	S	S	S	S	I	S
WT9	I	I	I	-	I	I	S	S	S	S	I	S
MDR1	R	R	R	R	R	R	R	R	R	R	R	R
MDR2	I	R	R	R	R	R	R	R	R	R	R	R
MDR3	I	R	R	R	R	R	S	R	R	R	R	I
MDR4	I	R	R	R	R	R	S	S	R	R	R	I
MDR5	R	R	R	R	R	R	S	S	R	R	R	R
MDR6	R	R	S	S	S	S	-	-	R	R	R	R
MDR7	R	R	R	R	R	R	S	S	R	R	R	R
MD8	R	R	R	R	R	R	S	R	S	S	R	I
MDR9	R	R	I	R	S	R	S	S	R	R	I	R
PDR2	R	R	R	I	R	R	R	R	R	R	R	R
PDR3	R	R	R	R	R	R	-	-	R	R	R	R
PDR4	R	R	R	R	R	R	R	R	R	R	R	R
PDR5	R	R	R	R	R	R	R	R	R	R	R	R
PDR6	R	R	R	R	R	R	R	R	R	R	R	R
PDR7	R	R	R	R	R	R	R	R	R	R	R	R
PDR8A	R	R	R	R	R	R	R	R	R	R	R	R

Antimicrobial susceptibility was performed according to the guidelines of EUCAST Clinical Breakpoint Tables v. 13.0 (valid from 1 January 2023). CIP: ciprofloxacin; LEV: levofloxacin; TZP: Piperacillin-tazobactam; ATM: aztreonam; CAZ: ceftazidime; FEP: cefepime; CZA: Ceftazidime/avibactam; C/T: ceftolozane/tazobactam; AK: amikacin; TOB: tobramycin; IM: imipenem; MRP: meropenem; S: sensitive, R: resistance; I: intermediate, -: not tested.

4.3. Biofilm Formation

The biofilm quantification was assessed by microtiter plate (MTP) biofilm assay [30]. An overnight bacterial culture was 1:100 diluted into BHI fresh medium and aliquoted in the wells of a sterile 96-well polystyrene flat base plate. The plates were overnight incubated at 37 °C under static conditions in aerobic and anaerobic conditions. After incubation, the supernatant containing planktonic cells were gently removed and the plates were washed with double-distilled water. Then the microtiter plates were patted dry in an inverted position. The staining was performed with 0.1% crystal violet for 15 min at room temperature. The excess of crystal violet was removed by washing the wells with double-distilled water. The microtiter plates were thoroughly dried. The remaining biofilm was dissolved with 20% (v/v) glacial acetic acid and 80% (v/v) ethanol, and spectrophotometrically measured at 590 nm. Each experiment was performed in 6-replicates, and each data point was composed of four independent experiments.

4.4. Pyocyanin Assay

Pyocyanin production was determined as previously described [31]. Briefly, bacterial cells were inoculated in BHI broth and aerobically incubated for different times at 37 °C. The cell-free supernatant, recovered after centrifugation at 10,000 rpm for 15 min, was used for pyocyanin extraction. The supernatant was mixed in a 1:1 ratio with chloroform. The mixture was inverted and then decanted at room temperature up to the separation between the two phases. Pyocyanin (lower phase) was transferred into a new tube and the same volume of 0.2 M HCl was added. The mixture was inverted and decanted to allow the separation of the two phases. The upper pink layer, containing pyocyanin, was recovered and quantified at 520 nm. Pyocyanin content was normalized for the optical density of the corresponding bacterial culture.

4.5. Pyoverdine Assay

For pyoverdine quantification, bacteria were grown in King's B Medium supplemented with 0.5% (*w*/*v*) of Casamino Acid (CAA) at 37 °C in aerobic and anaerobic conditions [31]. Pyoverdine was quantified by reading 100 µL of each *P. aeruginosa* cell-free supernatant into a black 96-well plate (Greiner, Stonehouse, UK) at excitation and emission wavelengths of 400/460 nm, as previously reported [32], on an Infinite 200 PRO (Molecular Devices, San Jose, CA, USA) fluorescence microplate reader. The background level of fluorescence was measured using the same medium. Each measurement was normalized for the optical density detected in each bacterial culture.

4.6. Protease Assay

The extracellular proteolytic activity of *P. aeruginosa* was determined by azocasein assay [33]. Cell-free culture supernatant (150 µL) was added to a solution containing 500 µL of 0.3% *w*/*v* azocasein (Sigma, St. Louis, MO, USA) in 50 mM Tris–HCl, 0.5 mM $CaCl_2$ pH 7.5 and incubated at 37 °C for 30 min. To stop the reaction, 650 µL of 10% ice-cold trichloroacetic acid was added. The obtained mixture was incubated at 4 °C for 10 min. Then, the insoluble azocasein was removed by centrifugation at 10,000 rpm for 10 min and the supernatant was measured at OD 400 nm.

4.7. Zymography Assay

Assay was performed on culture supernatants of *P. aeruginosa* clinical strains. Unconcentrated culture supernatants (20 µL) were combined with Laemmli sample buffer without reducing agent or boiling the protein samples [33]; these were then separated on a 10% SDS-PAGE gel containing 0.2% gelatin (Sigma–Aldrich, Milan, Italy) with a 4% polyacrylamide in stacking gel. Molecular masses of protease bands were estimated by using pre-stained molecular mass markers (mPAGE Color Protein Standard, Millipore, Milan, Italy). After electrophoretic run, gels were incubated, at room temperature, with 2.5% Triton X-100 for 1 h, thus removing SDS and renaturing the proteins. After two washes in distilled water, gels were incubated overnight at 37 °C in a development buffer, as previously described [33]. Gels were stained for 45 min with 0.5% Coomassie blue R-250 in glacial acetic/methanol/distilled water (1:3:6), and destained in distilled water. The zymogram images, containing clear bands on a blue background, were acquired by using ChemiDoc XRS+ System (Biorad, Segrate, Italy). Zymogram experiments were repeated at least twice.

4.8. Motility Assays

4.8.1. Swarming Assay

The swarming assay was performed as previously published by Yang and coworkers [34], with some modifications [31]. Plates after the inoculation, were incubated at 37 °C for 24 h, 48 h and 72 h in normoxia and anoxia. After the incubation period, plates were photographed, and halos were measured. Swarming assays were repeated three times.

4.8.2. Swimming Assay

The swimming assay was conducted in accordance with previous research [34], with some modifications [31]. Plates after the inoculation, were incubated at 37 °C for 24 h, 48 h and 72 h in normoxia and anoxia. After the incubation period, plates were photographed, and halos were measured. Swarming assays were repeated three times.

5. Conclusions

The novelty of this work is the systematic analysis of the virulence factors displayed by clinical strains in growth conditions that reflect the complexity of the hostile lung environment in which they grow and cause tissue damage. Our data show the variability of virulence factors expressed by *P. aeruginosa* during CF progression. They seem to be unpredictable so far, probably because of the number of interacting variables that modulate bacterial phenotype in such extreme microenvironments. Further studies would be useful to understand the real-world lifestyle of the pathogens isolated during disease progression in each individual patient. The identification of the most significant bacterial biomarkers, and their analysis/integration (possibly also with biomarkers of the patient), as the input for Artificial Intelligence algorithms could allow the design of therapeutic protocols of personalized medicine. This kind of approach is increasingly demanded by clinicians, mainly for chronic and multifactorial diseases, such as tumors and chronic infections, where therapy failure is a frustrating experience.

Supplementary Materials: The following supporting information can be downloaded at: https://www.mdpi.com/article/10.3390/antibiotics13010001/s1, Figure S1: Pyocyanin produc-on of *P. aeruginosa* clinical and reference strains; Figure S2: Pyoverdine produc-on of WT, MDR and PDR strains at 24 h, 48 h and 72 h in normoxia; Figure S3. Biofilm forma-on of *P. aeruginosa* clinical and reference strains; Figure S4: Protease produc-on of different clinical strains and PA14 reference strain at 24 h and 48 h in normoxia and anoxia. Table S1: Diameters of swimming moGlity of WT, MDR and PRD strains art 24, 48 and 72 h in normoxia and anoxia, respecGvely. Table S2: Diameters of swarming moGlity of WT, MDR and PRD strains art 24, 48 and 72 h in normoxia and anoxia, respecGvely

Author Contributions: Conceptualization, R.P. and G.V.; methodology, R.P., M.T., I.P., G.V. and E.I.; software, R.P., M.T., I.P., G.V. and E.I.; validation, R.P., M.T., I.P., G.V., E.I., M.A. and L.S.; formal analysis, R.P., M.T., I.P., G.V. and E.I.; investigation, R.P., M.T., I.P., G.V. and E.I.; resources, R.P., G.V., M.A. and L.S.; data curation, R.P., M.T., I.P., G.V. and E.I.; writing—original draft preparation, R.P. and E.I.; writing—review and editing, R.P., M.T., I.P., G.V., E.I., M.A. and L.S.; visualization, R.P., M.T., I.P., G.V., E.I., M.A. and L.S.; supervision, R.P., M.T., I.P., G.V., E.I., M.A. and L.S.; project administration, R.P., M.A. and L.S.; funding acquisition, R.P., M.A. and L.S. All authors have read and agreed to the published version of the manuscript.

Funding: This research received no external funding.

Institutional Review Board Statement: Not applicable.

Informed Consent Statement: Not applicable.

Data Availability Statement: Data may be obtained from the appropriate author depending on which data is requested.

Conflicts of Interest: The authors declare no conflict of interest.

References

1. Elborn, J.S. Cystic fibrosis. *Lancet* **2016**, *388*, 2519–2531. [CrossRef] [PubMed]
2. Castellani, C.; Assael, B.M. Cystic fibrosis: A clinical view. *Cell Mol. Life Sci.* **2017**, *74*, 129–140. [CrossRef] [PubMed]
3. Cabrini, G. Innovative Therapies for Cystic Fibrosis: The Road from Treatment to Cure. *Mol. Diagn. Ther.* **2019**, *23*, 263–279. [CrossRef] [PubMed]
4. Imperlini, E.; Papa, R. Clinical Advances in Cystic Fibrosis. *J. Clin. Med.* **2022**, *11*, 6306. [CrossRef] [PubMed]
5. Schick, A.; Kassen, R. Rapid diversification of *Pseudomonas aeruginosa* in cystic fibrosis lung-like conditions. *Proc. Natl. Acad. Sci. USA* **2018**, *115*, 10714–10719. [CrossRef]
6. Qin, S.; Xiao, W.; Zhou, C.; Pu, Q.; Deng, X.; Lan, L.; Liang, H.; Song, X.; Wu, M. *Pseudomonas aeruginosa*: Pathogenesis, virulence factors, antibiotic resistance, interaction with host, technology advances and emerging therapeutics. *Signal Transduct. Target. Ther.* **2022**, *7*, 199. [CrossRef] [PubMed]
7. Jurado-Martín, I.; Sainz-Mejías, M.; McClean, S. *Pseudomonas aeruginosa*: An Audacious Pathogen with an Adaptable Arsenal of Virulence Factors. *Int. J. Mol. Sci.* **2021**, *22*, 3128. [CrossRef] [PubMed]
8. Riquelme, S.A.; Ahn, D.; Prince, A. *Pseudomonas aeruginosa* and *Klebsiella pneumoniae* Adaptation to Innate Immune Clearance Mechanisms in the Lung. *J. Innate Immun.* **2018**, *10*, 442–454. [CrossRef]
9. Bajire, S.K.; Jain, S.; Johnson, R.P.; Shastry, R.P. 6-Methylcoumarin attenuates quorum sensing and biofilm formation in *Pseudomonas aeruginosa* PAO1 and its applications on solid surface coatings with polyurethane. *Appl. Microbiol. Biotechnol.* **2021**, *105*, 8647–8661. [CrossRef]
10. Yaeger, L.N.; Coles, V.E.; Chan, D.C.K.; Burrows, L.L. How to kill *Pseudomonas*-emerging therapies for a challenging pathogen. *Ann. N. Y. Acad. Sci.* **2021**, *1496*, 59–81. [CrossRef]

11. Yoon, S.S.; Hennigan, R.F.; Hilliard, G.M.; Ochsner, U.A.; Parvatiyar, K.; Kamani, M.C.; Allen, H.L.; DeKievit, T.R.; Gardner, P.R.; Schwab, U.; et al. *Pseudomonas aeruginosa* anaerobic respiration in biofilms: Relationships to cystic fibrosis pathogenesis. *Dev. Cell.* **2002**, *3*, 593–603. [CrossRef] [PubMed]
12. Hassett, D.J.; Sutton, M.D.; Schurr, M.J.; Herr, A.B.; Caldwell, C.C.; Matu, J.O. *Pseudomonas aeruginosa* hypoxic or anaerobic biofilm infections within cystic fibrosis airways. *Trends Microbiol.* **2009**, *17*, 130–138. [CrossRef] [PubMed]
13. Hall-Stoodley, L.; McCoy, K.S. Biofilm aggregates and the host airway-microbial interface. *Front. Cell Infect. Microbiol.* **2022**, *12*, 969326. [CrossRef] [PubMed]
14. Martin, L.W.; Gray, A.R.; Brockway, B.; Lamont, I.L. *Pseudomonas aeruginosa* is oxygen-deprived during infection in cystic fibrosis lungs, reducing the effectiveness of antibiotics. *FEMS Microbiol. Lett.* **2023**, *370*, fnad076. [CrossRef] [PubMed]
15. Rada, B.; Leto, T.L. Pyocyanin effects on respiratory epithelium: Relevance in *Pseudomonas aeruginosa* airway infections. *Trends Microbiol.* **2013**, *21*, 73–81. [CrossRef] [PubMed]
16. Jayaseelan, S.; Ramaswamy, D.; Dharmaraj, S. Pyocyanin: Production, applications, challenges and new insights. *World J. Microbiol. Biotechnol.* **2014**, *30*, 1159–1168. [CrossRef] [PubMed]
17. Visca, P.; Imperi, F.; Lamont, I.L. Pyoverdine siderophores: From biogenesis to biosignificance. *Trends Microbiol.* **2007**, *15*, 22–30. [CrossRef]
18. Winstanley, C.; O'Brien, S.; Brockhurst, M.A. *Pseudomonas aeruginosa* evolutionary adaptation and diversification in cystic fibrosis chronic lung infections. *Trends Microbiol.* **2016**, *24*, 327–337. [CrossRef]
19. La Rosa, R.; Rossi, E.; Feist, A.M.; Johansen, H.K.; Molin, S. Compensatory evolution of *Pseudomonas aeruginosa*'s slow growth phenotype suggests mechanisms of adaptation in cystic fibrosis. *Nat. Commun.* **2021**, *12*, 3186. [CrossRef]
20. Montemari, A.L.; Marzano, V.; Essa, N.; Levi Mortera, S.; Rossitto, M.; Gardini, S.; Selan, L.; Vrenna, G.; Onetti Muda, A.; Putignani, L.; et al. A Shaving Proteomic Approach to Unveil Surface Proteins Modulation of Multi-Drug Resistant *Pseudomonas aeruginosa* Strains Isolated From Cystic Fibrosis Patients. *Front. Med.* **2022**, *9*, 818669. [CrossRef]
21. Li, D.Y.; Han, J.T.; Zhang, M.Y.; Yan, X.; Zhang, N.; Ge, H.; Wang, Z.; He, Y.X. The Two-Component System RstA/RstB Regulates Expression of Multiple Efflux Pumps and Influences Anaerobic Nitrate Respiration in *Pseudomonas fluorescens*. *mSystems* **2021**, *6*, e0091121. [CrossRef] [PubMed]
22. Pallett, R.; Leslie, L.J.; Lambert, P.A.; Milic, I.; Devitt, A.; Marshall, L.J. Anaerobiosis influences virulence properties of *Pseudomonas aeruginosa* cystic fibrosis isolates and the interaction with *Staphylococcus aureus*. *Sci. Rep.* **2019**, *9*, 6748. [CrossRef] [PubMed]
23. Britigan, B.E.; Roeder, T.L.; Rasmussen, G.T.; Shasby, D.M.; McCormick, M.L.; Cox, C.D. Interaction of the *Pseudomonas aeruginosa* secretory products pyocyanin and pyochelin generates hydroxyl radical and causes synergistic damage to endothelial cells. Implications for *Pseudomonas*-associated tissue injury. *J. Clin. Investig.* **1992**, *90*, 2187–2196. [CrossRef] [PubMed]
24. Hunter, R.C.; Klepac-Ceraj, V.; Lorenzi, M.M.; Grotzinger, H.; Martin, T.R.; Newman, D.K. Phenazine content in the cystic fibrosis respiratory tract negatively correlates with lung function and microbial complexity. *Am. J. Respir. Cell Mol. Biol.* **2012**, *47*, 738–745. [CrossRef] [PubMed]
25. Schalk, I.J.; Perraud, Q. *Pseudomonas aeruginosa* and its multiple strategies to access iron. *Environ. microbiol.* **2023**, *25*, 811–831. [CrossRef] [PubMed]
26. Touati, D. Iron and oxidative stress in bacteria. *Arch. Biochem. Biophys.* **2000**, *373*, 1–6. [CrossRef] [PubMed]
27. Jeong, G.J.; Khan, F.; Khan, S.; Tabassum, N.; Mehta, S.; Kim, Y.M. *Pseudomonas aeruginosa* virulence attenuation by inhibiting siderophore functions. *Appl. Microbiol. Biotechnol.* **2023**, *107*, 1019–1038. [CrossRef] [PubMed]
28. Konovalova, A.; Søgaard-Andersen, L.; Kroos, L. Regulated proteolysis in bacterial development. *FEMS Microbiol. Rev.* **2014**, *38*, 493–522. [CrossRef]
29. Dong, S.; Chen, H.; Zhou, Q.; Liao, N. Protein degradation control and regulation of bacterial survival and pathogenicity: The role of protein degradation systems in bacteria. *Mol. Biol. Rep.* **2021**, *48*, 7575–7585. [CrossRef]
30. Artini, M.; Imperlini, E.; Buonocore, F.; Relucenti, M.; Porcelli, F.; Donfrancesco, O.; Tuccio Guarna Assanti, V.; Fiscarelli, E.V.; Papa, R.; Selan, L. Anti-Virulence Potential of a Chionodracine-Derived Peptide against Multidrug-Resistant *Pseudomonas aeruginosa* Clinical Isolates from Cystic Fibrosis Patients. *Int. J. Mol. Sci.* **2022**, *23*, 13494. [CrossRef]
31. Artini, M.; Vrenna, G.; Trecca, M.; Tuccio Guarna Assanti, V.; Fiscarelli, E.V.; Papa, R.; Selan, L. Serratiopeptidase Affects the Physiology of *Pseudomonas aeruginosa* Isolates from Cystic Fibrosis Patients. *Int. J. Mol. Sci.* **2022**, *23*, 12645. [CrossRef] [PubMed]
32. Wilderman, P.J.; Vasil, A.I.; Johnson, Z.; Wilson, M.J.; Cunliffe, H.E.; Lamont, I.L.; Vasil, M.L. Characterization of an endoprotease (PrpL) encoded by a PvdS-regulated gene in *Pseudomonas aeruginosa*. *Infect. Immun.* **2001**, *69*, 5385–5394. [CrossRef] [PubMed]
33. Kessler, E.; Safrin, M. Elastinolytic and proteolytic enzymes. *Methods Mol. Biol.* **2014**, *1149*, 135–169. [PubMed]
34. Yang, R.; Guan, Y.; Zhou, J.; Sun, B.; Wang, Z.; Chen, H.; He, Z.; Jia, A. Phytochemicals from *Camellia nitidissima* chi flowers reduce the pyocyanin production and motility of *Pseudomonas aeruginosa* PAO1. *Front. Microbiol.* **2018**, *8*, 2640. [CrossRef]

antibiotics

Review

Cefiderocol and Sulbactam-Durlobactam against Carbapenem-Resistant *Acinetobacter baumannii*

Arta Karruli [1,†], Antonella Migliaccio [2,†], Spyros Pournaras [3,*], Emanuele Durante-Mangoni [1,*] and Raffaele Zarrilli [2,*]

1 Department of Precision Medicine, University of Campania "L. Vanvitelli", 80138 Naples, Italy; arta.karruli@unicampania.it
2 Department of Public Health, University of Naples Federico II, 80131 Naples, Italy; antonella.migliaccio10@gmail.com
3 Clinical Microbiology Laboratory, Medical School, "Attikon" University General Hospital, National and Kapodistrian University of Athens, 1 Rimini Street, 12462 Athens, Greece
* Correspondence: spournaras@med.uoa.gr (S.P.); emanuele.durante@unicampania.it (E.D.-M.); rafzarri@unina.it (R.Z.)
† These authors contributed equally to this work and share first authorship.

Abstract: Infections caused by carbapenem-resistant *Acinetobacter baumannii* (CRAB) remain a clinical challenge due to limited treatment options. Recently, cefiderocol, a novel siderophore cephalosporin, and sulbactam-durlobactam, a bactericidal β-lactam–β-lactamase inhibitor combination, have been approved by the Food and Drug Administration for the treatment of *A. baumannii* infections. In this review, we discuss the mechanisms of action of and resistance to cefiderocol and sulbactam-durlobactam, the antimicrobial susceptibility of *A. baumannii* isolates to these drugs, as well as the clinical effectiveness of cefiderocol and sulbactam/durlobactam-based regimens against CRAB. Overall, cefiderocol and sulbactam-durlobactam show an excellent antimicrobial activity against CRAB. The review of clinical studies evaluating the efficacy of cefiderocol therapy against CRAB indicates it is non-inferior to colistin/other treatments for CRAB infections, with a better safety profile. Combination treatment is not associated with improved outcomes compared to monotherapy. Higher mortality rates are often associated with prior patient comorbidities and the severity of the underlying infection. Regarding sulbactam-durlobactam, current data from the pivotal clinical trial and case reports suggest this antibiotic combination could be a valuable option in critically ill patients affected by CRAB infections, in particular where no other antibiotic appears to be effective.

Keywords: carbapenem-resistant *Acinetobacter baumannii*; cefiderocol; sulbactam-durlobactam; antimicrobial susceptibility; mechanisms of resistance; antimicrobial therapy

check for updates

Citation: Karruli, A.; Migliaccio, A.; Pournaras, S.; Durante-Mangoni, E.; Zarrilli, R. Cefiderocol and Sulbactam-Durlobactam against Carbapenem-Resistant *Acinetobacter baumannii*. *Antibiotics* **2023**, *12*, 1729. https://doi.org/10.3390/antibiotics12121729

Academic Editor: Marc Maresca

Received: 24 November 2023
Revised: 11 December 2023
Accepted: 12 December 2023
Published: 14 December 2023

1. Introduction

Bacteria belonging to the genus *Acinetobacter* are non-fermentative Gram-negative coccobacilli that emerged as an increasingly frequent cause of healthcare-associated infections and hospital outbreaks [1–3]. *Acinetobacter baumannii* is the most clinically relevant species and is responsible for various healthcare-associated infections, including pneumonia, bloodstream infections, urinary tract infections, and wound infections. Also, *A. baumannii* causes community-acquired infections, although to a lesser extent [2,4].

A. baumannii strains responsible for epidemic spread are resistant to carbapenems and show intermediate resistance to tigecycline, but usually retain susceptibility to colistin and are classified as multi-drug-resistant (MDR) or extensively drug-resistant (XDR) [1,2,4]. Therefore, *A. baumannii* infections show limited treatment options and high mortality, especially in critically ill patients [2,4], and this has led the World Health Organization to classify carbapenem-resistant *A. baumannii* (CRAB) as a "critical" pathogen among the antibiotic-resistant bacteria of global priority for the development of new antibiotics [5].

Recently, cefiderocol (FDC; S-649266), a novel siderophore cephalosporin, which possesses a broad activity against CRAB in vitro and in vivo [6–8], has been approved by the Food and Drug Administration for the treatment of serious infections caused by carbapenem-resistant Gram-negative bacteria (CR-GNB) [9]. Also, sulbactam-durlobactam (SUL/DUR), a bactericidal β-lactam–β-lactamase inhibitor combination [10], has been demonstrated to be active against CRAB in vitro [11–13] and in vivo [14,15]. SUL/DUR, XACDURO, was approved in the USA for the treatment of hospital-acquired bacterial pneumonia and ventilator-associated bacterial pneumonia (HABP/VABP) caused by susceptible isolates of *A. baumannii* [16].

The aim of this review is to discuss: (i) chemical structures and pharmacological data of FDC and SUL/DUR; (ii) in vitro and in vivo activity of FDC and SUL/DUR against CRAB and mechanisms of resistance; (iii) utilization and efficacy of FDC and SUL/DUR in therapeutic regimens against CRAB.

2. Cefiderocol Is a Novel Siderophore Cephalosporin

2.1. Chemical Structure and Pharmacological Data

FDC is a novel catechol-substituted siderophore cephalosporin that is structurally different from other recently developed siderophore-conjugated molecules, showing a high stability against various serine-type and metallo-type carbapenemases and ESBLs [6]. FDC consists of a 4-membered β-lactam ring bound to a 6-membered dihydrothiazine ring, which is covalently bound in the 3-position to a catechol 2-chloro3,4-dihydroxybenzoic acid moiety. The quaternized N-methyl-pyrrolidine is identical to the pyrrolidinium group found in cefepime and confers zwitterionic properties that enable it to rapidly penetrate the outer cell membrane of Gram-negative bacteria (Figure 1A) [6]. Furthermore, FDC presents in the 7-position an aminothiadiazole ring bound to oxime and dimethyl groups, which improves the stability to β-lactamases and the overall antimicrobial activity (Figure 1A) [6]. The catechol moiety is important for the siderophore function and "Trojan horse" strategy of FDC; indeed, the molecule is able to chelate ferric iron and cross the outer membrane of Gram-negative organisms using the ferric iron transport system (Figure 1). The transport of FDC from the outer membrane to the periplasmic space is mediated by passive diffusion via porin channels or active transport linked to TonB-dependent siderophore receptors PiuA and PirA [6] (Figure 1A). Thus, a positive correlation has been demonstrated between FDC susceptibility and the expression of PiuA and PirA TonB-dependent receptors (TBDRs) in *A. baumannii* [17,18]. Also, the high affinity of PiuA and PirA transporters for siderophores allows the transport of FDC in the periplasmic space even at low concentrations (Figure 1B) [19,20]. The high affinity binding of FDC to penicillin-binding proteins (PBPs), primarily PBP3, in the periplasmic space (Figure 1B) results in the inhibition of peptidoglycan synthesis and cell death [6,17,21,22].

2.2. In Vitro and In Vivo Activity against CRAB

FDC susceptibility testing may be performed with disk diffusion or broth microdilution (BMD).

Disk diffusion is performed with an FDC 30 μg disk according to the EUCAST standard recommendations for non-fastidious organisms [23].

As the FDC requires low levels of iron to have optimal activity, BMD MICs need to be determined in iron-depleted Mueller–Hinton broth (MHB). As the iron concentration in the MHB affects reproducibility, broth preparation requires particular care (https://www.eucast.org/eucastguidancedocuments/, accessed on 10 November 2023). Indeed, chelators used to remove iron will also remove zinc, calcium, and magnesium, which have to be added back [24].

The EUCAST issued specific recommendations for reading the BMD MICs of FDC (EUCAST guidance document on FDC BMD, December 2020). According to this document, the MIC corresponds to the first well where the reduction of growth is represented by a button of <1 mm or by the presence of light haze/faint turbidity.

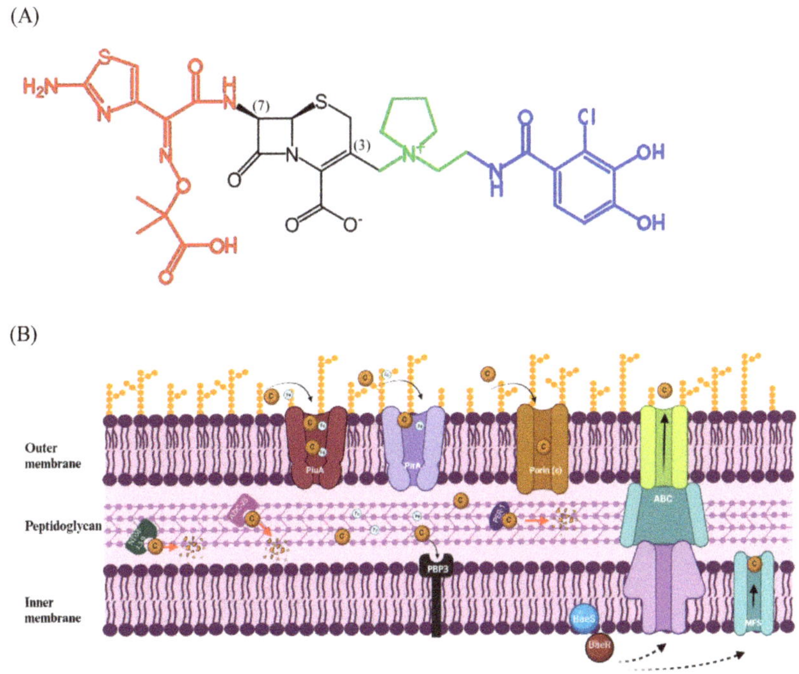

Figure 1. Cefiderocol: structure (**A**), mechanism of action and resistance (**B**). (**A**) Chemical structure of FDC: the C-7 side chain, N-methyl-pyrrolidine group, and the catechol moiety in C-3 chain are highlighted in red, green, and blue, respectively. (**B**) The active transport through PiuA and PirA iron transport systems, the passive diffusion via porin channels, the efflux through ABC and MSF efflux systems, and their modulation by BaeSR two component regulator are shown. The degradation of FDC by NDM 1-5-9, ADC-25, and PER-1 beta-lactamases is displayed. The binding of FDC to penicillin-binding protein 3 (PBP3) target molecule is also shown. Figure created with Biorender.

The inhibition zone diameters and MICs of FDC may be interpreted according to the EUCAST non-species-specific pharmacokinetic/pharmacodynamic (PK/PD) breakpoints (zone diameter $\geq$17 mm, corresponding to MIC below the PK-PD susceptible breakpoint $\leq$ 2 mg/L, MIC > 2 mg/L as resistant; EUCAST Clinical Breakpoint Tables v. 13.1, accessed 10 November 2023) or to CLSI breakpoints (zone diameter $\geq$15 mm or MIC $\leq$ 4 mg/L as susceptible, MIC $\geq$ 16 mg/L as resistant) [25].

The EUCAST has evaluated several commercial tests, all of which had problems of accuracy, reproducibility, bias, and/or skipped wells, and noted that it is difficult to interpret the FDC susceptibility of isolates in the area of technical uncertainty (ATU). For this reason, a warning has been issued (www.eucast.org/ast-of-bacteria/warnings, accessed on 10 November 2023) and it is recommended that laboratories start testing FDC with disk diffusion, which, when correctly performed and calibrated, is predictive of susceptibility and resistance outside the ATU. Within the ATU, the EUCAST recommended ignorance of the ATU and interpretation using the zone diameter breakpoints in the breakpoint table. Despite the investigation of several products by the EUCAST, the existing issues have not been resolved, and it was decided that the warning should still remain.

Some recent evaluations of FDC susceptibility tests for *Acinetobacter baumannii* isolates will be presented herein. Liu et al. [26] observed that disk diffusion was difficult to assess, but the limited isolates that actually exhibited resistance by BMD (CLSI breakpoint) were categorized as susceptible by disk diffusion. Jeannot et al. [27] evaluated two commercial BMD methods and discs from three manufacturers, compared with the reference

BMD. They showed that none of the tested methods met the accuracy requirements. Both BMD methods exhibited acceptable categorical agreement rates. MIC gradient testing was strongly discouraged, and disc diffusion could be used to screen for susceptibility, setting a critical diameter of 22 mm. Finally, a quick test, the rapid FDC *Acinetobacter baumannii* NP test for the detection of FDC susceptibility/resistance in *A. baumannii*, was recently proposed [28], and evaluated in comparison with the BMD reference method. The test showed very high sensitivity and specificity, obtained within 4:30–4:45 h of incubation, and had only a single very major error, using an isolate with the MIC of 8 mg/L. Shortridge et al. [7] reported that the susceptibility of meropenem-resistant *Acinetobacter* spp. (306 isolates) to FDC was 97.7% using CLSI criteria. Despite the susceptibility testing used, FDC showed excellent in vitro activity against *Acinetobacter* spp., ABC, and CARB [6,7,26–28].

In keeping with the in vitro activity of FDC, Matsumoto et al. [8] demonstrated the efficacy of FDC against CRAB in an immunocompetent rat respiratory tract infection model, recreating human plasma pharmacokinetics. Also, Gill et al. showed effective in vivo bactericidal activity of cefiderocol in combination with ceftazidime/avibactam or ampicillin/sulbactam using human simulated regimens against all the cefiderocol non-susceptible *A. baumannii* isolates tested [29].

2.3. Mechanisms of Resistance

FDC resistance has been associated with the reduced expression of the siderophore receptor *pirA* and *piuA* genes in several *A. baumannii* strains [17,18]. Also, Malik et al. showed that two mutations in the midst of a beta strand of PirA diminished the functionality of this receptor protein in two FDC-resistant *A. baumannii* strains [17]. Moreover, efflux-pump systems may underlie resistance to FDC. Liu et al. showed that mutations of the BaeS (D89V) and BaeR (S104N) two-component system regulator increased FDC MIC, and this effect was mediated by the up-regulation of MFS and the MacAB-TolC efflux pumps (Figure 1B) [30].

Mounting evidence indicates that resistance to FDC in *A. baumannii* is mediated by the production of PER- and NDM- β-lactamases [31,32]. Specifically, Poirel et al. demonstrated that: (i) *A. baumannii* isolates positive for PER-1-7 or NDM-1-5-9 β-lactamases showed increased MIC for FDC; (ii) the transformation of *E. coli* and *A. baumannii* CIP70.10 with recombinant plasmids producing PER-1 or NDM-1,5,9 β-lactamases increased FDC MIC; (iiI) β-lactamase crude enzymatic extracts from PER-1-producing recombinant *E. coli* strains and to a lesser extent from NDM-1-producing recombinant *E. coli* strains showed a significant hydrolysis rate of FDC [31]. Also, Liu et al. demonstrated that inactivation of the $bla_{\text{PER-1}}$ gene through allelic replacement restored the susceptibility to FDC in *A. baumannii* XH740, and that the phenomenon was reverted by the introduction of PER-1 into the knockout strain [32]. In addition, Asrat et al. demonstrated that the over-expression of bla_{ADC} subtypes β-lactamases correlated with elevated FDC resistance, and that site-specific insertional inactivation of blaADC–25 or blaADC-33 increased FDC susceptibility in *A. baumannii* strains (Figure 1B) [18]. Finally, Ile236Asn and His370Tyr mutations were found in PBP3 from one FDC-resistant *A. baumannii* isolate [17].

FDC has also been reported to exhibit heteroresistance [33–35]. The population analysis profiling (PAP) of CRAB isolates showed the change to the non-susceptible phenotype after exposure to FDC in 1 of 10 isolates and the occurrence of heteroresistance in 8 of 10 isolates [33]. The occurrence of heteroresistance after FDC exposure and the relationship between heteroresistance and clinical outcomes were recently evaluated in CRAB isolates from the CREDIBLE-CR study [34]. By using the PAP, only 7/38 CRAB isolates were susceptible, 18/38 were heteroresistant, and 13/38 were resistant. Heteroresistance, however, was not related with worse clinical or microbiological outcomes compared to non-heteroresistant isolates [34].

In contrast with FDC, which is generally active against CRAB, none of the newer b-lactam/b-lactamase inhibitor combinations, including ceftazidime–avibactam, imipenem–relebactam, meropenem–vaborbactam, and ceftolozane–tazobactam, retain activity [7]. For this reason, FDC has not generally been reported to exhibit cross-resistance with these antimicrobial classes against CRAB.

2.4. Therapy against CRAB

2.4.1. Studies Evaluating the Clinical Efficacy of FDC

Randomized, Phase III Studies

The efficacy of FDC has been evaluated in two pivotal, randomized, multi-center, phase 3 clinical studies:

- APEKS-NP—enrolling patients with nosocomial pneumonia;
- CREDIBLE-CR—focusing on severe infections caused by CR Gram-negatives.

The APEKS-NP [36] was a double-blind, randomized, active-controlled, non-inferiority study, enrolling adult patients with documented nosocomial pneumonia due to Gram-negative bacteria. Exclusion criteria were CR pathogens, Gram-positive or anaerobic pathogens, viral, atypical, chemical, or community-onset pneumonia. A total of 292 patients were included in the study after modifying the randomization group (comprising a total of 300 patients) for exclusion criteria, of whom 148 patients received treatment with FDC and 145 with meropenem, both at a dose of 2 g every 8 h intravenously. All patients received concomitant treatment with linezolid. Infections due to *A. baumannii* were present in 47 patients (16%), equally divided among the two study groups.

Although CR pathogens were excluded from the study, 56 patients (19%) were eventually found to have a CR infection after randomization, mostly due to *A. baumannii* and *Acinetobacter* spp.; these patients were still included in the study analyses.

A clinical cure was achieved in 65% of patients in the FDC group compared to 67% of those in the meropenem group.

This trial also analyzed endpoints based on individual pathogens: FDC was non-inferior to meropenem in infections due to *A. baumannii*, with clinical cure rates of 52% and 58%, respectively. The microbiological eradication of *A. baumannii* was achieved in 39% of cases treated with FDC compared to 33% of those treated with meropenem. Also, similar mortality rates were seen in the two arms of treatment for *A. baumannii*, with mortality at 14 days being 19% in the FDC group and 22% in the meropenem one [36].

These findings appear substantially different from those obtained in the other study with FDC, the CREDIBLE-CR trial [37]. This was a multicenter, randomized, open-label clinical study of FDC versus the "best available therapy" (BAT). Inclusion criteria were the identification of a carbapenem-resistant Gram-negative organism in a clinical sample or prior antimicrobial treatment failure. In complicated urinary tract infections (cUTI), FDC was used solely as monotherapy, whereas for other infection syndromes, another anti-Gram-negative antibiotic along with FDC was allowed. For the BAT arm, up to three antibiotics combined were allowed, with only 29% of the BAT group patients receiving monotherapy. Most treatment regimens were based on colistin, which was administered to 66% of patients overall. FDC was given at a dose of 2 g every 8 h and was mostly used as monotherapy (83% of patients).

A total of 152 patients were randomized in CREDIBLE, of whom 101 received FDC and 49 BAT. *A. baumannii* was the causative pathogen in 54 of 118 carbapenem-resistant microbiological intention-to-treat (ITT) subjects (46%). Indeed, *Acinetobacter* was the most prevalent CR pathogen at baseline in the FDC group (65% in FDC vs. 53% in the BAT group). Considering all infection syndromes, clinical cure rates at the test of cure (TOC) for patients who received FDC treatment compared to those who received BAT were 53% versus 50%, respectively, with clinical failure rates equal to 34% versus 37%, respectively. Clinical cure rates at TOC were higher for cUTI compared to other infection syndromes. Clinical cure and microbiological eradication rates along with the counterpart negative outcome based on different types of infections are described in detail in Table 1.

Table 1. Clinical and microbiological outcomes at the test of cure based on types of infections in the CREDIBLE study [36].

Types of Infections	Outcomes							
	Clinical Cure at TOC		Clinical Failure at TOC		Microbiological Eradication at TOC		Microbiological Persistence at TOC	
	FDC	BAT	FDC	BAT	FDC	BAT	FDC	BAT
Nosocomial pneumonia	20 (50)	10 (53)	16 (40)	6 (32)	9 (23)	4 (21)	8 (20)	7 (37)
Bloodstream infections	10 (43)	6 (43)	9 (39)	7 (50)	7 (30)	4 (29)	3 (13)	2 (14)
Complicated urinary tract infections	12 (71)	3 (60)	2 (12)	1 (20)	9 (53)	1 (20)	5 (29)	1 (20)
Overall	42 (53)	19 (50)	27 (34)	14 (37)	25 (31)	9 (24)	16 (20)	10 (26)

Data are N (%). Abbreviations: BAT, best available treatment; TOC, test of cure.

Interestingly, all-cause mortality in the subgroup of patients with *A. baumannii* infections at the end of treatment with FDC was 49%, compared to only 18% with BAT. Most of the deaths occurred either early, within the first 3 days, or late, from day 29 until the last follow-up. In the period between these two time points, deaths were similar between the two groups, suggesting that factors related to the underlying condition at randomization for the first 3 days and other complications developed after day 29 could have influenced the outcome.

It should be underscored that patients with infections due to *Acinetobacter* spp. who received treatment with FDC were overall more complex, showing a higher prevalence of baseline moderate/severe renal dysfunction, ICU admission at randomization, ongoing shock, or shock within 31 days before randomization, compared to those with the same pathogen treated with BAT [37]. Moreover, as the authors stated in their data discussion, the 28-day mortality rate in the FDC group of the CREDIBLE study was similar to that of other studies comparing the efficacy of colistin monotherapy vs. combination therapy for *A. baumannii* infections [38–41].

Clinical Studies Comparing the Efficacy of FDC with That of Colistin

Clinical trials provide the first key insights into the effectiveness of new drugs and represent the gold standard in evaluating the efficacy and safety of novel antimicrobials. However, clinical trials may be limited by the selected population studied; with all the exclusion criteria, they may leave out most of the patient subgroups encountered in daily clinical practice.

While formal clinical trials achieve mostly internal validity, other clinical/real-world studies are the ones that reach external/generalized validity [42]. Various studies, not fulfilling criteria for randomized controlled trials, have assessed the efficacy and safety of FDC compared to other treatments. Six single-center and one multi-center [43–49] retrospective and prospective observational studies evaluated the efficacy of FDC compared to a colistin-based regimen or other antibiotics, with a range of 73–124 patients included, and mostly ventilator-associated pneumonia (VAP) or bloodstream infections (BSI) due to carbapenem-resistant *Acinetobacter baumannii* (CRAB). FDC and colistin were used both as monotherapy and in combination regimens in six studies [43–47,49], whereas Pascale et al. [48] used FDC monotherapy.

Mortality

The FDC treatment was non-inferior compared to colistin in terms of mortality, which ranged from 10% to 55% (Table 2). Only in the Mazzitelli et al. study, FDC had a mortality which was higher compared to colistin (51% vs. 37%) [46]. As in the CREDIBLE-CR study, mortality or clinical failure was higher in more comorbid patients or those with sepsis/septic shock, along with higher SOFA scores [43–48,50–52]. Indeed, mortality associated with *A. baumannii* infections was shown to be higher in patients whose infections were complicated by septic shock [53–56].

Table 2. Clinical studies comparing the efficacy of FDC to other treatments in CRAB infections.

Author	Study Design	Site of Infection and Pathogens	No. of Patients	Treatment	Clinical Outcome and Adverse Event (AKI)	Microbiological Outcome	Mortality
Falcone et al. [43]	Single-center observational retrospective study	CRAB: BSI 63.7% VAP 28.5% Other sites 8.1%	124	FDC 47 (37.9%): 15 (31.9%) monotherapy 32 (69.1%) combination	Clinical cure-NA	Microbiological failure FDC 17.4% Colistin 6.8%	All infections 30-day mortality FDC 34% Colistin 55.8% BSI 14-day mortality: FDC 7.4% Colistin 42.3% 30-day mortality: FDC 25.9% Colistin 57.7% VAP no difference [Septic shock, SOFA score associated with mortality]
				colistin-based regimens 77 (62.1%): 12 (15.6%) monotherapy 55 (84.4%) combination	Adverse event FDC 2.1% Colistin 21.1%		
Russo et al. [44]	Single-center, retrospective, observational study	VAP and bacteremia due to CRAB in COVID-19 in ICU	73	FDC 19 (26%) 100% in combination	NA	Microbiological clearance higher in survivors	FDC regimen and FDC+ fosfomycin associated with survival compared to colistin
				colistin-based regimen 54 (74%) 12 (22.2%) monotherapy 32 (77.8%) combination			
Dalfino et al. [45]	Single-center prospective observational study	VAP due to CRAB in COVID-19 and non-COVID-19	90	FDC 40 patients 19 (47.5%) monotherapy 21 (52.5%) + fosfomycin	Clinical failure FDC 25% Colistin 48%	Microbiological failure FDC 30% Colistin 60%	14-day mortality FDC 10% Colistin 38%
				colistin 50 patients all combination Inhaled colistin in both groups	AKI FDC 45% Colistin 47%		
Mazzitelli et al. [46]	Single-center, retrospective, observational study	CRAB: BSI 53 (47.7%) VAP 58 (52.3%)	111	FDC 60 patients 50% monotherapy	Clinical cure FDC 73% Colistin 67% FDC Monotherapy 76.7% Combination 70%	Microbiological eradication FDC 43% Colistin 41% FDC Monotherapy 50% Combination 36.7%	Death FDC 51% Colistin 37% FDC Monotherapy 33.3% Combination 53.3%
				colistin 51 patients all combination	AKI FDC 10% Colistin 25.5%		
Pascale et al. [48]	Multi-center retrospective observational	CRAB BSI 58% LRTI 41%	107	FDC 42 patients monotherapy all cases	Clinical cure 14 days FDC 40% Colistin 36%	Microbiological cure 14 days FDC 28% Colistin 21%	28-day mortality FDC 55% Colistin 58% (High SOFA score risk factor for mortality)
				colistin 65 patients 82% combination	AKI FDC 9.5% Colistin 9.2%		
Rando et al. [47]	Single-center, prospective, observational	CRAB VAP	121	55 FDC monotherapy 21.8% 66 other	Clinical cure-NA	Microbiological failure FDC 53% Other 31%	28-day mortality FDC 44% Other 67% (mortality higher in septic shock and higher SOFA score)
					AKI FDC 9.1% Other 17%		
Bavaro et al. [49]	Single-center, retrospective, observational	CRAB-BSI	118	−43 FDC combination 63%	Clinical cure FDC 60.5% Colistin 41.3%	NA	30-day all-cause mortality FDC 40% Colistin 59%
				−75 colistin combination 96%	Adverse event FDC 2% Colistin 16%		

Abbreviations: AKI, acute kidney injury; BSI, bloodstream infections; CRAB, carbapenem-resistant *Acinetobacter baumannii*; CR GN, carbapenem-resistant Gram-negative; EOT, end of treatment; Cefiderocol, FDC; LRTI, lower respiratory tract infections; IT, urinary tract infection; VAP, ventilator-associated pneumonia.

Mortality Based on Type of Infection

Falcone et al., Russo et al., Dalfino et al., Rando et al., and Bavaro et al. showed lower mortality in the FDC group compared to colistin [43–45,47,49]. This difference was seen also in patients with BSI [43], with the FDC regimen translating into lower 14-day and 30-day mortality rates of 7.4% and 25.9% vs. 42.3% and 57.7% with the colistin-based regimen. Interestingly, there was no difference in 14- and 30-day mortality for the two groups among patients with VAP. Indeed, in another study [51], FDC was shown to be associated with a higher rate of clinical success in BSI (75%) compared to respiratory infections (45.8%). Similarly, Palermo et al. [57] observed a mortality of 61.5% in VAP compared with 46.7% in BSI due to CRAB.

Other Outcomes

In the Falcone et al. and Rando et al. studies, microbiological failure was higher with FDC compared to colistin (17.4% vs. 6.8% and 53% vs. 31%, respectively) [43,47]. In contrast, in the other three studies, microbiological failure was similar or higher in the colistin group [45,46,48], and in the Russo et al. study, microbiological failure was associated with higher mortality in the colistin group [44]. Clinical cure rates were also higher in the FDC group compared to colistin [45,46,48,49].

Adverse events were mostly seen in colistin-treated subjects [43,49]. Regarding treatment-emergent acute kidney injury (AKI), Pascale et al. and Dalfino et al. showed similar rates in both groups, whereas Rando et al. and Mazzitelli et al. observed this adverse event mostly in the colistin group [45–49], which is one of the major concerns for this antibiotic [58,59].

Clinical Studies and Case Series Assessing the Efficacy of FDC in Treating CRAB Infections

Two multi-center and several single-center studies (Table 3) have assessed the real-life efficacy of FDC in treating CRAB infection [50–52,60–65], generating similar data to the comparative studies described above. Most of the studies included multiple carbapenem-resistant Gram-negatives [52,57,60–63,65]. One study, the largest in size, focused only on *Acinetobacter* spp. infections [51], including 147 patients, of whom 146 had *A. baumannii* infections. One study focused on CRAB infections only [50]. Case series included 8 to 16 patients (with *A. baumannii*). Infection syndromes included in these studies were mostly BSI and respiratory infections. Clinical cure rates ranged from 32.5% to 77.8% [50,51,57,60–66]. Microbiological failure occurred in 0% to 25%. One case series on burn patients by Smoke et al. reported a microbiological failure rate as high as 88% [64]. However, in this study, the reported resistance to FDC was 60% of the tested isolates. Mortality ranged widely, from 12.5% to 51% [50–52,57,60–65].

Table 3. Clinical studies and case series evaluating efficacy and safety of FDC in CRAB infections.

Study	Study Design	Type of Infection and Pathogens	Patients	Clinical Outcome and Adverse Events	Microbiological Outcome	Mortality
			Clinical studies			
Palermo et al. [57]	Single-center, retrospective observational	CR GN For CRAB: BSI 48.4% HAP 41.9% SSTI 25.8% cIAIs 19.3% cUTI 9.7% Other 12.8%	41 total patients: 31 CRAB	CRAB: Clinical cure at EOT 64.5% Adverse events 4.9%	CRAB: Microbiological eradication at EOT 80.6%	CRAB: 30-day mortality 35.5% VAP 61.5% BSI 46.7%
Calò et al. [50]	Multi-center, retrospective/prospective, observational study	CRAB All types of infections (mostly): BSI 45% Respiratory 40%	38 patients	Clinical failure at EOT 32.5% Monotherapy 27.6% Combination 45.5% (non-significant) Adverse event none	Microbiological failure at EOT 10% Monotherapy 13.8% Combination 0 (non-significant)	30-day mortality 47.5% Monotherapy: 48.3% Combination 45.5%

Table 3. *Cont.*

Study	Study Design	Type of Infection and Pathogens	Patients	Clinical Outcome and Adverse Events	Microbiological Outcome	Mortality
Giannella et al. [51]	Multi-center, retrospective observational study	*Acinetobacter* spp: Respiratory 65.3% BSI 26.5% Other 8.2%	147 patients 146 *A. baumannii* 1 other *Acinetobacter*	Resolution of infection 39.5% Improved symptoms 12.2% Failure 38.1% Clinical success: BSI 52.2%-75% Respiratory 45.8% Monotherapy 61.2% Combination 49% Adverse event 4.8%	NA	28-day mortality Overall 51% Monotherapy 53.1% Combination 40.8% Survival: septic shock 24.3% Without Septic shock 60.7%
Piccica et al. [52]	Multi-center, retrospective observational	CR GN LRTI 57% IAI 9.2% UTI 8.5% BSI 13.4% ABSSSI 6.3% OTHERS 2.1%	142 patients: 70 monotherapy 72 combination (89 *A. baumannii*)	Clinical cure-NA AKI 38%	Microbiological eradication All pathogens 48.9% Monotherapy 45.8% Combination 52.3%	Survived All pathogens 36.6% Monotherapy 32.9% Combination 40.3% Mortality higher in septic shock (Survival in *A. baumannii* 62.9%)
			CASE SERIES			
Bavaro et al. [61]	Case series	XDR GN For CRAB: BSI 70% BSI+ VAP 10% VAP 10% Perihepatic abscess 10%	13 patients: CRAB in 10	CRAB: Clinical cure 70% No adverse event	CRAB: Microbiological eradication 100%	CRAB: Mortality 30%
Corcione et al. [62]	Case series	CR GN VAP + BSI 61.2% BSI 16.7% VAP 11.1% Other 11.1%	18 patients: CRAB in 16: 4 monotherapy 11 combination	All pathogens: Clinical cure 66.7% Monotherapy 75% Combination 64.29% No serious adverse event	All pathogens: Microbiological failure 22.2% Monotherapy 25% Combination 21.43%	All pathogens: 30-day mortality 27.8% Monotherapy 25% Combination 28.57%
Gavaghan et al. [63]	Case series	CR GN For CRAB: Pneumonia 10 UTI 1 Pneumonia + BSI 2 Wound + BSI 1	24 patients: *A. baumannii* in 14	CRAB: Clinical success 35.7% No adverse event	NA	CRAB: Mortality 42.8% 62% of isolates susceptible to FDC
Falcone et al. [60]	Case series	CR GN For CRAB: 6 patients BSI 2 patients VAP	10 patients *A. baumannii* in 8: 7 Monotherapy 1 Combination	CRAB: Clinical success 62.5% No severe adverse event	CRAB: Microbiological failure 25%	CRAB: Mortality 12.5%
Wicky et al. [65]	Case series	DTR GN For CRAB: Mostly VAP	16 patients CRAB in 9: 1 Monotherapy 8 Combination	CRAB: Clinical cure 77.8% Adverse events 66.6% 66.7% (mostly) encephalopathy		CRAB: Death 22.2%
Smoke et al. [64]	Case series	CRAB burn patients 5 BSI 4 BSI/VAP 1 VAP 1 VAT	11 patients: 3 Combination	Clinical cure 36% (FDC resistance 60% of tested isolates)	Microbiological failure 90 days 88% of 8 patients who completed initial treatment	Mortality 27.3%

Abbreviations: ABSSSI, acute bacterial skin and skin structure infection; BSI, bloodstream infections; cIAIs, complicated intra-abdominal infections; CR GN, carbapenem-resistant Gram-negative; CRAB, carbapenem-resistant *Acinetobacter baumannii*; CR GN, carbapenem-resistant Gram-negative; cUTI, complicated urinary tract infections; DTR GN, difficult to treat Gram-negative; EOT, end of treatment; FDC, FDC; LRTI, low respiratory tract infections; UTI, urinary tract infection; SSTI, skin and soft tissue infections; VAP, ventilator-associated pneumonia; VAT, ventilator-associated tracheobronchitis; XDR GN, extensively resistant Gram-negative.

Adverse events associated with FDC administration were very infrequent, being described in none to less than 5% of treated subjects. However, in the study by Piccica et al. [52], acute kidney injury occurred in 38% of cases, and in the case series by Wicky et al. [65], 6 of 9 patients experienced an adverse event, mostly encephalopathy.

Regarding the efficacy of combination treatment compared to FDC monotherapy, no significant difference was found in these studies in terms of mortality [46,50–52,58,62], microbiological eradication [46,50,52,62], and clinical cure [46,50,51,62].

In the case series by Gavaghan et al., mortality was 60%, with 62% of the analyzed isolates that were susceptible to FDC. In this study, the explanation for such a high mortality rate could actually be related with the observed high resistance rates to FDC [63]. Also, the study by Smoke et al. on 11 burn patients with infection due to CRAB showed not only a clinical cure rate of 36% with a high microbiological failure rate, but also a 60% rate of resistance prior to/after FDC treatment [64]. In a multi-center study by Piccica et al. on 142 CR GN infections treated with FDC, among 28 strains of *A. baumannii* with available microbiological data, 10 (35.7%) were resistant to FDC. However, in this study, resistance was not associated with a higher risk of mortality [52].

Reduced FDC susceptibility was also seen in the CREDIBLE and APEKS-NP clinical trials [36,37], with an FDC MIC increase of fourfold or more during/after treatment observed in 12 of 106 isolates in the CREDIBLE study and 7 of 159 isolates in the APEKS-NP study [67]. However, despite increasing more than fourfold, the MIC remained $\leq$4 μg/mL for most of the isolates. Mutations were found in only three isolates.

The in vivo emergence of resistance or reduced susceptibility to FDC was addressed also by other clinical studies. In the Falcone et al. study, 8.5% of isolates developed resistance to FDC during treatment and 50% (4/8) of patients who showed microbiological failure developed resistance to FDC during treatment [43]. In another case series by Falcone et al. [60] on 10 patients, of whom eight had an *Acinetobacter baumannii* infection, one patient with available repeated MIC values (of two who experienced microbiological failure) had a 16-fold increase in MIC from 0.25 μg/mL to 4 μg/mL [60]. The occurrence of heteroresistance in CRAB isolates has been demonstrated in CRAB isolates [33–35]. The presence of heteroresistance in Gram-negative isolates susceptible to FDC was found also in another study, which observed a 59% heteroresistance rate (64 of 108 *A. baumannii* isolates) with full resistance detected only in 8% [68]. Heteroresistance implies the presence of minor subpopulations of cells resistant to a specific antibiotic which remain undetectable by most antibiotic susceptibility tests and may predominate after exposure to that antibiotic, causing treatment failure or microbiological persistence. It was suggested as a possible mechanism underlying the MIC increase of isolates in the CREDIBLE study [68]. However, despite the high prevalence of heteroresistance to FDC in *A. baumannii*, its clinical impact is yet to be understood. In vivo, the emergence of resistance during treatment remains uncommon and is mostly described in low percentages/sporadic cases of the published studies [69].

Consistent with these data, the Infectious Disease Society of America (IDSA) recommended that FDC should be used with caution for CRAB infections and as part of a combination treatment regimen [70].

Other studies on a few peculiar cases in terms of site of infection described the efficacy of FDC in treating XDR/PDR infections, including one case of spondylodiscitis [71], one case of osteomyelitis, and one of prosthetic joint infection [72,73]. FDC was shown to reach optimal cerebrospinal fluid concentrations above MIC levels, achieving a clinical cure in two patients with central nervous system infections [74].

In conclusion, the growing wealth of data from several, mostly small, retrospective and single-center studies, provides somehow conflicting real-life data that do not add further robust evidence, and do not help solving the doubts raised by the CREDIBLE-CR study.

3. Sulbactam-Durlobactam: A Novel Combination of a Beta-Lactamase Inhibitor with Beta-Lactam Activity and a Non-Beta-Lactam/Beta-Lactamase Inhibitor

3.1. Chemical Structure and Pharmacological Data

Durlobactam (DUR), previously identified as ETX2514, is a non-beta lactam diazabicyclooctanone (DBO) inhibitor of class A, C, and D serine β-lactamase, but not class B metallo-β –lactamases [75]. In particular, DUR exhibits variable activity against class D beta-lactamases, with the highest activity being found for OXA-48 with respect to OXA-10, OXA-23, and OXA-24 [75]. DUR is active in its carbamylated state, i.e., when its active site serine nucleophile reacts with a β-lactamase, opening the cyclic urea in a reversible manner (Figure 2A) [10]. Sulbactam (SUL) is a semi-synthetic penicillanic acid which exhibits weak intrinsic inhibitory activity against CTX-M-15, SHV-5, TEM-1, and KPC-2 class A serine β-lactamases, but not class C and D beta-lactamases (Figure 2A) [76]. The combination of DUR with SUL restores the susceptibility to SUL in *A. baumannii* strains producing class A, C, and D β-lactamases [10,75].

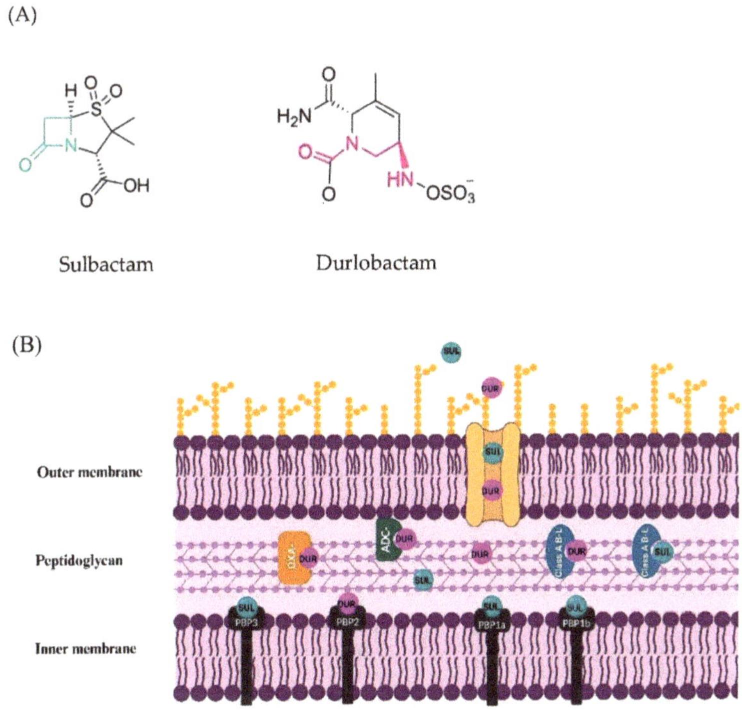

Figure 2. Sulbactam and durlobactam: Structures (**A**), mechanism of action and resistance (**B**). (**A**) Chemical structures of SUL (left) and DUR (right). The beta-lactam ring of SUL and the carbamoylated active site of DUR are highlighted in green and magenta, respectively. The active form of DUR with the cyclic urea opened is shown. (**B**) The passive diffusion of SUL and DUR via porin channels is shown. The binding of SUL to class A β-lactamase (A B-L) and DUR to class A B-L, AmpC-type of class C B-L, and OXA-type of class D B-L is shown. The binding of SUL to penicillin-binding protein (PBP) 1a, 1b, and 3 and DUR to PBP 2 target molecules is also shown. Figure created with Biorender.

The passive diffusion via porin channels (e.g., the most common OmpA porin) mediates the transport of DUR and SUL across the outer membrane. Subsequently, within the periplasmic space, SUL exerts its intrinsic antibacterial activity through the inhibition of PBP1a, PBP1b, and PBP3, but not PBP2, whilst DUR inhibits PBP2 in *A. baumannii* (Figure 2) [10]. A recent study also demonstrated that, in the presence of DUR, the FDC MICs decreased in *A. baumannii* strains producing PER-1 beta-lactamase and provided an in silico structural modeling of PER-1 binding with both FDC and DUR [32]. Based on the structural model of PER-1 binding, the authors suggested that the FDC and DUR combination might be an effective therapeutic approach against *A. baumannii* strains producing PER-1 enzymes [32].

3.2. In Vitro and In Vivo Activity against CRAB

Sulbactam/durlobactam (SUL/DUR) was approved in May 2023 by the USA FDA to treat hospital-acquired and ventilator-associated bacterial pneumonia (HABP/VABP) caused by susceptible isolates of the *Acinetobacter baumannii-calcoaceticus* complex (ABC) in patients 18 years of age and older [77].

For this reason, the CLSI Subcommittee on Antimicrobial Susceptibility Testing performed a breakpoint revision meeting, and new breakpoints were introduced for SUL/DUR by an ad hoc working group, and subsequently approved by the subcommittee. The proposed MIC breakpoints for ABC were susceptible (S; $\leq$4/4 µg/mL, $\geq$17 mm), intermediate (I; 8/4 µg/mL, 14–16 mm), and resistant (R; $\geq$16/4 µg/mL, $\leq$13 mm). DUR was tested at a fixed concentration of 4 µg/mL.

These breakpoints aligned with the epidemiological cut-off value, that included 98.3% of isolates tested in a large representative collection of 5,032 clinical isolates of ABC collected in 33 countries across the Asia/South Pacific region, Europe, Latin America, the Middle East, and North America between 2016 and 2021 [11]. In this study, SUL alone had a MIC_{90} of 64 µg/mL, whereas the SUL/DUR combination had a MIC_{90} of 2/4 µg/mL.

In most studies, all showing excellent activity of SUL/DUR against representative collections of ABC, MICs for SUL-DUR were determined by the CLSI standard BMD using a cation-adjusted Mueller–Hinton broth, with SUL-DUR tested in 2-fold dilutions of SUL in combination with a fixed concentration of 4 mg/mL of DUR [11,12,78]. In keeping with the in vitro data, treatment with SUL/DUR resulted in a dose-dependent reduction in XDR *A. baumannii* in both the neutropenic mice thigh abscess and pneumonia infection models [14].

3.3. Mechanisms of Resistance

The frequency of spontaneous in vitro resistance to SUL/DUR was low and occurred at 7.6×10^{-10} to $<9.0 \times 10^{-10}$ at $4 \times$ MIC [79]. Most frequent mutations S390T, V505L, and T511A identified in stable mutants occurred in the *ftsI* gene that encoded the target of SUL PBP3 and was near the active site serine (Ser336) [79]. Mutations in tRNA synthetases, *aspS*, and *gltX* genes were also identified, and were linked to the induction of the stringent response, which rendered PBP2 dispensable [79]. Mutations A515V and less frequently T526S were identified also in the PBP3 encoding gene in the proximity to the SUL-binding site in SUL/DUR-resistant isolates from Greece, Switzerland, and France [12,13]. Additional mutations occurred in PBP1a, PBP1b, and PBP2 encoding genes in SUL/DUR resistant isolates from Switzerland and France (Figure 2) [13].

3.4. Therapy against CRAB

3.4.1. Clinical Trials Assessing Efficacy and Safety of SUL-DUR

The efficacy of SUL-DUR in treating *A. baumannii* infections was assessed in a phase III clinical trial [80]. This was a multinational, randomized, active-controlled, non-inferiority trial. A total of 125 patients were included in the efficacy analysis, of whom 63 were treated with SUL-DUR at the dose of 2 g every 6 h, and 62 with colistin, with a maintenance dose of 2.5 mg/kg after a loading dose of 5 mg/kg of colistin base activity. Imipenem/cilastatin was used as a combination agent in both groups. Infections included BSI, hospital-acquired pneumonia (HAP), and VAP. SUL–DUR was non-inferior to colistin in terms of 28-day all-cause mortality in the microbiologically (CRAB) modified ITT population.

The 28-day all-cause mortality was 19% (12 patients) in the SUL–DUR group compared with 32.3% (20 patients) in the colistin group, with an observed treatment difference of –13.2% (95% C.I. –30 to 3.5). The 14-day all-cause mortality rate in the microbiologically (CRAB) modified ITT population was 6% (4 of 64) with SUL–DUR versus 19% (12 of 63) with colistin.

The authors performed a Kaplan–Meier analysis showing patients treated with SUL-DUR had a higher survival probability compared to those treated with colistin, with the difference becoming evident after the 6th day of treatment.

As for the microbiological eradication at the test of cure, a more favorable outcome was observed in the SUL-DUR group compared to colistin (43 of 63 treated with SUL-DUR [68%] vs. 26 of 62 patients treated with colistin [42%]).

Nephrotoxicity assessed by RIFLE criteria was significantly lower in the SUL-DUR group compared to colistin, (12 of 91 (13%) vs. 31 of 85 (38%), respectively). The overall incidence of any adverse event was 88% in the SUL-DUR and 94% in the colistin group, whereas for serious adverse events, it was 40% in the SUL-DUR and 49% in the colistin group.

3.4.2. Clinical Studies/Case Reports of SUL-DUR Efficacy in Real Life

At present, there is no published real-life experience that has assessed SUL-DUR efficacy. However, a few case reports have been published (Table 4) [81–83].

Table 4. Case description of real-life SUL-DUR use in CRAB infection.

Case Report	Patient	Comorbidities and History Prior to *A. baumannii* Isolation	*A. baumannii* Isolations and Clinical Characteristics	Prior Treatment Failure for *A. baumannii* Infection	Treatment Success	MIC	Outcome
Zaidan et al. [81]	55-year-old female	Comorbidities: Diabetes mellitus hypertension gastric bypass for obesity History prior to *A. baumannii* isolation: COVID-19 pneumonia with RF, mechanical ventilation → clinical improvement, extubation	During hospitalization, after improvement of COVID-19 pneumonia: → refractory hypoxia, intubation, vasopressor support → respiratory culture PDR *A. baumannii* Wild-type *P. aeruginosa*	Empiric meropenem+vancomycin After AST results meropenem + ampicillin/SUL	SUL-DUR 2 gr q6h + FDC 2 gr q8h	SUL-DUR 4 mg/L FDC 0.5 mg/L	Improvement after 72 h 14-day course treatment → tracheostomy- → discharged long-term care facility → discharged home
Holger et al. [82]	50-year-old male	Pulmonary embolism + infarction → intubation, thoracotomy, partial decortication	Bronchoalveolar lavage *A. baumannii* meropenem S → thoracotomy, RLL resection (abscess), decortication → Pleural tissue *A. baumannii* I to colistin → 45 days later, persistence of *A. baumannii* which developed resistance to FDC, eravacycline	Empiric piperacillin/tazobactam + vancomycin After AST results meropenem tigecycline colistin + meropenem FDC eravacycline FDC + tigecycline	SUL-DUR 2 gr q8 + meropenem 1 gr q6	SUL-DUR MIC 8 mg/L alone 4 mg/L with meropenem	13-day treatment Clinical cure Discharged home
Tiseo et al. [83]	Young female	Burn injury 45% of total body surface area → intubated	Central line BSI CRAB R to FDC, S to colistin → skin lesion CRAB + CR *P. aeruginosa* → BAL: CRAB R to FDC and colistin + CR *P. aeruginosa*	Colistin + tigecycline	SUL-DUR + colistin (for *P. aeruginosa*)	SUL-DUR MIC 1.5 mg/L	12-day regimen Clinical improvement Microbiological eradication of CRAB Alive at 30 days

Abbreviations: AST, antibiotic susceptibility testing; RF, respiratory failure; BSI, bloodstream infection; BAL, bronchoalveolar lavage; CR, carbapenem-resistant; CRAB, carbapenem-resistant *A. baumannii*; S, susceptible; I, intermediate; RLL, right lower lobe; R, resistant; SUL-DUR, SUL/DUR; FDC, FDC.

Zaidan et al. [81] described a 55-year-old woman with pneumonia due to COVID-19, complicated by respiratory failure and on mechanical ventilation, with sputum cultures negative after extubation and clinical improvement. She later developed hypoxia again requiring mechanical ventilation. Respiratory cultures then became positive for PDR *A. baumannii* and pan-susceptible *Pseudomonas aeruginosa*. Treatment with meropenem and ampicillin-SUL (after an empirical course with meropenem and vancomycin) did not improve her clinical conditions (respiratory failure and septic shock requiring vasopressor support). Therefore, she was switched to FDC plus SUL-DUR. This treatment appeared to be successful, as her clinical condition improved within 72 h, and after a 14-day course regimen, her respiratory conditions also improved, allowing a simple tracheostomy. The patient was subsequently transferred to a long-term care facility and discharged home several weeks later.

Holger et al. [82] described a 50-year-old male patient hospitalized for pulmonary embolism with pulmonary infarction. After intubation, thoracotomy, partial decortication, and right thoracoscopy with the placement of three tubes, a bronchoalveolar aspirate was positive for *A. baumannii*. Being the strain susceptible to meropenem, the empirical treatment with piperacillin/tazobactam was switched to meropenem. The patient underwent a second thoracotomy, with right lower lobe resection (which showed necrosis, hemorrhage, and abscess) and complete decortication due to multiloculated pleural effusions. *A. baumannii* was again isolated in the pleural tissue, but with a worse antimicrobial susceptibility pattern: intermediate to colistin and with a tigecycline MIC of 2 mg/L. Treatment was initially switched to tigecycline and subsequently to colistin plus meropenem due to deteriorating clinical conditions. In light of the side effects of colistin and defined susceptibility to FDC, treatment was switched to FDC. Respiratory samples remained positive for *A. baumannii* over 45 days, after which the isolate became resistant to FDC. The patient was then switched to eravacycline. Clinical conditions worsened again, with increasing eravacycline MIC, and treatment was switched to FDC + tigecycline. Due to the persistence of purulent secretions from the chest tube, the MIC of SUL-DUR was assessed, being 8 mg/L and decreased to 4 mg/L after meropenem addition. After a 2-week treatment course with SUL-DUR at a dose of 2 g every 6 h plus meropenem, no further output from the chest tube was seen. SUL-DUR maintained activity after the last debridement and no adverse events were observed. The patient was subsequently discharged.

Tiseo et al. [83] described treatment with SUL-DUR in a young woman who suffered a severe burn injury with 45% of body involvement. She developed a central line BSI due to CRAB isolate, which showed resistance to FDC and susceptibility to colistin. She was treated with colistin and tigecycline. CRAB and *P. aeruginosa* were isolated from skin lesions. She developed a kidney injury and started continuous renal replacement therapy. Her respiratory function worsened with bilateral multiple consolidations, and a bronchoalveolar lavage fluid became positive for CRAB resistance to FDC and colistin and for carbapenem-resistant *P. aeruginosa*. The MIC for SUL/DUR was 1.5 mg/L. Therefore, she was switched to SUL-DUR combined with colistin for the carbapenem-resistant *P. aeruginosa*. After 12 days of treatment, her respiratory function improved, and CRAB was eradicated from all sites of isolation. The patient was alive at 30 days.

In conclusion, with the possible selection bias of positively evolving cases, SUL-DUR appeared as an effective salvage treatment in a very limited number of clinical case reports.

4. Conclusions and Future Directions

Appropriate antibiotic administration is lifesaving in critically ill patients, and its judicious use is vital to keep the antimicrobial resistance risk as low as possible in our communities. This is particularly true for novel antibiotics that are being specifically developed to tackle CR Gram-negatives, including *A. baumannii*. The data we presented clearly urge clinicians not to use FDC and SUL/DUR for non-severe infections or when less potent options are appropriate.

In the frame of strategies to control MDR infection spread, the reduced use of antibiotics appears crucial. Combination antimicrobial regimens are very commonly used in an attempt to control DTR infections and reduce mortality, and this is particularly relevant for CRAB infections [84]. However, combination therapy inherently increases antibiotic usage, possibly perpetuating antimicrobial resistance spread. In addition to several clinical studies showing a combination of antibiotics to colistin does not improve mortality in CRAB infections [38–41], we have summarized in this article the current evidence suggesting the combination of other antibiotics to FDC also does not improve clinical outcomes. Data are still too limited to provide any suggestion as to whether SUL/DUR should be used in combination with other antimicrobials active against CRAB or as active monotherapy.

The literature review from in vitro and in vivo clinical studies suggests that FDC has potent antimicrobial activity against CRAB. Resistance to FDC in *A. baumannii* has been associated with the reduced expression and/or mutations in siderophore receptor *pirA* and *piuA*, and the production of PER- and NDM- β-lactamases. The literature review of clinical studies evaluating the efficacy of FDC therapy against CRAB indicated that FDC is non-inferior to colistin/other treatments for CRAB infections, with a possibly better safety profile. Combination treatment was not associated with improved outcomes compared to monotherapy. Higher mortality rates were often associated with prior patient comorbidities and the severity of the underlying infection.

SUL-DUR, a novel combination of a beta-lactamase inhibitor and non-beta-lactam beta-lactamase inhibitor, restores the susceptibility to SUL and shows excellent antimicrobial activity against CRAB. Although there are limited data regarding clinical studies in the real-life setting on the efficacy of SUL-DUR, current data from the pivotal clinical trial and case reports suggest this antibiotic could be a valuable option in critically ill patients affected by CRAB infections, especially where no other options appear to be effective.

Several open questions remain for the future. Among them are which of the novel antibiotics we presented should be used for diverse types of patients. The evidence is still very limited. With both FDC and SUL-DUR being very well-tolerated agents [67,80], the severity of illness would not be a major determinant. In contrast, the infection syndrome could play a role in clinical practice decision-making. Considering the reduced efficacy of FDC in respiratory infections, SUL/DUR could be a better option in this setting. Indeed, the large majority of patients enrolled in the ATTACK trial [80] had CRAB-related pneumonia. On the other hand, very little evidence has been generated so far on SUL/DUR in the setting of bloodstream infections, where indeed FDC could be preferred. Finally, in light of the broad antimicrobial spectrum, FDC would also be the agent of choice when polymicrobial Gram-negative infections exist or when suspected or documented infection with other XDR microorganisms complicates the clinical course.

Surely, larger clinical studies and/or clinical trials showing adequate power to answer such important clinical questions are strongly awaited.

Author Contributions: A.K., A.M., S.P., E.D.-M. and R.Z. have equally contributed to the conception and design of the work and have approved the submitted version of the manuscript. All authors have read and agreed to the published version of the manuscript.

Funding: The authors acknowledge funding support from the Italian Ministry of University and Research (MUR), grant PRIN 2022 (no. 202288EJ8B to R.Z.), and grant PRIN 2022 PNRR (no. P2022YCH8R to E.D.M and R.Z.).

Institutional Review Board Statement: Not applicable.

Informed Consent Statement: Not applicable.

Data Availability Statement: No new data were created or analyzed in this study. Data sharing is not applicable to this article.

Acknowledgments: We apologize to those colleagues whose work could not be cited in this paper due to space limitations.

Conflicts of Interest: The authors declare no conflict of interest. The funders had no role in the design of the study; in the collection, analyses, or interpretation of data; in the writing of the manuscript; or in the decision to publish the results.

References

1. Zarrilli, R.; Pournaras, S.; Giannouli, M.; Tsakris, A. Global evolution of multidrug-resistant *Acinetobacter baumannii* clonal lineages. *Int. J. Antimicrob. Agents* **2013**, *41*, 11–19. [CrossRef] [PubMed]
2. Wong, D.; Nielsen, T.B.; Bonomo, R.A.; Pantapalangkoor, P.; Luna, B.; Spellberg, B. Clinical and pathophysiological overview of *Acinetobacter* infections: A century of challenges. *Clin. Microbiol. Rev.* **2017**, *30*, 409–447. [CrossRef] [PubMed]
3. Migliaccio, A.; Bray, J.; Intoccia, M.; Stabile, M.; Scala, G.; Jolley, K.A.; Brisse, S.; Zarrilli, R. Phylogenomics of Acinetobacter species and analysis of antimicrobial resistance genes. *Front. Microbiol.* **2023**, *14*, 1264030. [CrossRef] [PubMed]
4. Dickstein, Y.; Lellouche, J.; Amar, M.B.D.; Schwartz, D.; Nutman, A.; Daitch, V.; Yahav, D.; Leibovici, L.; Skiada, A.; Antoniadou, A.; et al. Treatment outcomes of colistin- and carbapenem-resistant *Acinetobacter baumannii* infections: An exploratory subgroup analysis of a randomized clinical trial. *Clin. Infect. Dis.* **2019**, *69*, 769–776. [CrossRef] [PubMed]
5. WHO. *WHO Global Priority List of Antibiotic-Resistant Bacteria to Guide Research, Discovery, and Development of New Antibiotics*; World Health Organization: Geneva, Switzerland, 2017.
6. El-Lababidi, R.M.; Rizk, J.G. Cefiderocol: A Siderophore Cephalosporin. *Ann. Pharmacother.* **2020**, *54*, 1215–1231. [CrossRef] [PubMed]

7. Shortridge, D.; Streit, J.M.; Mendes, R.; Castanheira, M. *In Vitro* Activity of Cefiderocol against U.S. and European Gram-Negative Clinical Isolates Collected in 2020 as Part of the SENTRY Antimicrobial Surveillance Program. *Microbiol. Spectr.* **2022**, *10*, e0271221. [CrossRef] [PubMed]
8. Matsumoto, S.; Singley, C.M.; Hoover, J.; Nakamura, R.; Echols, R.; Rittenhouse, S.; Tsuji, M.; Yamano, Y. Efficacy of Cefiderocol against carbapenem-resistant Gram-negative bacilli in immunocompetent-rat respiratory tract infection models recreating human plasma pharmacokinetic. *Antimicrob. Agents Chemother.* **2017**, *61*, e00700-17. [CrossRef] [PubMed]
9. US Food and Drug Administration. Highlights of Prescribing Information: Fetroja (Cefiderocol). 2019. Available online: https://www.accessdata.fda.gov/drugsatfda_docs/label/2019/209445s000lbl.pdf (accessed on 12 November 2023).
10. Shapiro, A.B.; Moussa, S.H.; McLeod, S.M.; Durand-Réville, T.; Miller, A.A. Durlobactam, a New Diazabicyclooctane β-Lactamase Inhibitor for the Treatment of Acinetobacter Infections in Combination With Sulbactam. *Front. Microbiol.* **2021**, *12*, 709974. [CrossRef]
11. Karlowsky, J.A.; Hackel, M.A.; McLeod, S.M.; Miller, A.A. *In Vitro* Activity of Sulbactam-Durlobactam against Global Isolates of *Acinetobacter baumannii-calcoaceticus* Complex Collected from 2016 to 2021. *Antimicrob. Agents Chemother.* **2022**, *66*, e0078122. [CrossRef]
12. Petropoulou, D.; Siopi, M.; Vourli, S.; Pournaras, S. Activity of Sulbactam-Durlobactam and Comparators Against a National Collection of Carbapenem-Resistant *Acinetobacter baumannii* Isolates from Greece. *Front. Cell. Infect. Microbiol.* **2022**, *11*, 814530. [CrossRef]
13. Findlay, J.; Poirel, L.; Bouvier, M.; Nordmann, P. In vitro activity of sulbactam-durlobactam against carbapenem-resistant *Acinetobacter baumannii* and mechanisms of resistance. *J. Glob. Antimicrob. Resist.* **2022**, *30*, 445–450. [CrossRef]
14. O'donnell, J.P.; Bhavnani, S.M. The Pharmacokinetics/Pharmacodynamic Relationship of Durlobactam in Combination With Sulbactam in In Vitro and In Vivo Infection Model Systems Versus *Acinetobacter baumannii-calcoaceticus* Complex. *Clin. Infect. Dis.* **2023**, *76* (Suppl. 2), S202–S209. [CrossRef] [PubMed]
15. Papp-Wallace, K.M.; McLeod, S.M.; A Miller, A. Durlobactam, a Broad-Spectrum Serine β-lactamase Inhibitor, Restores Sulbactam Activity Against *Acinetobacter* Species. *Clin. Infect. Dis.* **2023**, *76* (Suppl. 2), S194–S201. [CrossRef] [PubMed]
16. Keam, S.J. Sulbactam/Durlobactam: First Approval. *Drugs* **2023**, *83*, 1245–1252. [CrossRef] [PubMed]
17. Malik, S.; Kaminski, M.; Landman, D.; Quale, J. Cefiderocol resistance in *Acinetobacter baumannii*: Roles of β-lactamases, siderophore receptors, and penicillin binding protein 3. *Antimicrob. Agents Chemother.* **2020**, *64*, e01221-20. [CrossRef] [PubMed]
18. Asrat, H.; Samaroo-Campbell, J.; Ata, S.; Quale, J. Contribution of Iron-Transport Systems and β-Lactamases to Cefiderocol Resistance in Clinical Isolates of *Acinetobacter baumannii* Endemic to New York City. *Antimicrob. Agents Chemother.* **2023**, *67*, e0023423. [CrossRef] [PubMed]
19. Moynié, L.; Luscher, A.; Rolo, D.; Pletzer, D.; Tortajada, A.; Weingart, H.; Braun, Y.; Page, M.G.P.; Naismith, J.H.; Köhler, T. Structure and Function of the PiuA and PirA Siderophore-Drug Receptors from Pseudomonas aeruginosa and *Acinetobacter baumannii*. *Antimicrob. Agents Chemother.* **2017**, *61*, e02531-16. [CrossRef]
20. Ito, A.; Nishikawa, T.; Matsumoto, S.; Yoshizawa, H.; Sato, T.; Nakamura, R.; Tsuji, M.; Yamano, Y. Siderophore Cephalosporin Cefiderocol Utilizes Ferric Iron Transporter Systems for Antibacterial Activity against Pseudomonas aeruginosa. *Antimicrob. Agents Chemother.* **2016**, *60*, 7396–7401. [CrossRef]
21. Hackel, M.A.; Tsuji, M.; Yamano, Y.; Echols, R.; Karlowsky, J.A.; Sahm, D.F. *In Vitro* Activity of the Siderophore Cephalosporin, Cefiderocol, against Carbapenem-Nonsusceptible and Multidrug-Resistant Isolates of Gram-Negative Bacilli Collected Worldwide in 2014 to 2016. *Antimicrob. Agents Chemother.* **2018**, *62*, e01968-17. [CrossRef]
22. Zhanel, G.G.; Golden, A.R.; Zelenitsky, S.; Wiebe, K.; Lawrence, C.K.; Adam, H.J.; Idowu, T.; Domalaon, R.; Schweizer, F.; Zhanel, M.A.; et al. Cefiderocol: A siderophore cephalosporin with activity against carbapenem- resistant and multidrug-resistant Gram-negative bacilli. *Drugs* **2019**, *79*, 271–289. [CrossRef]
23. EUCAST Disk Diffusion Manual v 11.0 (2 January 2023). Available online: https://www.eucast.org/fileadmin/src/media/PDFs/EUCAST_files/Disk_test_documents/2022_manuals/Cefiderocol_disk_diffusion_training.pdf (accessed on 10 November 2023).
24. Hackel, M.A.; Tsuji, M.; Yamano, Y.; Echols, R.; Karlowsky, J.A.; Sahm, D.F. Reproducibility of broth microdilution MICs for the novel siderophore cephalosporin, cefiderocol, determined using iron-depleted cation-adjusted Mueller-Hinton broth. *Diagn. Microbiol. Infect. Dis.* **2019**, *94*, 321–325. [CrossRef] [PubMed]
25. Clinical and Laboratory Standards Institute (CLSI). *Performance Standards for Antimicrobial Susceptibility Testing M100 S*, 33rd ed; Clinical and Laboratory Standards Institute: Wayne, PA, USA, 2023.
26. Liu, Y.; Ding, L.; Han, R.; Zeng, L.; Li, J.; Guo, Y.; Hu, F. Assessment of cefiderocol disk diffusion versus broth microdilution results when tested against *Acinetobacter baumannii* complex clinical isolates. *Microbiol. Spectr.* **2023**, *11*, e0535522. [CrossRef] [PubMed]
27. Jeannot, K.; Gaillot, S.; Triponney, P.; Portets, S.; Pourchet, V.; Fournier, D.; Potron, A. Performance of the Disc Diffusion Method, MTS Gradient Tests and Two Commercially Available Microdilution Tests for the Determination of Cefiderocol Susceptibility in *Acinetobacter* spp. *Microorganisms* **2023**, *11*, 1971. [CrossRef] [PubMed]
28. Raro, O.H.F.; Bouvier, M.; Kerbol, A.; Decousser, J.-W.; Poirel, L.; Nordmann, P. Rapid detection of cefiderocol susceptibility/resistance in *Acinetobacter baumannii*. *Eur. J. Clin. Microbiol. Infect. Dis.* **2023**, *42*, 1511–1518. [CrossRef] [PubMed]

29. Gill, C.M.; Santini, D.; Takemura, M.; Longshaw, C.; Yamano, Y.; Echols, R.; Nicolau, D.P. *In vivo* efficacy & resistance prevention of cefiderocol in combination with ceftazidime/avibactam, ampicillin/sulbactam or meropenem using human-simulated regimens versus *Acinetobacter baumannii*. *J. Antimicrob. Chemother.* **2023**, *78*, 983–990. [CrossRef] [PubMed]
30. Liu, X.; Chang, Y.; Xu, Q.; Zhang, W.; Huang, Z.; Zhang, L.; Weng, S.; Leptihn, S.; Jiang, Y.; Yu, Y.; et al. Mutation in the two-component regulator BaeSR mediates cefiderocol resistance and enhances virulence in *Acinetobacter baumannii*. *mSystems* **2023**, *8*, e0129122. [CrossRef] [PubMed]
31. Poirel, L.; Sadek, M.; Nordmann, P. Contribution of PER-type and NDM-type b-lactamases to Cefiderocol resistance in *Acinetobacter baumannii*. *Antimicrob. Agents Chemother.* **2021**, *65*, e00877-21. [CrossRef]
32. Liu, X.; Lei, T.; Yang, Y.; Zhang, L.; Liu, H.; Leptihn, S.; Yu, Y.; Hua, X. Structural Basis of PER-1-Mediated Cefiderocol Resistance and Synergistic Inhibition of PER-1 by Cefiderocol in Combination with Avibactam or Durlobactam in *Acinetobacter baumannii*. *Antimicrob. Agents Chemother.* **2022**, *66*, e0082822. [CrossRef]
33. Stracquadanio, S.; Bonomo, C.; Marino, A.; Bongiorno, D.; Privitera, G.F.; Bivona, D.A.; Mirabile, A.; Bonacci, P.G.; Stefani, S. *Acinetobacter baumannii* and Cefiderocol, between Cidality and Adaptability. *Microbiol. Spectr.* **2022**, *10*, e0234722. [CrossRef]
34. Longshaw, C.; Henriksen, A.S.; Dressel, D.; Malysa, M.; Silvestri, C.; Takemura, M.; Yamano, Y.; Baba, T.; Slover, C.M. Heteroresistance to cefiderocol in carbapenem-resistant *Acinetobacter baumannii* in the CREDIBLE-CR study was not linked to clinical outcomes: A post hoc analysis. *Microbiol. Spectr.* **2023**, *15*, e0237123. [CrossRef]
35. Choby, J.E.; Ozturk, T.; Satola, S.W.; Jacob, J.T.; Weiss, D.S. Does cefiderocol heteroresistance explain the discrepancy between the APEKS-NP and CREDIBLE-CR clinical trial results? *Lancet Microbe* **2021**, *2*, e648–e649. [CrossRef] [PubMed]
36. Wunderink, R.G.; Matsunaga, Y.; Ariyasu, M.; Clevenbergh, P.; Echols, R.; Kaye, K.S.; Kollef, M.; Menon, A.; Pogue, J.M.; Shorr, A.F.; et al. Cefiderocol versus high-dose, extended-infusion meropenem for the treatment of Gram-negative nosocomial pneumonia (APEKS-NP): A randomised, double-blind, phase 3, non-inferiority trial. *Lancet Infect. Dis.* **2020**, *21*, 213–225. [CrossRef] [PubMed]
37. Bassetti, M.; Echols, R.; Matsunaga, Y.; Ariyasu, M.; Doi, Y.; Ferrer, R.; Lodise, T.P.; Naas, T.; Niki, Y.; Paterson, D.L.; et al. Efficacy and safety of cefiderocol or best available therapy for the treatment of serious infections caused by carbapenem-resistant Gram-negative bacteria (CREDIBLE-CR): A randomised, open-label, multicentre, pathogen-focused, descriptive, phase 3 trial. *Lancet Infect. Dis.* **2021**, *21*, 226–240. [CrossRef] [PubMed]
38. Paul, M.; Daikos, G.L.; Durante-Mangoni, E.; Yahav, D.; Carmeli, Y.; Benattar, Y.D.; Skiada, A.; Andini, R.; Eliakim-Raz, N.; Nutman, A.; et al. Colistin alone versus colistin plus meropenem for treatment of severe infections caused by carbapenem-resistant Gram-negative bacteria: An open-label, randomised controlled trial. *Lancet Infect. Dis.* **2018**, *18*, 391–400. [CrossRef] [PubMed]
39. Durante-Mangoni, E.; Signoriello, G.; Andini, R.; Mattei, A.; De Cristoforo, M.; Murino, P.; Bassetti, M.; Malacarne, P.; Petrosillo, N.; Galdieri, N.; et al. Colistin and rifampicin compared with colistin alone for the treatment of serious infections due to extensively drug-resistant *Acinetobacter baumannii*: A multicenter, randomized clinical trial. *Clin. Infect. Dis.* **2013**, *57*, 349–358. [CrossRef] [PubMed]
40. Sirijatuphat, R.; Thamlikitkul, V. Preliminary Study of Colistin versus Colistin plus Fosfomycin for Treatment of Carbapenem-Resistant *Acinetobacter baumannii* Infections. *Antimicrob. Agents Chemother.* **2014**, *58*, 5598–5601. [CrossRef] [PubMed]
41. Kaye, K.S.; Marchaim, D.; Thamlikitkul, V.; Carmeli, Y.; Chiu, C.-H.; Daikos, G.; Dhar, S.; Durante-Mangoni, E.; Gikas, A.; Kotanidou, A.; et al. Colistin Monotherapy versus Combination Therapy for Carbapenem-Resistant Organisms. *NEJM Évid.* **2023**, *2*, EVIDoa2200131. [CrossRef]
42. Ho, Y.; Hu, F.; Lee, P. The Advantages and Challenges of Using Real-World Data for Patient Care. *Clin. Transl. Sci.* **2020**, *13*, 4–7. [CrossRef]
43. Falcone, M.; Tiseo, G.; Leonildi, A.; Della Sala, L.; Vecchione, A.; Barnini, S.; Farcomeni, A.; Menichetti, F. Cefiderocol- Compared to Colistin-Based Regimens for the Treatment of Severe Infections Caused by Carbapenem-Resistant *Acinetobacter baumannii*. *Antimicrob. Agents Chemother.* **2022**, *66*, e0214221. [CrossRef]
44. Russo, A.; Bruni, A.; Gullì, S.; Borrazzo, C.; Quirino, A.; Lionello, R.; Serapide, F.; Garofalo, E.; Serraino, R.; Romeo, F.; et al. Efficacy of cefiderocol- vs colistin-containing regimen for treatment of bacteraemic ventilator-associated pneumonia caused by carbapenem-resistant *Acinetobacter baumannii* in patients with COVID-19. *Int. J. Antimicrob. Agents* **2023**, *62*, 106825. [CrossRef]
45. Russo, A.; Bruni, A.; Gullì, S.; Borrazzo, C.; Quirino, A.; Lionello, R.; Torti, C. Effectiveness of First-Line Therapy with Old and Novel Antibiotics in Ventilator-Associated Pneumonia Caused by Carbapenem-Resistant *Acinetobacter baumannii*: A Real Life, Prospective, Observational, Single-Center Study. *Antibiotics* **2023**, *12*, 1048. [CrossRef]
46. Mazzitelli, M.; Gregori, D.; Sasset, L.; Trevenzoli, M.; Scaglione, V.; Lo Menzo, S.; Marinello, S.; Mengato, D.; Venturini, F.; Tiberio, I.; et al. Cefiderocol -Based versus Colistin-Based Regimens for Severe Carbapenem-Resistant *Acinetobacter baumannii* Infections: A Propensity Score-Weighted, Retrospective Cohort Study during the First Two Years of the COVID-19 Pandemic. *Microorganisms* **2023**, *11*, 984. [CrossRef] [PubMed]
47. Rando, E.; Cutuli, S.L.; Sangiorgi, F.; Tanzarella, E.S.; Giovannenze, F.; De Angelis, G.; Murri, R.; Antonelli, M.; Fantoni, M.; De Pascale, G. Cefiderocol-containing regimens for the treatment of carbapenem-resistant *A. baumannii* ventilator-associated pneumonia: A propensity-weighted cohort study. *JAC-Antimicrobial Resist.* **2023**, *5*, dlad085. [CrossRef] [PubMed]
48. Pascale, R.; Pascale, R.; Pasquini, Z.; Pasquini, Z.; Bartoletti, M.; Bartoletti, M.; Caiazzo, L.; Caiazzo, L.; Fornaro, G.; Fornaro, G.; et al. Cefiderocol treatment for carbapenem-resistant *Acinetobacter baumannii* infection in the ICU during the COVID-19 pandemic: A multicentre cohort study. *JAC-Antimicrobial Resist.* **2021**, *3*, dlab174. [CrossRef] [PubMed]

49. Bavaro, D.F.; Papagni, R.; Belati, A.; Diella, L.; De Luca, A.; Brindicci, G.; De Gennaro, N.; Di Gennaro, F.; Romanelli, F.; Stolfa, S.; et al. Cefiderocol Versus Colistin for the Treatment of Carbapenem-Resistant *Acinetobacter baumannii* Complex Bloodstream Infections: A Retrospective, Propensity-Score Adjusted, Monocentric Cohort Study. *Infect. Dis. Ther.* **2023**, *12*, 2147–2163. [CrossRef] [PubMed]
50. Calò, F.; Onorato, L.; De Luca, I.; Macera, M.; Monari, C.; Durante-Mangoni, E.; Massa, A.; Gentile, I.; Di Caprio, G.; Pagliano, P.; et al. Outcome of patients with carbapenem-resistant *Acinetobacter baumannii* infections treated with cefiderocol: A multicenter observational study. *J. Infect. Public Health* **2023**, *16*, 1485–1491. [CrossRef] [PubMed]
51. Giannella, M.; Verardi, S.; Karas, A.; Hadi, H.A.; Dupont, H.; Soriano, A.; Henriksen, A.S.; Cooper, A.; Falcone, M.; Viale, P.; et al. Carbapenem-Resistant *Acinetobacter* spp Infection in Critically Ill Patients with Limited Treatment Options: A Descriptive Study of Cefiderocol Therapy During the COVID-19 Pandemic. *Open Forum Infect. Dis.* **2023**, *10*, ofad329. [CrossRef]
52. Piccica, M.; Spinicci, M.; Botta, A.; Bianco, V.; Lagi, F.; Graziani, L.; Faragona, A.; Parrella, R.; Giani, T.; Bartolini, A.; et al. Cefiderocol use for the treatment of infections by carbapenem-resistant Gram-negative bacteria: An Italian multicentre real-life experience. *J. Antimicrob. Chemother.* **2023**, *78*, 2752–2761. [CrossRef]
53. Inchai, J.; Pothirat, C.; Bumroongkit, C.; Limsukon, A.; Khositsakulchai, W.; Liwsrisakun, C. Prognostic factors associated with mortality of drug-resistant *Acinetobacter baumannii* ventilator-associated pneumonia. *J. Intensiv. Care* **2015**, *3*, 9. [CrossRef]
54. Silva, D.; Lima, C.; Magalhães, V.; Baltazar, L.; Peres, N.; Caligiorne, R.; Moura, A.; Fereguetti, T.; Martins, J.; Rabelo, L.; et al. Fungal and bacterial coinfections increase mortality of severely ill COVID-19 patients. *J. Hosp. Infect.* **2021**, *113*, 145–154. [CrossRef]
55. Ballouz, T.; Aridi, J.; Afif, C.; Irani, J.; Lakis, C.; Nasreddine, R.; Azar, E. Risk Factors, Clinical Presentation, and Outcome of *Acinetobacter baumannii* Bacteremia. *Front. Cell. Infect. Microbiol.* **2017**, *7*, 156. [CrossRef] [PubMed]
56. Lee, C.M.; Kim, C.-J.; Kim, S.E.; Park, K.-H.; Bae, J.Y.; Choi, H.J.; Jung, Y.; Lee, S.S.; Choe, P.G.; Park, W.B.; et al. Risk factors for early mortality in patients with carbapenem-resistant *Acinetobacter baumannii* bacteraemia. *J. Glob. Antimicrob. Resist.* **2022**, *31*, 45–51. [CrossRef] [PubMed]
57. Palermo, G.; Medaglia, A.A.; Pipitò, L.; Rubino, R.; Costantini, M.; Accomando, S.; Giammanco, G.M.; Cascio, A. Cefiderocol Efficacy in a Real-Life Setting: Single-Centre Retrospective Study. *Antibiotics* **2023**, *12*, 746. [CrossRef] [PubMed]
58. Durante-Mangoni, E.; Andini, R.; Signoriello, S.; Cavezza, G.; Murino, P.; Buono, S.; De Cristofaro, M.; Taglialatela, C.; Bassetti, M.; Malacarne, P.; et al. Acute kidney injury during colistin therapy: A prospective study in patients with extensively-drug resistant *Acinetobacter baumannii* infections. *Clin. Microbiol. Infect.* **2016**, *22*, 984–989. [CrossRef] [PubMed]
59. Wagenlehner, F.; Lucenteforte, E.; Pea, F.; Soriano, A.; Tavoschi, L.; Steele, V.R.; Henriksen, A.S.; Longshaw, C.; Manissero, D.; Pecini, R.; et al. Systematic review on estimated rates of nephrotoxicity and neurotoxicity in patients treated with polymyxins. *Clin. Microbiol. Infect.* **2021**, *27*, 671–686. [CrossRef] [PubMed]
60. Falcone, M.; Tiseo, G.; Nicastro, M.; Leonildi, A.; Vecchione, A.; Casella, C.; Forfori, F.; Malacarne, P.; Guarracino, F.; Barnini, S.; et al. Cefiderocol as Rescue Therapy for *Acinetobacter baumannii* and Other Carbapenem-resistant Gram-negative Infections in Intensive Care Unit Patients. *Clin. Infect. Dis.* **2021**, *72*, 2021–2024. [CrossRef] [PubMed]
61. Bavaro, D.F.; Belati, A.; Diella, L.; Stufano, M.; Romanelli, F.; Scalone, L.; Saracino, A. Cefiderocol-Based Combination Therapy for "Difficult-to-Treat" Gram-Negative Severe Infections: Real-Life Case Series and Future Perspectives. *Antibiotics* **2021**, *10*, 652. [CrossRef]
62. Corcione, S.; De Benedetto, I.; Pinna, S.M.; Vita, D.; Lupia, T.; Montrucchio, G.; Brazzi, L.; De Rosa, F.G. Cefiderocol use in Gram negative infections with limited therapeutic options: Is combination therapy the key? *J. Infect. Public Health* **2022**, *15*, 975–979. [CrossRef]
63. Gavaghan, V.; Miller, J.L.; Dela-Pena, J. Case series of cefiderocol for salvage therapy in carbapenem-resistant Gram-negative infections. *Infection* **2023**, *51*, 475–482. [CrossRef]
64. Smoke, S.M.; Brophy, A.; Reveron, S.; Iovleva, A.; Kline, E.G.; Marano, M.; Miller, L.P.; Shields, R.K. Evolution and Transmission of Cefiderocol-Resistant *Acinetobacter baumannii* during an Outbreak in the Burn Intensive Care Unit. *Clin. Infect. Dis.* **2023**, *76*, e1261–e1265. [CrossRef]
65. Wicky, P.-H.; Poiraud, J.; Alves, M.; Patrier, J.; D'humières, C.; Lê, M.; Kramer, L.; de Montmollin, É.; Massias, L.; Armand-Lefèvre, L.; et al. Cefiderocol Treatment for Severe Infections due to Difficult-to-Treat-Resistant Non-Fermentative Gram-Negative Bacilli in ICU Patients: A Case Series and Narrative Literature Review. *Antibiotics* **2023**, *12*, 991. [CrossRef] [PubMed]
66. Karruli, A.; Massa, A.; Andini, R.; Marrazzo, T.; Ruocco, G.; Zampino, R.; Durante-Mangoni, E. Clinical efficacy and safety of cefiderocol for resistant Gram-negative infections: A real-life, single-centre experience. *Int. J. Antimicrob. Agents* **2023**, *61*, 106723. [CrossRef] [PubMed]
67. Takemura, M.; Yamano, Y.; Matsunaga, Y.; Ariyasu, M.; Echols, R.; Nagata, T.D. 1266. Characterization of Shifts in Minimum Inhibitory Concentrations During Treatment with Cefiderocol or Comparators in the Phase 3 CREDIBLE-CR and APEKS-NP Studies. *Open Forum Infect. Dis.* **2020**, *7*, S649–S650. [CrossRef]
68. E Choby, J.; Ozturk, T.; Satola, S.W.; Jacob, J.T.; Weiss, D.S. Widespread cefiderocol heteroresistance in carbapenem-resistant Gram-negative pathogens. *Lancet Infect. Dis.* **2021**, *21*, 597–598. [CrossRef] [PubMed]
69. Karakonstantis, S.; Rousaki, M.; Kritsotakis, E.I. Cefiderocol: Systematic Review of Mechanisms of Resistance, Heteroresistance and In Vivo Emergence of Resistance. *Antibiotics* **2022**, *11*, 723. [CrossRef] [PubMed]

70. Tamma, P.D.; Aitken, S.L.; Bonomo, R.A.; Mathers, A.J.; van Duin, D.; Clancy, C.J. Infectious Diseases Society of America 2023 Guidance on the Treatment of Antimicrobial Resistant Gram-Negative Infections. *Clin Infect Dis.* **2023**, ciad428. [CrossRef] [PubMed]
71. Oliva, A.; Ceccarelli, G.; De Angelis, M.; Sacco, F.; Miele, M.C.; Mastroianni, C.M.; Venditti, M. Cefiderocol for compassionate use in the treatment of complicated infections caused by extensively and pan-resistant *Acinetobacter baumannii. J. Glob. Antimicrob. Resist.* **2020**, *23*, 292–296. [CrossRef]
72. Dagher, M.; Ruffin, F.; Marshall, S.; Taracila, M.; A Bonomo, R.; Reilly, R.; Fowler, V.G.; Thaden, J.T. Case Report: Successful Rescue Therapy of Extensively Drug-Resistant *Acinetobacter baumannii* Osteomyelitis with Cefiderocol. *Open Forum Infect. Dis.* **2020**, *7*, ofaa150. [CrossRef]
73. Mabayoje, A.D.; NicFhogartaigh, C.; Cherian, B.P.; Tan, M.G.M.; Wareham, D.W. Compassionate use of cefiderocol for carbapenem-resistant *Acinetobacter baumannii* prosthetic joint infection. *JAC-Antimicrobial Resist.* **2021**, *3*, dlab109. [CrossRef]
74. Kufel, W.D.; Abouelhassan, Y.; Steele, J.M.; Gutierrez, R.L.; Perwez, T.; Bourdages, G.; Nicolau, D.P. Plasma and cerebrospinal fluid concentrations of cefiderocol during successful treatment of carbapenem-resistant *Acinetobacter baumannii* meningitis. *J. Antimicrob. Chemother.* **2022**, *77*, 2737–2741. [CrossRef]
75. Durand-Reville, T.F.; Guler, S.; Comita-Prevoir, J.; Chen, B.; Bifulco, N.; Huynh, H.; Lahiri, S.; Shapiro, A.B.; McLeod, S.M.; Carter, N.M.; et al. ETX2514 is a broad-spectrum b-lactamase inhibitor for the treatment of drug-resistant Gram-negative bacteria including *Acinetobacter baumannii. Nat. Microbiol.* **2017**, *2*, 17104. [CrossRef] [PubMed]
76. Shapiro, A.B. Kinetics of SUL hydrolysis by β-lactamases, and kinetics of β-lactamase inhibition by Sulbactam. *Antimicrob. Agents Chemother.* **2017**, *61*, e01612-17. [CrossRef] [PubMed]
77. FDA Approves New Treatment for Pneumonia Caused by Certain Difficult-to-Treat Bacteria. News Release. 23 May 2023. Available online: https://www.fda.gov/news-events/press-announcements/fda-approves-new-treatment-pneumonia-caused-certain-difficult-treat-bacteria (accessed on 12 November 2023).
78. Seifert, H.; Müller, C.; Stefanik, D.; Higgins, P.G.; Miller, A.; Kresken, M. In vitro activity of sulbactam/durlobactam against global isolates of carbapenem-resistant *Acinetobacter baumannii. J. Antimicrob. Chemother.* **2020**, *75*, 2616–2621. [CrossRef] [PubMed]
79. McLeod, S.M.; Shapiro, A.B.; Moussa, S.H.; Johnstone, M.; McLaughlin, R.E.; de Jonge, B.L.M.; Miller, A.A. Frequency and mechanism of spontaneous resistance to SUL combined with the novel β-lactamase inhibitor ETX2514 in clinical isolates of *Acinetobacter baumannii. Antimicrob. Agents Chemother.* **2018**, *62*, e01576-17. [CrossRef] [PubMed]
80. Kaye, K.S.; Kaye, K.S.; Shorr, A.F.; Shorr, A.F.; Wunderink, R.G.; Wunderink, R.G.; Du, B.; Du, B.; E Poirier, G.; E Poirier, G.; et al. Efficacy and safety of sulbactam–durlobactam versus colistin for the treatment of patients with serious infections caused by *Acinetobacter baumannii*–calcoaceticus complex: A multicentre, randomised, active-controlled, phase 3, non-inferiority clinical trial (ATTACK). *Lancet Infect. Dis.* **2023**, *23*, 1072–1084. [CrossRef] [PubMed]
81. Zaidan, N.; Hornak, J.P.; Reynoso, D. Extensively Drug-Resistant *Acinetobacter baumannii* Nosocomial Pneumonia Successfully Treated with a Novel Antibiotic Combination. *Antimicrob. Agents Chemother.* **2021**, *65*, e0092421. [CrossRef] [PubMed]
82. Holger, D.J.; Coyne, A.J.K.; Zhao, J.J.; Sandhu, A.; Salimnia, H.; Rybak, M.J. Novel Combination Therapy for Extensively Drug-Resistant *Acinetobacter baumannii* Necrotizing Pneumonia Complicated by Empyema: A Case Report. *Open Forum Infect. Dis.* **2022**, *9*, ofac092. [CrossRef]
83. Tiseo, G.; Giordano, C.; Leonildi, A.; Riccardi, N.; Galfo, V.; Limongi, F.; Nicastro, M.; Barnini, S.; Falcone, M. Salvage therapy with sulbactam/durlobactam against cefiderocol-resistant *Acinetobacter baumannii* in a critically ill burn patient: Clinical challenges and molecular characterization. *JAC-Antimicrobial Resist.* **2023**, *5*, dlad078. [CrossRef]
84. Durante-Mangoni, E.; Utili, R.; Zarrilli, R. Combination therapy in severe *Acinetobacter baumannii* infections: An update on the evidence to date. *Futur. Microbiol.* **2014**, *9*, 773–789. [CrossRef]

antibiotics

Article

Probable Three-Species In Vivo Transfer of bla_{NDM-1} in a Single Patient in Greece: Occurrence of NDM-1-Producing *Klebsiella pneumoniae*, *Proteus mirabilis*, and *Morganella morganii*

Georgios Meletis [1,*], Andigoni Malousi [2], Areti Tychala [1], Angeliki Kassomenaki [1], Nikoletta Vlachodimou [1], Paraskevi Mantzana [1], Simeon Metallidis [3], Lemonia Skoura [1] and Efthymia Protonotariou [1]

[1] Department of Microbiology, AHEPA University Hospital, School of Medicine, Aristotle University of Thessaloniki, 54636 Thessaloniki, Greece; aretich@gmail.com (A.T.); angelikikasso@gmail.com (A.K.); nikoletta.vlachodimou@gmail.com (N.V.); vimantzana@gmail.com (P.M.); mollyskoura@gmail.com (L.S.); protonotariou@gmail.com (E.P.)

[2] Laboratory of Biological Chemistry, School of Medicine, Aristotle University of Thessaloniki, 54124 Thessaloniki, Greece; andigoni@auth.gr

[3] First Department of Internal Medicine, Infectious Diseases Division, AHEPA University Hospital, School of Medicine, Aristotle University of Thessaloniki, 54636 Thessaloniki, Greece; metallidissimeon@yahoo.gr

* Correspondence: meletisg@hotmail.com; Tel.: +30-697-428-2575

Abstract: NDM carbapenemase-encoding genes disseminate commonly among Enterobacterales through transferable plasmids carrying additional resistance determinants. Apart from the intra-species dissemination, the inter-species exchange of plasmids seems to play an additional important role in the spread of bla_{NDM}. We here present the genetics related to the isolation of three species (*Klebsiella pneumoniae, Proteus mirabilis*, and *Morganella morganii*) harboring the bla_{NDM-1} gene from a single patient in Greece. Bacterial identification and antimicrobial susceptibility testing were performed using the Vitek2. Whole genome sequencing and bioinformatic tools were used to identify resistance genes and plasmids. Bla_{NDM-1} harboring plasmids were found in all three isolates. Moreover, the plasmid constructs of the respective incomplete or circular contigs showed that the bla_{NDM-1} and its neighboring genes form a cluster that was found in all isolates. Our microbiological findings, together with the patient's history, suggest the in vivo transfer of the bla_{NDM-1}-containing cluster through three different species in a single patient.

Keywords: *Klebsiella pneumoniae*; carbapenemases; NDM; plasmid; *Proteus mirabilis*; *Morganella morganii*

check for updates

Citation: Meletis, G.; Malousi, A.; Tychala, A.; Kassomenaki, A.; Vlachodimou, N.; Mantzana, P.; Metallidis, S.; Skoura, L.; Protonotariou, E. Probable Three-Species In Vivo Transfer of bla_{NDM-1} in a Single Patient in Greece: Occurrence of NDM-1-Producing *Klebsiella pneumoniae, Proteus mirabilis*, and *Morganella morganii*. *Antibiotics* **2023**, *12*, 1206. https://doi.org/10.3390/antibiotics12071206

Academic Editor: Masafumi Seki

Received: 18 June 2023
Revised: 11 July 2023
Accepted: 18 July 2023
Published: 20 July 2023

1. Introduction

The emergence and spread of carbapenem-resistant Gram-negative nosocomial pathogens, especially those producing one or more of the major carbapenemases, are of great concern to public health [1]. Carbapenemases are enzymes that are able to hydrolyze carbapenems and the majority of β-lactams. *Klebsiella pneumoniae* carbapenemase (KPC), Verona integron-encoded metallo-β-lactamase (VIM), Imipenemase (IMP), Oxacillinase-48 (OXA-48), and the later New Delhi metallo-β-lactamase (NDM) are clinically the most important carbapenemases because they often confer high-level resistance to carbapenems and have successfully spread since they are mainly located on a variety of plasmids [2].

The existence of plasmids harboring resistance genes, initially called R-factors, has been known for decades, and nowadays, the correlation of resistance plasmids with the spread of resistance genes to new organisms is widely accepted. More light was shed on the understanding of antimicrobial resistance dissemination with the later discovery of smaller and more versatile mobile genetic elements such as transposons and gene cassettes accompanied by integrons. Using such elements as vehicles, resistance genes are able to relocate among different DNA molecules, including the bacterial chromosome and various plasmids. In such cases, and when the plasmids are conjugative or mobilizable, resistance genes can be transferred to other bacterial cells of the same, and in some cases, of different species.

In particular for NDMs, their extensive dissemination among different Enterobacterales is mainly due to the presence of the *bla*$_{NDM}$ gene in transferable plasmids of various incompatibility groups that propagate within each species, including incompatibility group A/C-type (IncA/C-type), incompatibility group N (IncN), and incompatibility group F (IncF) types [3–5]. Moreover, these plasmids commonly carry additional resistance determinants to antimicrobial categories other than the β-lactams, thus contributing to multi-drug resistance (MDR), leaving very few therapeutic options. Interspecies dissemination is deemed to play an important role as well, even though it is not frequently documented. At the clinical level, the presence of NDM is important for yet another reason. New antibiotic combinations with novel β-lactamase inhibitors such as ceftazidime/avibactam and meropenem/vaborbactam have been made available lately. Unfortunately, however, these are not active for a specific category of carbapenemases called the metallo-β-lactamases. NDM enzymes, together with VIM and IMP, belong to this category for which no clinically useful inhibitors are available for the time being. Due to the aforementioned reasons, and since carbapenems are the safest last line treatment option for Gram-negative bacterial infections, the detection of NDM in hospital settings is considered of high importance, and measures are needed to tackle its dissemination.

This study reports the isolation of three *bla*$_{NDM-1}$–harboring Enterobacterales (*Klebsiella pneumoniae*, *Proteus mirabilis*, and *Morganella morganii*) from a single patient in Greece together with whole genome sequencing analysis that suggest in vivo transfer of the NDM-1-encoding gene.

2. Results

2.1. Patient History

In April 2021, a 67-year-old male was admitted to AHEPA University Hospital (Thessaloniki, Greece) due to fever, cough, and shortness of breath. After testing positive for severe acute respiratory syndrome coronavirus 2 (SARS-CoV-2) by PCR, the patient was hospitalized for coronavirus infectious disease 2019 (COVID-19). From his medical record, the patient had a history of pre-existing and well-managed type-2 diabetes mellitus. During his hospitalization, fecal surveillance samples taken for infection control purposes identified a carbapenem-resistant (CR) *K. pneumoniae*. The patient's clinical condition soon deteriorated, and he was transferred to the intensive care unit (ICU) where the patient was intubated. During the ICU stay, the patient was additionally colonized with an extensively drug-resistant *Acinetobacter baumannii* strain. The patient underwent a series of co-infections during the hospitalization in the ICU. These included a carbapenem-resistant *K. pneumoniae* blood infection (isolate D730) on 5 May 2021 and *P. mirabilis* and *A. baumannii* lower respiratory tract infections on May 5 and 10, respectively. Susceptibility testing was performed upon isolation for each strain. The patient's treatment included colistin, tigecycline, ceftazidime/avibactam, and aminoglycosides. The patient was discharged from the ICU on 14 July 2021, hospitalized in the Internal Medicine Department, and left the hospital on 19 July 2021.

A few months later, however, on 8 October 2021, he was again admitted to the hospital presenting with respiratory infection and final stage renal disease. Blood cultures taken upon admission gave positive results for a carbapenem-resistant *P. mirabilis* (isolate D1633) and a carbapenem-resistant *M. morganii* (isolate D1644). CR *P. mirabilis* was also recovered from the patient's urine culture. In the following days the patient's clinical condition deteriorated, necessitating his transfer to the ICU. On the seventh day of his hospitalization, blood cultures were repeated and showed a simultaneous infection by CR *P. mirabilis*, CR *M. morganii*, and *Enterococcus faecium*. The patient went into septic shock and eventually died of cardiac arrest on 15 October 2021.

2.2. Susceptibility Testing

All three study isolates were multi-drug-resistant, presenting high minimum inhibitory concentrations (MICs) to β-lactams, including carbapenems (Table 1).

Table 1. Minimum inhibitory concentrations (MICs) of various antimicrobials for the three study isolates tested with the Vitek2 and the Micronaut-S broth microdilution method (BMD).

	MIC (mg/L)/Interpretation											
Antimicrobials	*K. pneumoniae*				*P. mirabilis*				*M. morganii*			
	Vitek2		BMD		Vitek2		BMD		Vitek2		BMD	
Amikacin	–	–	>32	R	-	-	>32	R	8	S	4	S
Amoxicillin/clavulanic acid	≥32	R			≥32	R						
Ampicillin	≥32	R			≥32	R						
Aztreonam	≥64	R			32	R			≤1	S		
Cefepime	≥32	R			16	R			≥64	R		
Cefixime	≥4	R			≥4	R						
Cefotaxime	≥64	R			≥64	R					>2	R
Cefoxitin	≥64	R			≥64	R						
Ceftolozan/tazobactam	≥32	R	>2	R	≥32	R	>2	R			>2	R
Ceftazidime	≥64	R	>128	R	≥64	R	128	R	≥64	R	>128	R
Ceftazidime/avibactam	≥16	R	>4	R	≥16	R	>4	R			>4	R
Ceftriaxone	≥64	R			≥64	R						
Cefuroximeaxetil	≥64	R			≥64	R						
Cefuroxime	≥64	R			≥64	R						
Colistin	≥16	R	>8	R	≥16	R	>8	R	≥16	R	8	R
Ciprofloxacin	≥4	R	>2	R	≥4	R	>2	R	≥4	R	>2	R
Ertapenem	≥8	R			≥8	R						
Fosfomysin	≥256	R	>128	R	≥256	R	>128	R	128	R	128	R
Gentamicin	≥16	R			≥16				2	S		
Chloramphenicol	32	R	>16	R	≥64	R	>16	R			8	S
Piperacillin	–	–	>16	R	–	–	>16	R	≥128	R	>16	R
Piperacillin/tazobactam	≥128	R	>16	R	64	R	>4	I	≥128	R	>16	R
Ticarcillin/clavulanic acid	–	–			–	–			≥128	R		
Ticarcillin	–	–			–	–			≥128	R		
Tobramycin	≥16	R			≥16	R			8	R		
Levofloxacin	≥8	R	>2	R	≥8	R	>2	R			>2	R
Trimeth/sulfamethoxazole	≥320	R	>4/76	R	≥320	R	>4/76	R	≥320	R	>4/76	R
Cefotaxime	≥64	R	>2	R	≥64	R	>2	R				
Tigecycline	–		1	–	–	–	2	–			0.5	
Imipenem	–	–	>8	R	–	–	>8	R	≥16	R	>8	R
Meropenem	–	–	128	R	–	–	32	R	≥16	R	32	R

2.3. Genomic Characterization of the Isolates

Whole genome sequencing (WGS) was performed on the three isolates using the same sequencing Ion Torrent protocols and data analysis pipelines. The main WGS summary statistics are shown in Table 2. Overall, 0.7–1.7 M reads per isolate were produced. The read length was normally distributed across isolates with an average 305–311 bp and the duplication rate was 13.2%. The assembled genomes contained on average 90 contigs for *P. mirabilis* and *M. morganii*. *K. pneumoniae* assembly was significantly more fragmented due to the 35% larger genome size, yet the N50 value was similar to that of *P. mirabilis*.

Table 2. Summary statistics of the assembled genomes.

Isolate ID	Species	Genome Length	% Bacteria Abundance	N50	# Contigs
D730	*Klebsiella pneumoniae*	5,772,350	78	169,475	156
D1633	*Proteus mirabilis*	4,256,486	69	174,791	89
D1644	*Morganella morganii*	4,108,218	84	231,582	91

2.4. Taxonomic Assignment

The taxonomic classification of the isolates showed that the abundance of the *P. mirabilis*, *M. morganii*, and *K. pneumoniae* species was on average 77% among other bacteria (Table 2). Taxonomic assignments to *E. coli*-like sequences covered 7% of the total bacteria-derived *P. mirabilis* and *M. morganii* sequencing reads and 3% of the *K. pneumoniae*. Half of the *M. morganii* sequencing reads belonged to *morganii* subspecies, while the remaining did not have any subspecies assignments. *P. mirabilis* sequencing reads were further taxonomically assigned to the HI4320 (10% of the reads) and BB2000 (17% of the reads) strains, while the remaining had no further strain assignment. *K. pneumoniae* isolate was taxonomically assigned to *pneumoniae* subspecies (40% of the reads), while 54% of the *K. pneumoniae* reads had further taxonomic assignment. The remaining 6% classified to a wide range of *K. pneumoniae* strains and subspecies, including rhinoscleromatis SB3432 (*subsp*), *K. pneumoniae* ozaenae (*subsp*), *K. pneumoniae* MGH 39 (*str*), and *K. pneumoniae* 500_1420 (*str*).

2.5. Antimicrobial Resistance Genes

In silico analysis detected 48 antimicrobial resistance (AMR) genes (>98% identity). *K. pneumoniae* and *P. mirabilis* carried 16 common AMR genes that corresponded to almost half of the total genes (Figure 1). *OqxAB* genes conferring resistance to fluoroquinolones were detected in *K. pneumoniae*. *M. morganii* included 12 AMR genes. Among these, *aac(6')-Ib-cr*, *bla*$_{NDM-1}$, *bla*$_{OXA-10}$, *qacE*, and *sul1* were shared among all isolates.

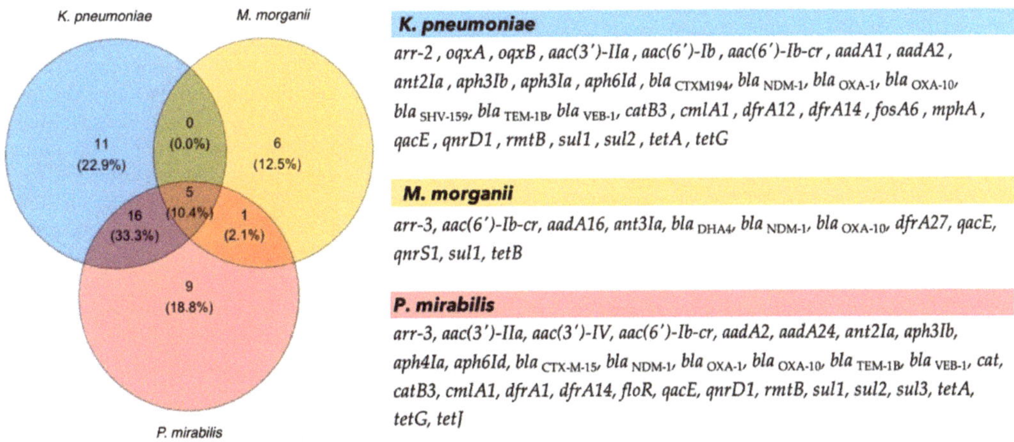

Figure 1. Number of shared (**left**) and list of (**right**) AMR genes detected in each isolate.

2.6. Replicon Types of Plasmids

The replicon types of the detected plasmids, according to PlasmidFinder, are shown in Table 3. The Inc plasmid family is very frequent in Enterobacterales. In this study, Inc plasmids were detected in all isolates in a single or in multiple subtypes. IncC and IncFIA(HI1) types were found in *K. pneumoniae* and *P. mirabilis* isolates. *M. morganii* included an IncN plasmid subtype. bla_{NDM-1} was identified in IncFIA(HI1) replicons for the *K. pneumoniae* and *P. mirabilis* isolates. The initial gene detection analysis of the IncN replicon type in *M. morganii* did not verify the presence of bla_{NDM-1} in the assembled 6.5 kb contig. Further analysis of the assembled *M. morganii* genome using MOB-suite revealed the presence of bla_{NDM-1} in a plasmid sequence that is classified to the AA552 plasmid type and that is identical to the closest cluster type of the IncN replicon. *M. morganii* includes a relaxase of the MOB_F type in a plasmid contig that is of the same AA552 cluster type. Both *K. pneumoniae* and *P. mirabilis* carried a relaxase of the MOB_H family on the IncC replicon. Overall, relaxases are essential components for conjugative DNA transfer, and both MOB_F and MOB_H types are prevalent in transmissible plasmids hosted in γ-Proteobacteria.

Table 3. List of plasmid types per isolate detected by PlasmidFinder. AMR gene: Genes conferring resistance in the corresponding contig; Conf: Confidence level [range: 0–1] of being a true plasmid; Cluster: Closest group of plasmids.

Species	Plasmid Types	Conf/Cluster	AMR Genes
M. morganii	IncN	0.555/AA552(I)	-
	Col3 M	0.745/AB434(C)	*qnrD1*
P. mirabilis	IncC	1.000/AA860(C)	-
	IncFIA(HI1)	1.000/AB187(I)	bla_{NDM-1}
	ColRNAI	1.000/AA941(I)	-
K. pneumoniae	IncC	1.000/AA860(C)	-
	IncFIA(HI1)	0.996/AA964(I)	bla_{NDM-1}
	IncFIB(K)	0.045/AA275(I)	-

2.7. Construction and Comparative Analysis of the Plasmids Harboring bla_{NDM-1}

Plasmids that carry bla_{NDM-1} were further analyzed in order to build the topological features of the plasmid constructs and to assess the level of gene co-localization. Overall, the plasmid constructs of the respective incomplete contigs showed that bla_{NDM-1} and its neighboring genes are found in all isolates. Table 4 lists relevant genomic features of the plasmids harboring the bla_{NDM-1} gene, including the confidence level of being a plasmid and respective cluster group, the putative origin of transfer (oriT) sequences, and distance of the nearest neighbor. During the conjugation process, oriT sequences are recognized by relaxases and initiate the transfer. *oriT* sequences of MOB_F type (NZ_CP016035) were found in both *K. pneumoniae* and *P. mirabilis*. In *M. morganii* no oriT sequences are identified; however, a strong homology with the same MOB_F type is detected, as shown in the corresponding track in Figure 2. Notably, all isolates lacked any membrane-associated mating pair formation (MPF) complex, implying the presence of mobilizable bla_{NDM-1}-containing plasmids for *K. pneumoniae* and *P. mirabilis*. Overall, all bla_{NDM-1} plasmids were identified as incomplete.

Table 4. Genomic features of the bla_{NDM-1} plasmids; Conf: Confidence level [range: 0–1] of being a plasmid; Cluster: Closest group of plasmids; OriT: MOB-group to oriT sequences; Type: Plasmid type, Neighbor: Nearest species, MASH dist: Genome-based pairwise distance.

Isolate	Conf	Cluster	OriT	Type	Neighbor	MASH Dist
K. pneumoniae	0.996	AA964	MOB_F	IncFIA	*K. pneumoniae*	0.036
P. mirabilis	1.000	AB187	MOB_F	IncFIA	*K. pneumoniae*	0.016
M. morganii	0.304	AA552	-	-	*K. pneumoniae*	0.006

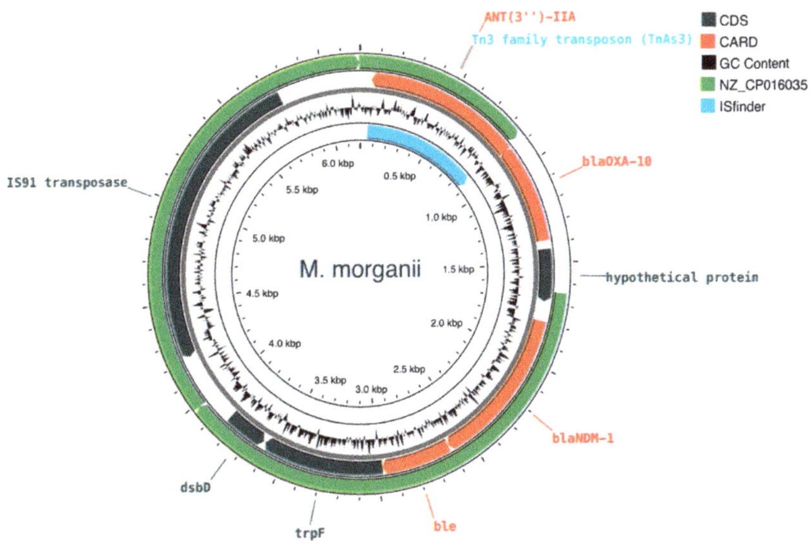

Figure 2. Circular map of the bla_{NDM-1}-containing plasmid of *M. morganii*. Features colored in red correspond to AMR genes identified by CARD. Grey colored tracks correspond to additional coding regions other than AMR genes, of the plasmid (CDS). The green track shows the homology level with the MOB_F (NZ_CP016035) oriT type. The hypothetical protein on the left side of the plasmid belongs to an integration/excision mobile element.

K. pneumoniae (ST11) was the most abundant isolate in plasmids and AMR genes (Figure 1, Table 3). The bla_{NDM-1} was found on a mobilizable 16,354 bp plasmid with a MOB_F oriT type closest to *K. pneumoniae* (Figure 3). bla_{NDM-1} colocalized with *ble*, and both were enclosed by the *trpF* and *dsbD* genes on the 3′ end, and by the *Tn3* family transposase element on the 5′ end.

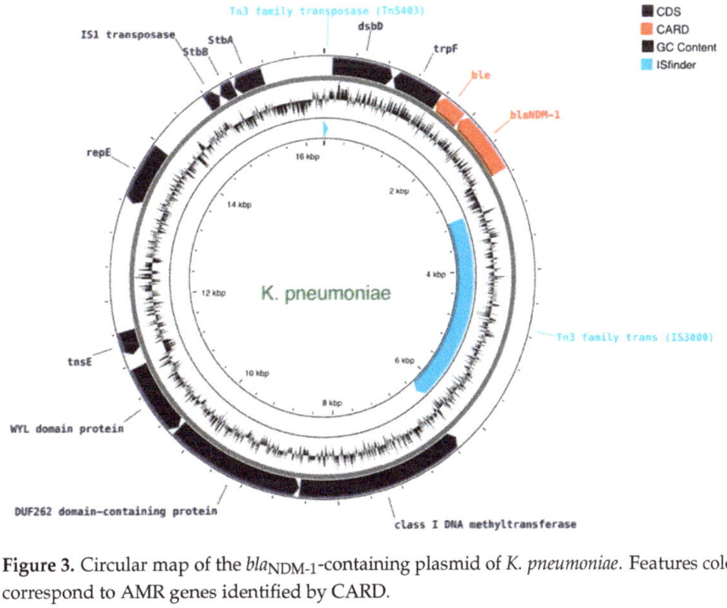

Figure 3. Circular map of the bla_{NDM-1}-containing plasmid of *K. pneumoniae*. Features colored in red correspond to AMR genes identified by CARD.

Similar organizational features are present on the bla_{NDM-1}-containing plasmid *P. mirabilis* isolate. The genome of *P. mirabilis* includes a mobilizable plasmid of 35,703 bp (57.54× depth of coverage) that contains bla_{NDM-1} and is identified as being closest to *K. pneumoniae* (MOB_F, MASH: 0.006). The plasmid was identified by both Deeplasmid and plasmidSPAdes (Figure 4).

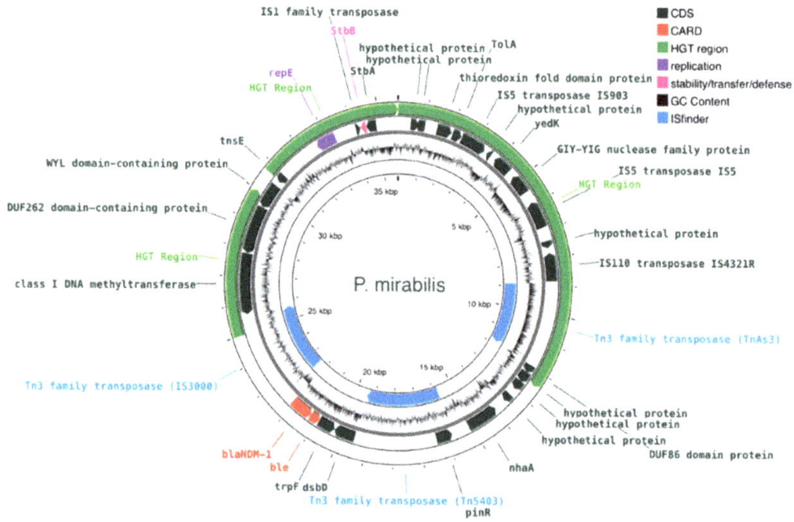

Figure 4. Circular map of the bla_{NDM-1}-containing plasmid of *P. mirabilis*. Features colored in red correspond to AMR genes identified by CARD. The HGT region colored in green corresponds to a horizontal gene transfer region.

In *M. morganii*, the bla_{NDM-1} was identified in a 6191 bp contig (162× depth) that was characterized as a plasmid-deriving sequence by both PlasmidSPAdes and MOB-suite. AA552 is the closest cluster type of the bla_{NDM-1}-containing plasmid, which is also the closest cluster type of the IncN replicon detected in the *M. morganii* isolate (Table 3). Besides bla_{NDM-1}, the plasmid contains three other AMR genes, bla_{OXA-10}, *ant(3″)-IIa*, and *ble* (Figure 2). In addition, the plasmid includes a mobile genetic element that is involved in integration/excision mechanisms (WP_000050481.1, mobileOG_000731050). Moreover, in all three isolates, the best scoring matches (e-value = 0) of the bla_{NDM-1}-containing plasmids with the insertion sequences of the ISfinder database belong to the Tn3 family of replicative transposons. The sequences producing high-confidence alignments are shared by at least two isolates, and include IS3000 (*K. pneumoniae*, *P. mirabilis*), Tn5403 (*K. pneumoniae*, *P. mirabilis*), and TnAS3 (*M. morganii*, *P. mirabilis*) insertion elements.

Overall, on the gene level (Figures 2–4), *M. morganii*, *K. pneumoniae*, and *P. mirabilis* include the same region covering four genes: *dsbD–trpF–ble–*bla_{NDM-1}. The relative position and orientation of these genes are aligned in the three isolates (same-stranded *trpF–ble–*bla_{NDM-1}, and *dsbD* on the opposite strand). bla_{NDM-1} and *ble* are systematically found in adjacent loci and are co-expressed as they share a common promoter upstream: the bla_{NDM-1} [6].

The genome similarity of the bla_{NDM-1}-containing plasmids was further assessed on the genomic level using the average nucleotide identity that was calculated by FastANI 1.3.3. Despite the highly variable lengths, bla_{NDM-1}-containing plasmids shared common genomic regions of significant identity. *K. pneumoniae* and *P. mirabilis* were less divergent compared with *M. morganii*. The latter was closer to *K. pneumoniae* considering both the overlapping genome length and the identity level. An important confounding factor for all isolates was the incomplete plasmids constructs.

3. Discussion

NDM-1 carbapenemase was first reported from Sweden in 2008 from a patient previously hospitalized in New Delhi, India [7]. The encoding gene, however, was already widely disseminated in the Indian subcontinent, and to a lesser extent, in other parts of the world. After its first identification, rapid worldwide spread occurred throughout all continents [8,9] thus placing NDM-1 and its variants among the most successful carbapenemases in terms of epidemiology and dissemination. This is mostly due to the incorporation of bla_{NDM-1} in transposable genetic elements and its horizontal transfer by plasmid replicons. Indeed, bla_{NDM-1} is commonly detected in conjugative plasmids of various incompatibility groups, through which rapid intra-species and even inter-species spread is possible. In Greece, NDM-1 was first reported in late 2011 [10]. Since then, it has been well-established, together with KPC, VIM, and OXA-48 carbapenemases, as also shown by recent data from our hospital [11]. Unfortunately, during the COVID-19 pandemic there was an increase in carbapenem resistance rates in our institution that was attributed to the increased number of COVID-19 patient admissions, their prolonged time of hospitalization, and the extensive antimicrobial treatment that these patients received. Even worse, the personnel dedicated to infection control were constrained [12].

The bacterial isolates described in this study were recovered by a COVID-19 patient hospitalized during the predominance of the Alpha SARS-CoV-2 variant of concern in our region [13] that caused increased pressure in the national healthcare system. For this reason, the case described here could be characterized as indicative of the outcomes of the concomitant presence of the pandemic with an endemic situation regarding carbapenemases among Gram-negative nosocomial opportunistic pathogens. Before the patient was admitted in the ICU, he was already colonized by a carbapenem-resistant *K. pneumoniae*, whereas, during his ICU stay, he was moreover colonized by a carbapenem-resistant *Acinetobacter baumannii* strain. Among other co-infections for which he received a series of last-line antibiotic treatment, the patient presented blood infection by an NDM-1-producing *K. pneumoniae*. Even though initially discharged from the hospital, he was admitted again after a few months. During this second hospitalization, he presented blood infection with an NDM-1-producing *P. mirabilis* and an NDM-1-producing *M. morganii* resulting in his death by septic shock. Since NDM-1-producing *K. pneumoniae* are common in our hospital but this specific type of carbapenemase is not frequently encountered in *P. mirabilis* and *M. morganii* in our institution, further laboratory investigation was needed to document in vivo transfer.

Plasmids are important "vehicles" for gene transfer among bacteria. Conjugation and mobilization of plasmids may confer clinically relevant traits to host cells, promoting their rapid evolution and adaptation to challenging environments [14]. This is particularly evident for plasmids carrying antimicrobial resistance genes in Enterobacterales like the IncF group (or MOB_F according to relaxase typing) [15]. These are the most frequently described conjugative plasmids from human and animal sources and are commonly related to antimicrobial resistance gene transfer and spread [15].

In this study, conjugative plasmids and numerous resistance genes were found in all isolates; interestingly, bioinformatics analyses showed that the bla_{NDM-1} and its neighboring genes built a cluster of ordered genes that was present in all three isolates. This cluster was found in IncFIA plasmids of *K. pneumoniae* and *P. mirabilis*, indicating in vivo transfer. The same cluster was found in a plasmid sequence of *M. morganii*, but unfortunately, this specific plasmid could not be characterized regarding its incompatibility group. Thus, a process of conjugation, mobilization of the cluster to another plasmid in *M. morganii*, and then loss of the first plasmid due to fitness costs cannot be excluded. The preservation of the bla_{NDM-1}-containing cluster was probably favored by the presence of last-line antimicrobials like ceftazidime/avibactam and the versatility of this carbapenemase-encoding gene to incorporate with different types of conjugative plasmid backbones. Indeed, the spread of bla_{NDM-1} through IncFIA plasmids is well documented in the literature [16–19].

A case of in vivo transfer of the *bla*$_{VIM}$ among three different bacterial species in Greece was published in 2012 [20]. To our knowledge, however, this is the first report of three Enterobacterales isolated from the same patient in Greece harboring the *bla*$_{NDM-1}$ with the same surrounding gene cluster located in transferable plasmids. In accordance with our study, recent similar reports from other countries indicate the occurrence of inter-species NDM-1 in vivo transfer. In Portugal, NDM-1-producing *M. morganii* and *P. mirabilis* were isolated from a single patient [21], and the same was reported in Italy for three Enterobacterales: *K. pneumoniae*, *P. mirabilis* and *Enterobacter cloacae*, suggesting a possible transfer of the *bla*$_{NDM-1}$ among clinical species [22]. This in vivo transfer of *bla*$_{NDM-1}$ in hospitalized patients is not limited to species frequently encountered in clinical environments, since it has been reported also among rather less common microorganisms. In the Republic of Korea, this transfer was documented from a *Klebsiella oxytoca* to a *Citrobacter freundii* isolate [23]. In China, an NDM-1-producing *Raoultella ornithinolytica* and an NDM-1-producing *E. cloacae* were isolated from a single patient, indicating in vivo conjugation [24].

This study has a certain limitation. The indications are drawn based on incomplete, non-circular constructs of the plasmids that contain *bla*$_{NDM-1}$. Since the complete reconstruction was not possible, we may have missed even more robust evidence of in vivo transfer or additional information about the structures of the three plasmids.

4. Materials and Methods

4.1. Hospital Setting and Patient Data

The study was conducted in AHEPA University Hospital, a 700-bed tertiary care hospital in Thessaloniki, Greece that served as one of the reference hospitals for COVID-19 patients in Northern Greece during the pandemic. The isolates described in the present study were collected as part of the standard of care protocol. Patient history data were retrieved by the hospital's electronic database.

4.2. Susceptibility Testing

Bacterial identification and antimicrobial susceptibility testing were performed using the Vitek2 semi-automated system (bioMérieux, Marcy l'Étoile, France). Confirmatory susceptibility testing using the gold standard broth microdilution method was performed where applicable by the MICRONAUT-S MDR MRGN-Screening system (Bruker Daltonics GmbH & Co. KG, Bremen, Germany). The interpretation of susceptibility testing results was done according to the European Committee on Antimicrobial Susceptibility Testing (EUCAST) 2021 clinical breakpoints for bacteria (v 11.0) available at: https://www.eucast.org/ast_of_bacteria/previous_versions_of_documents (accessed on: 30 July 2023).

4.3. Whole Genome Sequencing and Genomes Assembly

Libraries were prepared using Ion Torrent technology and Ion Chef workflows (Thermo Scientific, Waltham, MA, USA). Sequencing was performed in the S5XLS system, and analysis of the raw sequencing data was conducted by Ion Torrent Suite v.5.10.0, according to the manufacturer's instructions. Sequencing of the D1644 isolate performed twice to resolve ambiguities related to the presence of an IncN plasmid type harboring *bla*$_{NDM-1}$. The detection of the plasmid type and downstream analysis for D1644 performed on the merged sequencing products. Raw sequencing data were quality-checked and trimmed to keep only the reads that match the minimum length and quality criteria. Taxonomic assignment was performed as a quality step to filter out samples that may have been contaminated by foreign DNA during sample preparation and to verify the presence of the isolated species. Kraken2 was used for the taxonomy classification built on the standard database that includes NCBI taxonomic information, complete RefSeq microbe genomes, human genomes, and a vector collection [25]. Genomes were de novo assembled by AssemblerSPAdes and SPAdes using the default settings for the k-mers and the Ion Torrent parameter [26].

4.4. Functional Annotation of the Assembled Genomes

To identify and annotate genes on the assembled chromosomal and plasmid contigs, we used Prokka v.1.14.6, based on the *Enterococcus* genomic features [27]. Open reading frames (ORFs) were further annotated by manual comparative curation and determination of sequence similarity using the BLASTn searches. The antimicrobial resistant determinants and plasmid types were detected by Staramr version 0.9.1 [28] that integrates diverse molecular profiling tools, including CGE's Resfinder [29] and PlasmidFinder [30]. To classify contigs either as plasmids or chromosomal, we applied a deep learning methodology called Deeplasmid, that post-assembles plasmids using a combination of assembled sequences and sequence features [31]. In addition, to further verify the synthesis of the plasmid sequences, we used plasmidSPAdes [26] and MOB-Recon [32]. MOB-Recon applies an alternative approach that relies on clustered plasmid reference databases to reconstruct plasmids. MASH indices were calculated to assess the distance of the plasmids against plasmid clusters [33]. ISfinder was used to identify insertion sequence segments in the detected $bla_{\text{NDM-1}}$-containing plasmids. Nucleotide similarities between genomes were calculated by FastANI [34].

Proksee was used to build circular maps of the $bla_{\text{NDM-1}}$-containing plasmids [35]. Plasmid sequences were further annotated based on the presence of AMR genes by CARD Resistance Gene Identifier v.5.2.1 [36] and mobile genetic elements by mobileOG-db [37]. $bla_{\text{NDM-1}}$ genes in the three isolates were examined for the presence of genomic alterations and aligned to assess the evolutionary relationship among isolates. To detect horizontal gene transfer events, we applied an interpolated variable order motifs algorithm [38].

5. Conclusions

Based on our molecular microbiology data and the patient's history, it is probable that in vivo transfer of a $bla_{\text{NDM-1}}$-containing cluster among three different Enterobacterales may have occurred in our patient in Greece during the COVID-19 pandemic. The treatment options for infections due to carbapenemase-producing microorganisms are limited, and often include formerly abandoned antibiotics or antibiotic combinations. Even these choices, however, depend on the overall patient condition, the antimicrobial susceptibility testing results, and the site of infection. Patients hospitalized for long periods of time can serve as reservoirs of emergence and dissemination of resistance genes. Therefore, this finding highlights the need for close monitoring, patient isolation, implementation of infection control measures, and antimicrobial stewardship, even during difficult times for the national healthcare systems.

Author Contributions: G.M. contributed to study design, laboratory diagnosis, data acquisition, and data interpretation, and drafted the manuscript; A.M. contributed to laboratory investigation, data acquisition, and data interpretation, and drafted the manuscript; A.T. contributed to laboratory diagnosis and data acquisition; A.K. contributed to laboratory diagnosis and data acquisition; N.V. contributed to laboratory diagnosis and data acquisition; P.M. contributed to laboratory diagnosis and data acquisition; S.M. contributed to patient admission and data acquisition; L.S. contributed to supervision and critically revised the manuscript; E.P. contributed to conceptualization, study design, and supervision, and critically revised the manuscript. All authors have read and agreed to the published version of the manuscript.

Funding: This research received no external funding.

Institutional Review Board Statement: The publication of the present data was approved by the AHEPA University Hospital bioethics committee (protocol number: 36129/5 July 2022).

Informed Consent Statement: Not applicable.

Data Availability Statement: The data presented in this study are available in the article.

Conflicts of Interest: The authors declare no conflict of interest.

References

1. Jean, S.S.; Harnod, D.; Hsueh, P.R. Global Threat of Carbapenem-Resistant Gram-Negative Bacteria. *Front. Cell. Infect. Microbiol.* **2022**, *12*, 823684. [CrossRef] [PubMed]
2. Meletis, G. Carbapenem resistance: Overview of the problem and future perspectives. *Ther. Adv. Infect. Dis.* **2016**, *3*, 15–21. [CrossRef] [PubMed]
3. Dortet, L.; Poirel, L.; Nordmann, P. Worldwide dissemination of the NDM-type carbapenemases in Gram-negative bacteria. *Biomed. Res. Int.* **2014**, *2014*, 249856. [CrossRef] [PubMed]
4. Nordmann, P.; Poirel, L.; Walsh, T.R.; Livermore, D.M. The emerging NDM carbapenemases. *Trends Microbiol.* **2011**, *19*, 588–595. [CrossRef]
5. Poirel, L.; Dortet, L.; Bernabeu, S.; Nordmann, P. Genetic features of blaNDM-1-positive Enterobacteriaceae. *Antimicrob. Agents Chemother.* **2011**, *55*, 5403–5407. [CrossRef]
6. Dortet, L.; Girlich, D.; Virlouvet, A.L.; Poirel, L.; Nordmann, P.; Iorga, B.I.; Naas, T. Characterization of BRPMBL, the Bleomycin Resistance Protein Associated with the Carbapenemase NDM. *Antimicrob. Agents Chemother.* **2017**, *61*, e02413–e02416. [CrossRef]
7. Yong, D.; Toleman, M.A.; Giske, C.G.; Cho, H.S.; Sundman, K.; Lee, K.; Walsh, T.R. Characterization of a new metallo-beta-lactamase gene, bla(NDM-1), and a novel erythromycin esterase gene carried on a unique genetic structure in Klebsiella pneumoniae sequence type 14 from India. *Antimicrob. Agents Chemother.* **2009**, *53*, 5046–5054. [CrossRef]
8. Cantón, R.; Akóva, M.; Carmeli, Y.; Giske, C.G.; Glupczynski, Y.; Gniadkowski, M.; Livermore, D.M.; Miriagou, V.; Naas, T.; Rossolini, G.M.; et al. European Network on Carbapenemases. Rapid evolution and spread of carbapenemases among Enterobacteriaceae in Europe. *Clin. Microbiol. Infect.* **2012**, *18*, 413–431. [CrossRef]
9. Johnson, A.P.; Woodford, N. Global spread of antibiotic resistance: The example of New Delhi metallo-beta-lactamase (NDM)-mediated carbapenem resistance. *J. Med. Microbiol.* **2013**, *62*, 499–513. [CrossRef]
10. Voulgari, E.; Gartzonika, C.; Vrioni, G.; Politi, L.; Priavali, E.; Levidiotou-Stefanou, S.; Tsakris, A. The Balkan region: NDM-1-producing Klebsiella pneumoniae ST11 clonal strain causing outbreaks in Greece. *J. Antimicrob. Chemother.* **2014**, *69*, 2091–2097. [CrossRef]
11. Protonotariou, E.; Meletis, G.; Pilalas, D.; Mantzana, P.; Tychala, A.; Kotzamanidis, C.; Papadopoulou, D.; Papadopoulos, T.; Polemis, M.; Metallidis, S.; et al. Polyclonal Endemicity of Carbapenemase-Producing Klebsiella pneumoniae in ICUs of a Greek Tertiary Care Hospital. *Antibiotics* **2022**, *11*, 149. [CrossRef] [PubMed]
12. Protonotariou, E.; Mantzana, P.; Meletis, G.; Tychala, A.; Kassomenaki, A.; Vasilaki, O.; Kagkalou, G.; Gkeka, I.; Archonti, M.; Kati, S.; et al. Microbiological characteristics of bacteremias among COVID-19 hospitalized patients in a tertiary referral hospital in Northern Greece during the second epidemic wave. *FEMS Microbes.* **2021**, *2*, xtab021. [CrossRef] [PubMed]
13. Meletis, G.; Tychala, A.; Ntritsos, G.; Verrou, E.; Savvidou, F.; Dermitzakis, I.; Chatzidimitriou, A.; Gkeka, I.; Fyntanidou, B.; Gkarmiri, S.; et al. Variant-Related Differences in Laboratory Biomarkers among Patients Affected with Alpha, Delta and Omicron: A Retrospective Whole Viral Genome Sequencing and Hospital-Setting Cohort Study. *Biomedicines* **2023**, *11*, 1143. [CrossRef] [PubMed]
14. Shintani, M.; Sanchez, Z.K.; Kimbara, K. Genomics of microbial plasmids: Classification and identification based on replication and transfer systems and host taxonomy. *Front. Microbiol.* **2015**, *6*, 242. [CrossRef] [PubMed]
15. Rozwandowicz, M.; Brouwer, M.S.M.; Fischer, J.; Wagenaar, J.A.; Gonzalez-Zorn, B.; Guerra, B.; Mevius, D.J.; Hordijk, J. Plasmids carrying antimicrobial resistance genes in Enterobacteriaceae. *Antimicrob. Chemother.* **2018**, *73*, 1121–1137. [CrossRef]
16. Alcántar-Curiel, M.D.; Fernández-Vázquez, J.L.; Toledano-Tableros, J.E.; Gayosso-Vázquez, C.; Jarillo-Quijada, M.D.; López-Álvarez, M.D.R.; Giono-Cerezo, S.; Santos-Preciado, J.I. Emergence of IncFIA Plasmid-Carrying blaNDM-1 Among Klebsiella pneumoniae and Enterobacter cloacae Isolates in a Tertiary Referral Hospital in Mexico. *Microb. Drug Resist.* **2019**, *25*, 830–838. [CrossRef]
17. Mendes, G.; Ramalho, J.F.; Duarte, A.; Pedrosa, A.; Silva, A.C.; Méndez, L.; Caneiras, C. First Outbreak of NDM-1-Producing Klebsiella pneumoniae ST11 in a Portuguese Hospital Centre during the COVID-19Pandemic. *Microorganisms* **2022**, *10*, 251. [CrossRef]
18. Arend, L.N.V.S.; Bergamo, R.; Rocha, F.B.; Bail, L.; Ito, C.; Baura, V.A.; Balsanelli, E.; Pothier, J.F.; Rezzonico, F.; Pilonetto, M.; et al. Dissemination of NDM-producing bacteria in Southern Brazil. *Diagn. Microbiol. Infect. Dis.* **2023**, *106*, 115930. [CrossRef]
19. Zou, H.; Han, J.; Zhao, L.; Wang, D.; Guan, Y.; Wu, T.; Hou, X.; Han, H.; Li, X. The shared NDM-positive strains in the hospital and connecting aquatic environments. *Sci. Total Environ.* **2023**, *860*, 160404. [CrossRef]
20. Sianou, E.; Kristo, I.; Petridis, M.; Apostolidis, K.; Meletis, G.; Miyakis, S.; Sofianou, D. A cautionary case of microbial solidarity: Concurrent isolation of VIM-1-producing Klebsiella pneumoniae, Escherichia coli and Enterobacter cloacae from an infected wound. *J. Antimicrob. Chemother.* **2012**, *67*, 244–246. [CrossRef]
21. Aires-de-Sousa, M.; Ortiz de la Rosa, J.M.; Goncalves, M.L.; Costa, A.; Nordmann, P.; Poirel, L. Occurrence of NDM-1-producing Morganellamorganii and Proteus mirabilis in a single patient in Portugal: Probable in vivo transfer by conjugation. *J. Antimicrob. Chemother.* **2020**, *75*, 903–906. [CrossRef] [PubMed]
22. Piccirilli, A.; Meroni, E.; Mauri, C.; Perilli, M.; Cherubini, S.; Pompilio, A.; Luzzaro, F.; Principe, L. Analysis of Antimicrobial Resistance Genes (ARGs) in Enterobacteralesand A. baumannii Clinical Strains Colonizing a Single Italian Patient. *Antibiotics* **2023**, *12*, 439. [CrossRef] [PubMed]

23. Kim, J.S.; Jin, Y.H.; Park, S.H.; Han, S.; Kim, H.S.; Yu, J.K.; Jang, J.I.; Kim, J.; Hong, C.K.; Lee, J.H.; et al. Horizontal transfer of bla_{NDM-1}-carrying IncX3 plasmid between carbapenem-resistant Enterobacteriaceae in a single patient. *J. Infect.* **2020**, *81*, 816–846. [CrossRef]
24. Yu, C.; Wei, X.; Wang, Z.; Liu, L.; Liu, Z.; Liu, J.; Wu, L.; Guo, H.; Jin, Z. Occurrence of two NDM-1-producing Raoultellaornithinolytica and Enterobacter cloacae in a single patient in China: Probable a novel antimicrobial resistance plasmid transfer in vivo by conjugation. *J. Glob. Antimicrob. Resist.* **2020**, *22*, 835–841. [CrossRef]
25. Wood, D.E.; Salzberg, S.L. Kraken: Ultrafast Metagenomic Sequence Classification Using Exact Alignments. *Genome Biol.* **2014**, *15*, R46. [CrossRef]
26. Prjibelski, A.; Antipov, D.; Meleshko, D.; Lapidus, A.; Korobeynikov, A. Using SPAdes De Novo Assembler. *Curr. Protoc. Bioinform.* **2020**, *70*, e102. [CrossRef] [PubMed]
27. Seemann, T. Prokka: Rapid Prokaryotic Genome Annotation. *Bioinformatics* **2014**, *30*, 2068–2093. [CrossRef]
28. Bharat, A.; Petkau, A.; Avery, B.P.; Chen, J.; Folster, J.; Carson, C.A.; Kearney, A.; Nadon, C.; Mabon, P.; Thiessen, J.; et al. Correlation between Phenotypic and In Silico Detection of Antimicrobial Resistance in Salmonella Enterica in Canada Using Staramr. *Microorganisms* **2022**, *10*, 292. [CrossRef]
29. Florensa, A.F.; Kaas, R.S.; Clausen, P.T.L.C.; Aytan-Aktug, D.; Aarestrup, F.M. ResFinder—An Open Online Resource for Identification of Antimicrobial Resistance Genes in next-Generation Sequencing Data and Prediction of Phenotypes from Genotypes. *Microb. Genom.* **2022**, *8*, 748. [CrossRef]
30. Carattoli, A.; Zankari, E.; Garciá-Fernández, A.; Larsen, M.V.; Lund, O.; Villa, L.; Aarestrup, F.M.; Hasman, H. In Silico Detection and Typing of Plasmids Using Plasmidfinder and Plasmid Multilocus Sequence Typing. *Antimicrob. Agents Chemother.* **2014**, *58*, 3895–3903. [CrossRef]
31. Andreopoulos, W.B.; Geller, A.M.; Lucke, M.; Balewski, J.; Clum, A.; Ivanova, N.N.; Levy, A. Deeplasmid: Deep Learning Accurately Separates Plasmids from Bacterial Chromosomes. *Nucleic Acids Res.* **2022**, *50*, e17. [CrossRef] [PubMed]
32. Robertson, J.; Nash, J.H.E. MOB-Suite: Software Tools for Clustering, Reconstruction and Typing of Plasmids from Draft Assemblies. *Microb.Genom.* **2018**, *4*, e000206. [CrossRef] [PubMed]
33. Ondov, B.D.; Treangen, T.J.; Melsted, P.; Mallonee, A.B.; Bergman, N.H.; Koren, S.; Phillippy, A.M. Mash: Fast Genome and Metagenome Distance Estimation Using MinHash. *Genome Biol.* **2016**, *17*, 132. [CrossRef]
34. Jain, C.; Rodriguez-R, L.M.; Phillippy, A.M.; Konstantinidis, K.T.; Aluru, S. High Throughput ANI Analysis of 90K Prokaryotic Genomes Reveals Clear Species Boundaries. *Nat. Commun.* **2018**, *9*, 5114. [CrossRef]
35. Grant, J.R.; Enns, E.; Marinier, E.; Mandal, A.; Herman, E.K.; Chen, C.; Graham, M.; Van Domselaar, G.; Stothard, P. Proksee: In-Depth Characterization and Visualization of Bacterial Genomes. *Nucleic Acids Res.* **2023**. [CrossRef] [PubMed]
36. Alcock, B.P.; Raphenya, A.R.; Lau, T.T.Y.; Tsang, K.K.; Bouchard, M.; Edalatmand, A.; Huynh, W.; Nguyen, A.L.V.; Cheng, A.A.; Liu, S.; et al. CARD 2020: Antibiotic Resistome Surveillance with the Comprehensive Antibiotic Resistance Database. *Nucleic Acids Res.* **2020**, *48*, D517–D525. [CrossRef]
37. Brown, C.L.; Mullet, J.; Hindi, F.; Stoll, J.E.; Gupta, S.; Choi, M.; Keenum, I.; Vikesland, P.; Pruden, A.; Zhang, L. MobileOG-Db: A Manually Curated Database of Protein Families Mediating the Life Cycle of Bacterial Mobile Genetic Elements. *Appl. Environ. Microbiol.* **2022**, *88*, e0099122. [CrossRef]
38. Vernikos, G.S.; Parkhill, J. Interpolated Variable Order Motifs for Identification of Horizontally Acquired DNA: Revisiting the Salmonella Pathogenicity Islands. *Bioinformatics* **2006**, *22*, 2196–2203. [CrossRef]

antibiotics

MDPI

Communication

Efficacy of Novel Combinations of Antibiotics against Multidrug-Resistant—New Delhi Metallo-Beta-Lactamase-Producing Strains of *Enterobacterales*

Shamsi Khalid [1], Antonella Migliaccio [2], Raffaele Zarrilli [2,*] and Asad U. Khan [1,*]

[1] Medical Microbiology Laboratory, Interdisciplinary Biotechnology Unit, Aligarh Muslim University, Aligarh 202001, India; shamsi123khalid@gmail.com
[2] Department of Public Health, University of Naples Federico II, 80131 Naples, Italy; antonella.migliaccio10@gmail.com
* Correspondence: rafzarri@unina.it (R.Z.); akhan.cb@amu.ac.in (A.U.K.)

Abstract: The emergence of multidrug-resistance (MDR)—New Delhi metallo-beta-lactamase (NDM)-producing microorganisms—has become a serious concern for treating such infections. Therefore, we investigated the effective antimicrobial combinations against multidrug-resistant New Delhi metallo-beta-lactamase-producing strains of *Enterobacterales*. The tests were carried out using the 2D(two-dimensional) checkerboard method. Of 7 antimicrobials, i.e., doripenem (DRP), streptomycin (STR), cefoxitin (FOX), imipenem (IPM), cefotaxime (CTX), meropenem (MER), and gentamicin (GEN), 19 different combinations were used, and out of them, three combinations showed synergistic effects against 31 highly drug-resistant strains carrying bla_{NDM} and other associated resistance markers. Changes in the minimum inhibitory concentration (MIC) values were interpreted using the test fractional inhibitory concentration index (FIC Index). The FIC Index values of these combinations were found in the range of 0.1562 to 0.5, which shows synergy, whereas no synergism was observed in the remaining antimicrobial combinations. We conclude that these antibiotic combinations can be analyzed in in vivo and pharmacological studies to establish an effective therapeutic approach.

Keywords: antimicrobial resistance; antibiotics synergy; carbapenemase; NDM-beta-lactamase

check for updates

Citation: Khalid, S.; Migliaccio, A.; Zarrilli, R.; Khan, A.U. Efficacy of Novel Combinations of Antibiotics against Multidrug-Resistant—New Delhi Metallo-Beta-Lactamase-Producing Strains of *Enterobacterales*. *Antibiotics* **2023**, *12*, 1134. https://doi.org/10.3390/antibiotics12071134

Academic Editors: Ilias Karaiskos and Theodoros Karampatakis

Received: 12 June 2023
Revised: 26 June 2023
Accepted: 29 June 2023
Published: 30 June 2023

1. Introduction

The emergence of multidrug-resistant (MDR) bacterial infections is one of the major worldwide problems having difficulty being treated. Metallo-β-lactamases (MBLs) and extended β- lactamases (ESBLs) are the major causes of resistance in bacteria against available antibiotics [1]. Antibiotic efficacy against many Gram-negative pathogens is increasingly compromised by the spread and emergence of MDR strains that produce β-lactamases. These pathogens can confer resistance to one or more carbapenems, cephalosporins, monobactam penicillin, or drugs that are routinely used in clinical practice [2]. Carbapenems are considered the last resort of antibiotics against infections caused by MDR gram-negative microorganisms that show stability and high resistance values through bacterial outer membranes [1]. The common mechanisms for the members of the Enterobacteriaceae family include the production of β- lactamases, efflux pumps, and modification of penicillin-binding proteins (PBPs). In some bacterial species, the combination of these pathways can result in significant resistance to carbapenems [3]. In contrast to monobactam, New Delhi Metallo- β-lactamases (NDM) are a form of MBL that can hydrolyze most β-lactams, including carbapenems. The primary antimicrobial medicines of choice for treating severe infections caused by Gram-negative bacteria are carbapenems. Clinically used β-lactamase inhibitors, such as avibactam, clavulanate, sulbactam, and tazobactam, cannot stop the hydrolysis of β-lactams by NDM enzymes. In 2008, a Swedish patient who had been hospitalized in New Delhi, India, was infected with a strain of *Klebsiella pneumoniae* that included NDM-1 for the first time [4]. Since then, NDM-1 has been detected in numerous *Enterobacteriaceae*, *Acinetobacter*, and *Pseudomonas* species. These infections are the major challenge for clinicians in treating critically ill patients. The risk factors for NDM-1 strains include the lack of effective antibiotics, the failure to recognize highly prevalent asymptomatic carriers, the absence of a routine phenotypic test for the detection of Metallo-beta lactamases, and the presence of MBL on plasmids with the potential to rearrange and spread through horizontal gene transfer [3]. Therefore, to overcome the resistance problem, several studies for the treatment of MDR bacterial infections can be observed with the combination of ≥2 antimicrobial agents [5,6]. In the current scenario of the emergence of antibiotic resistance, where all classes of antibiotics fail to treat infections, a combination of these available antibiotics may be one of the most intelligent ideas to subside infections. Antimicrobial combination therapies can increase their efficacy over monotherapy and decrease MDR-based infections in clinical settings. Hence, we proposed to screen a large number of combinations of these available antibiotics against a set of MDR strains of clinical origin that have already been characterized in our previous studies.

2. Results and Discussion

2.1. Minimum Inhibitory Concentration

The MICs of cefoxitin, doripenem, imipenem, and streptomycin against MDR strains carrying blaNDM and other associated resistance markers on a plasmid, were reported in the range between 4096 µg/mL and 128 µg/mL (Table 1). All these MDR strains have already been characterized for the presence of resistance markers in our laboratory, and these isolates were collected from the neonatal intensive care unit (NICU) of an Indian hospital [7–13].

Table 1. Minimum Inhibitory concentrations and synergistic effects of doripenem, cefoxitin, imipenem, and streptomycin antibiotics against 31 MDR-NDM-producing clinical strains.

Strains	Resistance Markers	MIC µg/mL				MIC µg/mL FIC Index 5			Ref.
		DRP 1	FOX 2	IMP 3	STP 4	DRP + FOX	DRP + STP	IMP + FOX	
AK-33 *Escherichia coli*	NDM-4, OXA-1, CTX-M, Amp C	512	2048	1024	2048	128 + 512 0.5	128 + 256 0.375	128 + 256 0.25	[13]
AK-35 *Escherichia coli*	NDM-7, OXA-1	1024	2048	1024	2048	128 + 256 0.25	128 + 128 0.1875	256 + 512 0.5	[13]
AK-37 *Escherichia coli*	NDM-1, CMY-139, OXA-1, CTX-M	1024	1024	512	512	128 + 128 0.375	256 + 128 0.5	256 + 128 0.5	[13]
AK-83 *Escherichia coli*	NDM-7, OXA-1, SHV-1	512	4096	1024	2048	64 + 512 0.25	128 + 512 0.5	128 + 512 0.25	[8]
AK-66 *Klebsiella pneumoniae*	NDM-1, OXA-1, OXA-9, CMY-1	256	1024	1024	2048	128 + 256 0.375	256 + 256 0.25	128 + 512 0.1875	[8]
AK-102 *Klebsiella pneumoniae*	NDM-5, OXA-1, OXA-9, CMY-4	1024	2048	1024	2048	128 + 512 0.375	64 + 512 0.3125	64 + 256 0.1875	[8]
AK-121 *Klebsiella pneumoniae*	NDM-1	512	1024	512	1024	128 + 256 0.5	64 + 256 0.375	128 + 256 0.5	[10]
AK-125 *Klebsiella pneumoniae*	NDM-1	512	1024	1024	2048	128 + 256 0.5	64 + 256 0.25	128 + 256 0.375	[10]
AK-130 *Klebsiella pneumoniae*	NDM-1	512	256	512	1024	64 + 64 0.375	128 + 128 0.375	128 + 64 0.5	[10]
AK-140 *Klebsiella pneumoniae*	NDM-1, OXA-48	1024	1024	2048	2048	128 + 256 0.375	256 + 512 0.5	512 + 128 0.375	[10]

Table 1. *Cont.*

Strains	Resistance Markers	MIC µg/mL				MIC µg/mL FIC Index [5]			Ref.
		DRP [1]	FOX [2]	IMP [3]	STP [4]	DRP + FOX	DRP + STP	IMP + FOX	
AK-142 *Klebsiella pneumoniae*	NDM-1	512	1024	512	2048	128 + 526 0.5	64 + 512 0.375	128 + 128 0.25	[10]
AK-144 *Klebsiella pneumoniae*	NDM-1	512	2048	1024	1024	128 + 512 0.25	64 + 128 0.25	512 + 512 0.25	[10]
AK-147 *Klebsiella pneumoniae*	NDM-1, OXA-48	1024	2048	2048	1024	128 + 512 0.375	64 + 128 0.1875	512 + 512 0.5	[10]
AK-149 *Klebsiella pneumoniae*	NDM-1, OXA-48	1024	2048	2048	2048	128 + 512 0.375	512 + 128 0.375	256 + 512 0.375	[10]
AK-158 *Klebsiella pneumoniae*	NDM-5	1024	2048	2048	2048	128 + 256 0.25	64 + 512 0.3125	512 + 512 0.5	[10]
AK-100 *Klebsiella oxytoca*	NDM-4, OXA-1, OXA-9	1024	4096	1024	2048	256 + 256 0.3125	256 + 512 0.5	128 + 512 0.25	[8]
AK-67 *Enterobacter aerogenes*	NDM-1, OXA-1, SHV-2	1024	2048	1024	2048	64 + 128 0.1875	128 + 64 0.1562	32 + 64 0.3125	[8]
AK-93 *Enterobacter aerogenes*	NDM-4, OXA-1, OXA-9, SHV-1	512	1024	256	1024	64 + 64 0.185	128 + 128 0.375	32 + 64 0.25	[11]
AK-95 *Enterobacter aerogenes*	NDM-5, OXA-1, OXA-9, CMY-149	256	1024	256	1024	64 + 128 0.375	64 + 256 0.5	32 + 256 0.375	[11]
AK-96 *Enterobacter aerogenes*	NDM-7, OXA-1, OXA-9, CMY-145	256	1024	256	2048	64 + 128 0.375	32 + 256 0.375	64 + 256 0.5	[11]
AK-108 *Enterobacter cloacae*	NDM-4, OXA-1, OXA-9, CMY-149	512	2048	512	1024	128 + 256 0.375	128 + 256 0.5	64 + 256 0.25	[8]
AK-154 *Acinetobacter baumannii*	NDM-5	1024	2048	1024	2048	128 + 256 0.25	128 + 256 0.25	64 + 256 0.1875	[9]
AK-42 *Citrobacter freundii*	NDM-1, CMY-42, OXA-1, CTX-M, AmpC	1024	4096	1024	4096	256 + 1024 0.5	256 + 512 0.375	128 + 1024 0.375	[13]
AK-58 *Citrobacter freudii*	NDM-7, CMY-2, OXA-1, CTX-M	256	1024	128	2048	32 + 256 0.375	64 + 512 0.5	32 + 128 0.375	[13]
AK-82 *Citrobacter freundii*	NDM-4, OXA-9, SHV-1, CMY-149	512	4096	2048	2048	128 + 1024 0.5	64 + 256 0.25	128 + 512 0.1875	[8]
AK-48 *Citrobacter braakii*	NDM-4, CMY-4, OXA-48	512	1024	1024	4096	128 + 256 0.5	128 + 512 0.375	128 + 128 0.25	[13]
AK-49 *Citrobacter farmer*	NDM-4, CMY-4, OXA-48	256	1024	1024	2048	64 + 256 0.5	128 + 1024 0.5	128 + 128 0.25	[13]
AK-68 *Cedecea lapagei*	NDM-1, CTX-M, SHV, TEM	512	1024	512	1024	128 + 256 0.5	128 + 256 0.5	128 + 128 0.375	[12]

Table 1. *Cont.*

Strains	Resistance Markers	MIC µg/mL				MIC µg/mL FIC Index [5]			Ref.
		DRP [1]	FOX [2]	IMP [3]	STP [4]	DRP + FOX	DRP + STP	IMP + FOX	
AK-152 *Cedecea davisae*	NDM-1	1024	2048	1024	1024	128 + 256 0.25	128 + 256 0.375	64 + 256 0.1875	[9]
AK-65 *Shigella boydii*	NDM-5, CMY-42, OXA-1, CTX-M	1024	2048	128	2048	64 + 64 0.125	128 + 128 0.1875	32 + 128 0.3125	[13]
AK-92 *Moellerella wisconsensis*	NDM-1	512	1024	256	2048	128 + 256 0.5	64 + 256 0.25	64 + 128 0.375	[7]

[1] DRP—doripenem; [2] FOX—cefoxitin; [3] STP—streptomycin; [4] I MP—imipenem; [5] FIC Index—Fractional Inhibitory Concentration Index.

2.2. Synergistic Effect of Antibiotic Combinations

MDR clinical strains harboring *blaNDM* and other associated resistance markers like *blaOXA-1, blaCTX-M, blaAmpC, blaCMY-1,* and *blaSHV* showed resistance toward doripenem (MIC ranges 256 µg/mL to 1024 µg/mL), imipenem (MIC ranges 128 µg/mL to 2048 µg/mL), cefoxitin (MIC ranges 256 µg/mL to 4096 µg/mL) and streptomycin (MIC ranges 512 µg/mL to 4096 µg/mL) which were previously well characterized in our laboratory (Table 1). To check the effect of these antibiotics in combination, a total of 19 combinations of different classes of antibiotics were tested by a 2D checkerboard microdilution assay. Of them, only three combinations were found to be most effective against the set of clinical strains (Table 1). These combinations belong to the classes of carbapenems, cephamycin, and aminoglycosides (doripenem with cefoxitin, doripenem with streptomycin, and imipenem with cefoxitin), showing synergy. It was observed that the MICs of doripenem decreased from 1024 µg/mL to 64 µg/mL, cefoxitin 4096 µg/mL to 64 µg/mL, and streptomycin 4096 µg/mL to 32 µg/mL (Table 1) in combination, and their FICI values were in the range of 0.156 to 0.5 (Table 1) for all highly resistant strains tested, which were found in synergistic range. Previously, a synergistic effect of doripenem in combination with cefoxitin and tetracycline in inhibiting *blaNDM-1*-producing bacterial strains was reported by our research group [14]. The synergistic interaction of antimicrobials allows the use of lower doses. Another combination of imipenem and cefoxitin was showing a synergistic effect, with FICI values in the range of 0.187 to 0.5 (Table 1). It has been previously reported that synergy is observed when the FICI value is ≤0.5 [14]. In this study, three combinations showed synergy. The FICI values of combinations were found to be 0.5 to 0.1562 (Table 1). Although several mechanism-based studies were performed earlier on the interaction of antibiotics with specific markers using biophysical and biochemical approaches [14–16]. Antibiotic combination therapy with ceftazidime/avibactam (CAZ/AVI) and aztreonam (ATM) was previously investigated for the treatment of infection with NDM producer Enterobacterales. The majority of Enterobacterales that are ATM resistant and NDM positive shows significant efficacies of the CAZ/AVI+ATM combination against them [17]. Another study demonstrated the effectiveness of the double carbapenem combination against Gram-negative bacteria through the in vitro synergistic activity of ertapenem and meropenem [18].

Although β-lactamase inhibitors have been crucial to fighting against β-lactam resistance in Gram-negative bacteria, their potency has been dwindling as a result of the development of numerous severe β-lactamases. Though it has a unique synergistic mechanism of action, a triple combination of β- lactam antibiotics meropenem, piperacillin, and tazobactam has been demonstrated to be an effective method for killing Methicillin-resistant *Staphylococcus aureus* (MRSA) in vitro and in a mouse model [19].

The susceptibility pattern of bla_{NDM}-producing bacteria may vary geographically depending on specific strains over time. In this study, all microorganisms present were NDM producers and showed high resistance values due to which synergy of antibiotics was not able to restore susceptibilities, but the combination values were in synergistic range, showing this combination can be used for the therapeutic approach.

3. Materials and Methods

3.1. Strain, Antibiotics and Chemicals

This study included 31 MDR clinical strains carrying blaNDM and other associated resistance markers to determine the minimum inhibitory concentration (MIC) and fractional inhibitory concentration index (FICI). These strains were obtained from NICU of a North Indian Hospital. These are ESBL- and MBL-producing strains with different resistant markers, as reported previously by our group [7–13]. The antibiotic resistance markers and MIC of these strains are presented in Table 1. Doripenem, cefoxitin, and imipenem were purchased from Sigma-Aldrich (Sigma, Milan, Italy). Streptomycin was purchased from Himedia (Mumbai, India). Mueller–Hinton broth was purchased from Himedia (Mumbai, India).

3.2. Combination of Antibiotics and MIC

The antibiotics used in this study were doripenem (DRP), streptomycin (STR), cefoxitin (FOX), imipenem (IPM), cefotaxime (CTX), meropenem (MER), and gentamicin (GEN). These antibiotics were used to prepare all possible combinations against highly resistant clinical strains. To examine the MIC of antibiotics for a set of clinical strains, the overnight-grown colonies were collected using a sterile loop and transferred into a tube containing 5 mL of Mueller–Hinton broth. This broth was incubated at 37 °C to obtain a final turbidity equivalent to that of 0.5 McFarland standards (10^8 CFU/mL) and diluted to 1:100 for the broth microdilution procedure. The strains were treated with decreasing drug concentrations from 4096 µg/mL to 0.5 µg/mL according to Clinical Laboratory Standards Institute (CLSI) guidelines [20].

3.3. D-Checkerboard Microdilution Assay to Determine FICI

A total of 19 different combinations of antibiotics were taken against 31 MDR clinical strains. A 2D checkerboard microdilution assay was performed using a 96-well microtiter plate. Serially diluted antibiotics were taken in concentrations less than, equal to, or greater than their MICs. To check their effect against MDR clinical stains, fractional inhibitory concentration indexes (FICI) were calculated. The FICI value $\leq$0.5 was defined as synergy, <4, indifference, and >4, antagonism [21].

4. Conclusions

The study revealed three combinations: doripenem with cefoxitin, doripenem with streptomycin, and imipenem with cefoxitin, which are effective against highly drug-resistant clinical strains carrying blaNDM and other associated resistance markers. Hence, we propose these novel combinations against highly drug resistant clinical strains of bacteria for further in vivo and pharmacological studies in order to establish effective infection control therapy.

Author Contributions: Conceptualization, A.U.K. and R.Z.; methodology, S.K.; validation, S.K. and A.M.; formal analysis, S.K. and A.M.; investigation, S.K. and A.M.; resources, A.U.K.; data curation, S.K. and A.M.; writing—original draft preparation, S.K.; writing—review and editing, A.U.K. and R.Z.; visualization, S.K.; supervision, A.U.K. and R.Z.; project administration, A.U.K.; funding acquisition, A.U.K. All authors have read and agreed to the published version of the manuscript.

Funding: The DBT: Government of India, grant numbers BT/PR40148/BTIS/137/20/2021 and HRD-16012/6/2020-AFS-DBT, is highly acknowledged for its support. This work was supported also by grant from the Italian Ministry of University and Research (MUR): PRIN2017 (Grant nr. 2017SFBER to R.Z.).

Institutional Review Board Statement: Not applicable.

Informed Consent Statement: Not applicable.

Data Availability Statement: All data are provided in the main text of the manuscript.

Conflicts of Interest: The authors declared no conflict of interest.

References

1. Cherak, Z.; Loucif, L.; Moussi, A.; Bendjama, E.; Benbouza, A.; Rolain, J.M. Emergence of Metallo-β-Lactamases and OXA-48 Carbapenemase Producing Gram-Negative Bacteria in Hospital Wastewater in Algeria: A Potential Dissemination Pathway Into the Environment. *Microb. Drug Resist.* **2022**, *28*, 23–30. [CrossRef] [PubMed]
2. Bush, K. Past and present perspectives on β-lactamases. *Antimicrob. Agents Chemother.* **2018**, *62*, e01076-18. [CrossRef] [PubMed]
3. Khan, A.U.; Maryam, L.; Zarrilli, R. Structure, Genetics and Worldwide Spread of New Delhi Metallo-β-lactamase (NDM): A threat to public health. *BMC Microbiol.* **2017**, *17*, 101. [CrossRef] [PubMed]
4. Yong, D.; Toleman, M.A.; Giske, C.G.; Cho, H.S.; Sundman, K.; Lee, K.; Walsh, T.R. Characterization of a new metallo-β-lactamase gene, blaNDM-1, and a novel erythromycin esterase gene carried on a unique genetic structure in Klebsiella pneumoniae sequence type 14 from India. *Antimicrob. Agents Chemother.* **2009**, *53*, 5046–5054. [CrossRef] [PubMed]
5. Almutairi, M.M. Synergistic activities of colistin combined with other antimicrobial agents against colistin-resistant *Acinetobacter baumannii* clinical isolates. *PLoS ONE* **2022**, *17*, e0270908. [CrossRef] [PubMed]
6. Scudeller, L.; Righi, E.; Chiamenti, M.; Bragantini, D.; Menchinelli, G.; Cattaneo, P.; Giske, C.G.; Lodise, T.; Sanguinetti, M.; Piddock, L.J.; et al. Systematic review and meta-analysis of in vitro efficacy of antibiotic combination therapy against carbapenem-resistant Gram-negative bacilli. *Int. J. Antimicrob. Agents* **2021**, *57*, 106344. [CrossRef] [PubMed]
7. Ahmad, N.; Ali, S.M.; Khan, A.U. Co-existence of bla_{NDM-1} and bla_{VIM-1} producing *Moellerella wisconsensis* in NICU of North Indian Hospital. *J. Infect. Dev. Ctries.* **2020**, *14*, 228–231. [CrossRef] [PubMed]
8. Ahmad, N.; Khalid, S.; Ali, S.M.; Khan, A.U. Occurrence of bla_{NDM} variants among Enterobacteriaceae from a neonatal intensive care unit in a northern India hospital. *Front. Microbiol.* **2018**, *9*, 407. [CrossRef] [PubMed]
9. Khalid, S.; Ahmad, N.; Ali, S.M.; Khan, A.U. The outbreak of efficiently transferred carbapenem-resistant bla_{NDM} producing gram-negative bacilli isolated from the neonatal intensive care unit of an Indian hospital. *Microb. Drug Resist.* **2020**, *26*, 284–289. [CrossRef] [PubMed]
10. Ahmad, N.; Ali, S.M.; Khan, A.U. Molecular characterization of novel sequence type of carbapenem-resistant New Delhi Metallo-β-lactamase-1-producing *Klebsiella pneumoniae* in the neonatal intensive care unit of an Indian hospital. *Int. J. Antimicrob. Agents* **2019**, *53*, 525–529. [CrossRef] [PubMed]
11. Ahmad, N.; Ali, S.M.; Khan, A.U. Detection of New Delhi metallo-β-lactamase variants NDM-4, NDM-5, and NDM-7 in *Enterobacter aerogenes* isolated from a neonatal intensive care unit of a North India Hospital: A first report. *Microb. Drug Resist.* **2018**, *24*, 161–165. [CrossRef] [PubMed]
12. Ahmad, N.; Ali, S.M.; Khan, A.U. First reported New Delhi metallo-β-lactamase-1-producing *Cedecea lapagei*. *Int. J. Antimicrob. Agents* **2017**, *49*, 118–119. [CrossRef] [PubMed]
13. Parvez, S.; Khan, A.U. Hospital sewage water: A reservoir for variants of New Delhi metallo- β-lactamase (NDM) and extended-spectrum β-lactamase (ESBL)-producing Enterobacteriaceae. *Int. J. Antimicrob. Agents* **2018**, *51*, 82–88. [CrossRef] [PubMed]
14. Maryam, L.; Khalid, S.; Ali, A.; Khan, A.U. Synergistic effect of doripenem in combination with cefoxitin and tetracycline in inhibiting NDM-1-producing bacteria. *Future Microbiol.* **2019**, *14*, 671–689. [CrossRef] [PubMed]
15. Hasan, S.; Ali, S.Z.; Khan, A.U. Novel combinations of antibiotics to inhibit extended-spectrum β-lactamase and Metallo-β-lactamase producers in vitro: A synergistic approach. *Future Microbiol.* **2013**, *8*, 939–944. [CrossRef] [PubMed]
16. Shakil, S.; Khan, R.; Zarrilli, R.; Khan, A.U. Aminoglycosides versus bacteria–a description of the action, resistance mechanism, and nosocomial battleground. *J. Biomed. Sci.* **2008**, *15*, 5–14. [CrossRef] [PubMed]
17. Rawson, T.M.; Brzeska-Trafny, I.; Maxfield, R.; Almeida, M.; Gilchrist, M.; Gonzalo, X.; Moore, L.S.; Donaldson, H.; Davies, F. A practical laboratory method to determine ceftazidime-avibactam-aztreonam synergy in patients with New Delhi metallo-beta-lactamase (NDM)-producing Enterobacterales infection. *J. Glob. Antimicrob. Resist.* **2022**, *29*, 558–562. [CrossRef] [PubMed]
18. Lu, J.; Qing, Y.; Dong, N.; Liu, C.; Zeng, Y.; Sun, Q.; Shentu, Q.; Huang, L.; Wu, Y.; Zhou, H.; et al. Effectiveness of a double-carbapenem combinations against carbapenem-resistant Gram-negative bacteria. *Saudi Pharm. J.* **2022**, *30*, 849–855. [CrossRef] [PubMed]
19. Bush, K. A resurgence of β-lactamase inhibitor combinations effective against multidrug-resistant Gram-negative pathogens. *Int. J. Antimicrob. Agents* **2015**, *46*, 483–493. [CrossRef] [PubMed]
20. Clinical and Laboratory Standards Institute (CLSI). *Performance Standards for Antimicrobial Susceptibility Testing*; 28th Informational Supplement; CLSI document M100-S29; Clinical and Laboratory Standards Institute: Wayne, PA, USA, 2019.
21. Doern, C.D. When does 2 plus 2 equal 5? A review of antimicrobial synergy testing. *J. Clin. Microbiol.* **2014**, *52*, 4124–4128. [CrossRef] [PubMed]

antibiotics

MDPI

Article

Genetic Characterization of Carbapenem-Resistant *Klebsiella pneumoniae* Clinical Isolates in a Tertiary Hospital in Greece, 2018–2022

Charalampos Zarras [1,2], Theodoros Karampatakis [1,*], Styliani Pappa [1], Elias Iosifidis [3], Eleni Vagdatli [2], Emmanuel Roilides [3] and Anna Papa [1]

1 Department of Microbiology, Medical Faculty, School of Health Sciences, Aristotle University of Thessaloniki, 541 24 Thessaloniki, Greece; zarraschak6@gmail.com (C.Z.); s_pappa@hotmail.com (S.P.); annap@auth.gr (A.P.)
2 Microbiology Department, Hippokration General Hospital, 546 42 Thessaloniki, Greece; evagdatli@gmail.com
3 Infectious Disease Unit, 3rd Department of Pediatrics, Medical Faculty, School of Health Sciences, Hippokration General Hospital, 546 42 Thessaloniki, Greece; iosifidish@gmail.com (E.I.); roilides@auth.gr (E.R.)
* Correspondence: tkarampatakis@yahoo.com

Abstract: Background: Carbapenem-resistant *Klebsiella pneumoniae* (CRKP) is a serious public health issue. The study aimed to identify the antimicrobial resistance and accessory genes, the clonal relatedness, and the evolutionary dynamics of selected CRKP isolates recovered in an adult and pediatric intensive care unit of a tertiary hospital in Greece. Methods: Twenty-four CRKP isolates recovered during 2018–2022 were included in the study. Next-generation sequencing was performed using the Ion Torrent PGM Platform. The identification of the plasmid content, MLST, and antimicrobial resistance genes, as well as the comparison of multiple genome alignments and the identification of core genome single-nucleotide polymorphism sites, were performed using various bioinformatics software. Results: The isolates belonged to eight sequence types: 11, 15, 30, 35, 39, 307, 323, and 512. A variety of carbapenemases (KPC, VIM, NDM, and OXA-48) and resistance genes were detected. CRKP strains shared visually common genomic regions with the reference strain (NTUH-K2044). ST15, ST323, ST39, and ST11 CRKP isolates presented on average 17, 6, 16, and 866 recombined SNPs, respectively. All isolates belonging to ST15, ST323, and ST39 were classified into distinct phylogenetic branches, while ST11 isolates were assigned to a two-subclade branch. For large CRKP sets, the phylogeny seems to change approximately every seven SNPs. Conclusions: The current study provides insight into the genetic characterization of CRKP isolates in the ICUs of a tertiary hospital. Our results indicate clonal dispersion of ST15, ST323, and ST39 and highly diverged ST11 isolates.

Keywords: carbapenem-resistant *Klebsiella pneumoniae*; whole genome sequencing; core genome single-nucleotide polymorphism analysis; molecular epidemiology

Citation: Zarras, C.; Karampatakis, T.; Pappa, S.; Iosifidis, E.; Vagdatli, E.; Roilides, E.; Papa, A. Genetic Characterization of Carbapenem-Resistant *Klebsiella pneumoniae* Clinical Isolates in a Tertiary Hospital in Greece, 2018–2022. *Antibiotics* **2023**, *12*, 976. https://doi.org/10.3390/antibiotics12060976

Academic Editor: Michael J. McConnell

Received: 4 May 2023
Revised: 23 May 2023
Accepted: 25 May 2023
Published: 28 May 2023

1. Introduction

Klebsiella pneumoniae is recognized as an important Gram-negative opportunistic pathogen that causes community- and hospital-acquired infections globally [1]. It can cause ventilator-associated pneumonia among patients in intensive care units (ICUs), bloodstream infections, and urinary tract infections [2–4]. The emergence of multidrug-resistant (MDR) *K. pneumoniae* has become a major public health problem, causing a significant increase in morbidity and mortality worldwide [5]. The spread of carbapenem-resistant *K. pneumoniae* (CRKP) has generally been considered an increasingly serious issue for clinical practice due to the limitation of therapeutic options [6–8]. Greece is endemic for CRKP [9–13]; data from the European Antimicrobial Resistance Surveillance Network committee for the year 2020 indicate that the prevalence of CRKP in Greece was 66.3%, the highest among the European countries [14].

The rapid dissemination of CRKP is closely related to the antimicrobial resistance genes carried by plasmids and transferable genetic elements [15–18]. CRKP strains accumulate antimicrobial resistance genes due to inadequate implementation of infection control measures in healthcare settings and the irrational use of antimicrobials, resulting in the emergence of MDR, extremely drug-resistant, and pan-drug-resistant phenotypes [19]. The main mechanism of resistance to carbapenems is through the production of enzymes called carbapenemases, with *K. pneumoniae* carbapenemase (KPC) being the most prevalent, followed by Verona integron-encoded metallo-β-lactamase (VIM), New Delhi metallo-β-lactamase (NDM), and OXA-48 carbapenemases [20–22]. The contribution of outer membrane protein (Omp) deficiencies, such as *OmpK35*, *OmpK36*, and *OmpK37*, is considered a secondary mechanism in the emergence of CRKP [23–25]. The pathogenic potential of these strains is determined by virulence factors such as capsular polysaccharide synthesis, fimbriae, pili, outer membrane proteins, and iron acquisition systems [26].

The application of next-generation sequencing (NGS) has become a powerful tool for obtaining whole genome sequences (WGS), resulting in genetic characterization of the strains and a better understanding of the genomic diversity, providing fast and in-depth information about bacterial pathogenicity [27]. WGS analysis is recommended for genotyping and determining relatedness between clinical *K. pneumoniae* strains, as well as for the surveillance of virulence genes [28]. The aim of the present study was to identify the antimicrobial resistance and accessory genes of selected CRKP isolates recovered in an adult intensive care unit (ICU) and a pediatric ICU (PICU) of a tertiary hospital in Greece, to evaluate their clonal relatedness through core genome single-nucleotide polymorphism (cgSNP) analysis, and to calculate the evolutionary dynamics of this bacterial population.

2. Results

2.1. Antimicrobial Susceptibility Testing

All CRKP isolates were resistant to β-lactam antibiotics (including penicillins, cephalosporins, monobactams, and carbapenems). Nineteen isolates (79.2%) were resistant to at least one aminoglycoside, 23 isolates (95.8%) showed resistance to quinolones, eight isolates (33.3%) were resistant to fosfomycin and tigecycline, and 21 (87.5%) to cotrimoxazole. Four out of five isolates that carried the class A β-lactamase bla_{KPC} showed susceptibility to ceftazidime-avibactam (CAZ-AVI). The phenotypic resistance rates to antimicrobials of CRKP isolates are shown in the heatmap in Figure 1.

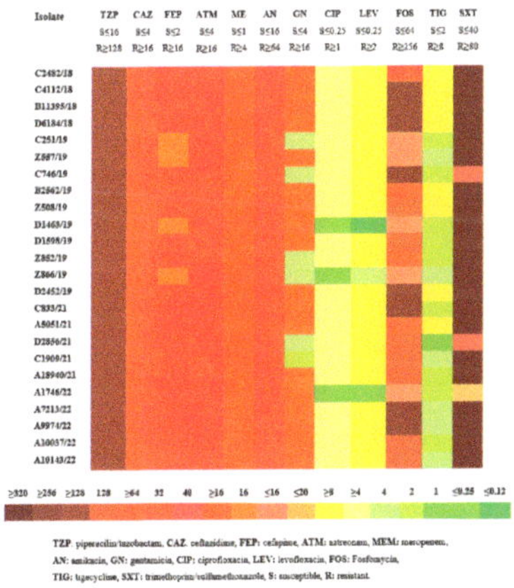

Figure 1. MICs (μg/mL) heatmap of the 24 clinical isolates of carbapenem-resistant *K. pneumoniae* (CRKP) according to CLSI.

2.2. MLST, Antimicrobial Resistance Genes, Plasmids

The CRKP isolates belonged to eight different STs: ST11 (7, 29.1%), ST39 (7, 29.1%), ST323 (3, 12.5%), ST15 (7, 29.1%), ST307 (1, 4.2%), ST30 (1, 4.2%), ST512 (1, 4.2%), and ST35 (1, 4.2%).

The resistance genes and the plasmid content per isolate are detailed in Table 1. The prevalence of genes conferring resistance to beta-lactams was 100%, to aminoglycosides 79.2%, to quinolones 95.8%, to fosfomycin and tigecycline 33.3%, and to cotrimoxazole 87.5%. The most prevalent plasmid was IncF, which is associated with the spread of several extended-spectrum β-lactamase (ESBL) genes or carbapenemases and was detected in all CRKP strains; additional plasmids were found in fifteen strains, such as Col and IncC.

Table 1. Genetic characteristics of the 24 carbapenem-resistant *K. pneumoniae* isolates of the study.

Isolates	ST	Plasmids	Antibiotic Resistance Genes Affecting					Virulence	Efflux/Regulator	Capsule
			Other Beta-Lactams	Carba-Penems	Amino-Glycosides	Quinolones	Other			
C2482/18	39	ColRNAI IncA/C2 IncFIB IncFIB(pQil) Col (pHAD28)	bla_{SHV-79} bla_{TEM-1B}	bla_{KPC-2} bla_{VIM-1}	*aac(3)-Iid aac(6′)-Ib aac(6′)-Im aant(2″)-Ia aph(2″)-Ib aph(3′)-Ia*	*oqxA, oqxB*	*fosA, sul1, Sul3, dfrA1*	*fyuA, mrkABCDFHIJ ybtASTX*	*acrB, marAR, soxSR, ramA, rob sdiA,* fis, *envR, oqxBR, rarA*	*wzc,*
C4112/18	11	ColRNAI IncFIA(HI1) IncFIB(K) IncFII(K)	$bla_{CTX-M-15}$ bla_{SHV-11}	bla_{NDM-1}	*aac(3)-Iia, aac(6′)-Ib aph(3′)-Ia*	*aac(6′)-Ib-cr, oqxA, oqxB*	*fosA, sul2 dfrA14, catB3, tet(D)*	*iutA, mrkABCDFHIJ*	*acrABR marAR soxSR ramAR rob sdiA, fis, envR, oqxR, rarA*	*wzc, wzi*
B11395/18	39	IncFIB (AP001918) IncFIB(pQil) IncFII(K)	bla_{SHV-79} bla_{TEM-1B}	bla_{KPC-2}	*aac(3)-Iid, aadA1, aadA2, aph(3′)-Ia*	*oqxA, oqxB*	*fosA, sul2, sul3, dfrA12, cmlA1*	*fyuA, mrk-ABCDFHIJ, ybtEQSTU*	*acrABR marAR soxSR, ramA, sdiA, fis, envR, oqxA, rarA*	*wzc, wzi*
D6184/18	39	ColRNAI IncFIB (AP001918) IncFIB(pQil)	bla_{SHV-11} bla_{SHV-40} bla_{TEM-1B}	bla_{KPC-2}	*aac(3)-Iid, aac(6′)-Ib, aph(3′)-Ia, aadA1, aadA2*	*aac(6′)-Ib-cr, oqxA, oqxB*	*fosA, sul2, sul3, tet(M), dfrA12, catA1, atB3,*	*fyuA, irp1, irp2, mrk-ABCDFHIJ, ybtAEPQS-TUX*	*acrABR marAR oxSR, ramA, rob, sdiA, fis, envR, qxABR, rarA*	*wzc, wzi*
C251/19	30	ColRNAI IncA/C2 IncFIB(K) IncFII(K) IncX3	bla_{SHV-12}, bla_{TEM-1A} bla_{OXA-9}	bla_{KPC-2} bla_{VIM-1}	*aac(6′)-Ib aph(3′)-Ia aadA2 aadA15*	*qnrS1, aac(6′)-Ib-cr, oqxA, oqxB*	*fosA, sul1, dfrA1, dfrA12, catA, tet(D)*	*fyuA, irp1, iutA mrkBCDFIJ, ybtAEPQSTU*	*acrABR, marAR, soxSR, ramAR, rob, sdiA, envR, oqxABR, rarA*	*wzc, wzi*
Z557/19	323	IncFIB(K) IncFIB(pQil) IncFII(K) IncFIB(Mar) IncFIB (pKPHS1) IncHI1B	bla_{SHV-1} bla_{SHV-99}	bla_{KPC-2} bla_{VIM-1}	*ant(2″)-Ia, aac(6′)-Il, aadA1*	*qnrA1 oqxA, oqxB*	*fosA, fosA7, sul1, dfrA1*	*mrkABCDFIJ*	*acrAR, marA, soxSR, ramA, rob, sdiA, fis, envR, oqxA, rarA*	*wzc, wzi*
C746/19	11	ColRNAI IncFIB(K) IncFIB(pQil) IncFII(K), IncR	bla_{SHV-1} bla_{OXA-1}	bla_{KPC-2}	*aac(6′)-Ib-cr aph(3′)-Ia aadA2*	*aac(6′)-Ib-cr, oqxA oqxB*	*fosA, sul1*	*fyuA, irp1, irp2 iutA, mrkABCDFIJ, ybtAEUX*	*acrAB, marAR, soxSR, ramAR, rob, sdiA, fis, envR, oqxABR,*	*wzi*
B2562/19	15	IncFIB(K) IncFII(K) IncFIA(HI1)	$bla_{CTX-M-15}$ bla_{TEM-1B} bla_{OXA-1}	bla_{NDM-1}	*aac(3)-Iia aac(6′)-Ib3 aph(3″)-Ib aph(6)-Id*	*aac(6′)-Ib-cr, oqxA, oqxB*	*fosA, sul2, dfrA14, catB3*	*fyuA, irp1, kfuABC, mrkABCDFIJ, ybtAPQSTUX*	*acrABR, marAR, soxSR, ramA, rob, sdiA,* fis, *envR, rarA*	*wzi*

Table 1. *Cont.*

Isolates	ST	Plasmids	Antibiotic Resistance Genes Affecting					Virulence	Efflux/Regulator	Capsule
			Other Beta-Lactams	Carba-Penems	Amino-Glycosides	Quinolones	Other			
Z508/19	39	ColRNAI IncA/C2 IncFIB IncFIB(pQil)	bla_{SHV-11} bla_{SHV-40} bla_{TEM-1B} $bla_{OXA-116}$ $bla_{OXA-366}$	bla_{KPC-2} bla_{VIM-1}	*aac(3)-Iid* *aac(6′)-Ib3* *aac(6′)-Im* *aph(2″)-Ib* *aph(3′)-Ia* *aph(3′)-Via* *aadA24*, *armA*, *aac(6′)-Il*	*qnrS1* *aac(6′)-Ib-cr*, *oqxA*, *oqxB*	*fosA*, *sul1*, *sul3*, *frA1*, *tet(M)*	*fyuA*, *irp1*, *irp2*, *mrk-ABCDFHIJ*, *ybtAEPQS-TUX*	*acrBR*, *marAR*, *soxSR*, *ramA*, *rob*, *sdiA*, fis, *envR*, *oqxABR*, *rarA*	*wzc*, *wzi*
D1463/19	323	IncFIB(K) IncFIB(Mar) IncFIB (pKPHS1) IncFIB(pQil) IncFII(K) IncHI1B	bla_{SHV-99}	bla_{KPC-2} bla_{VIM-1}	*ant(2″)-Ia* *aac(6′)-Il*	*oqxA* *oqxB*	*fosA*, *fosA7*, *dfrA1*	*mrkABCDFHJ*,	*acrAR*, *marA*, *soxSR*, *ramA*, *rob*, *sdiA*, fis, *envR*, *oqxA*, *rarA*	*wzc*, *wzi*
D1598/19	39	ColRNAI IncA/C2 IncFIB IncFIB(pQil)	bla_{SHV-40} bla_{TEM-1B}	bla_{KPC-2} bla_{VIM-1}	*aac(3)-Iid* *aac(6′)-Il* *aac(6′)-Im* *ant(3)-Ia* *aph(2″)-Ib* *aph(3′)-Ia*	*qnrS1*, *aac(6′)-Ib-cr*, *oqxA* *oqxB*	*fosA*, *sul1*, *sul3*, *dfrA1*	ND	ND	*wzi*
Z852/19	307	IncA/C2 IncFIB(pQil)	bla_{SHV-28}	bla_{KPC-2} bla_{VIM-1}	*aac(6′)-Il* *aadA24*	*oqxA* *oqxB*	*fosA*, *sul1*, *dfrA1*, *frA14*	*fyuA*, *irp1*, *irp2* *iutA*, *mrk-ABCDFHIJ*, *ybtAEPQSTX*	*acrABR*, *marAR*, *soxSR*, *ramAR*, *rob*, *sdiA*, fis, *envR*, *oqxABR*, *rarA*	*wzc*, *wzi*
Z866/19	323	IncFIB(K) IncFIB(Mar) IncFIB (pKPHS1) IncFIB(pQil) IncFII(K) IncHI1B	bla_{SHV-99}	bla_{KPC-2} bla_{VIM-1}	*aadA1*	*oqxA* *oqxB*	*fosA*, *fosA7*, *Sul1*	*mrkABCDFHIJ*	*acrAR*, *marA*, *soxSR*, *ramA*, *rob*, *sdiA*, fis, *envR*, *oqxAR*, *rarA*	*wzc*, *wzi*
D2452/19	39	ColRNAI IncA/C2 IncFIB(pQil)	bla_{SHV-79} bla_{TEM-1B}	bla_{KPC-2} bla_{VIM-1}	*aac(3)-Iid* *aac(6′)-Il* *aac(6′)-Im* *ant(3)-Ia* *aph(2″)-Ib* *aph(3′)-Ia*	*qnrS1*, *aac(6′)-Ib-cr*, *oqxA* *oqxB*	*fosA*, *sul1*, *Sul4*, *dfrA1*	*fyuA*, *irp1*, *irp2* *mrk-ABCDFHIJ*, *ybtAEPQSTX*	*acrBR*, *marAR*, *soxSR*, *ramA*, *rob*, *sdiA*, fis, *envR*, *oqxABR*, *rarA*	*wzc*, *wzi*
C833/21	11	Col(pHAD2) IncFIA(HI1) IncFIB(K) IncFII(K)	$bla_{SHV-182}$	bla_{OXA-48} bla_{NDM-1}	*aph(3′)-Ia* *aac(6′)-Ib-cr*	*aac(6′)-Ib-cr*, *oqxA*, *oqxB*	*fosA*, *sul2*, *dfrA14*, *catB3*	*mrkABCDFHIJ*	*acrABR*, *marA*, *soxSR*, *ramAR*, *rob*, *sdiA*, *fis*, *envR*, *rarA*	*wzc*, *wzi*
A5051/21	15	IncFIA(HI1) IncFIB(K) IncFII(K)	bla_{SHV-28} $bla_{SHV-106}$	bla_{NDM-1}	*aac(6′)-Ib-cr*	*aac(6′)-Ib-cr* *oqxA*	*fosA* *catB3*	*fyuA*, *kfuC*, *mrk-ABCDFHIJ* *ybtQU*	*acrABR*, *marA*, *soxSR*, *ramAR*, *rob*, *sdiA*, *fis*, *envR*, *rarA*	*wzcwzi*
D2856/21	11	Col440II IncFIA(HI1) IncFIB(K) IncFII(K) IncR	$bla_{SHV-182}$	bla_{OXA-48} bla_{NDM-1}	*aph(3′)-Ia* *aac(6′)-Ib-cr*	*oqxA*	*fosA*, *sul2*, *dfrA14*, *catB3*, *tet(A)*	*mrkABCDJ*, *ybtAPQS*	*acrABR*, *marAR*, *soxSR*, *ramAR*, *rob*, *sdiA*, *fis*, *envR*, *oqxAR*, *rarA*	*wzcwzi*
C1909/21	39	ColRNAI IncFII(K)	$bla_{SHV-40,\,-79,\,-85,\,-89}$	bla_{KPC-33}	*aac(6′)-Ib* *aac(6′)-Ib-cr*	*aac(6′)-Ib-cr*, *oqxA*, *oqxB*	*fosA*, *dfrA12* *tet(A)*	*mrkABCDFHIJ* *ybtEPQSTU*	*acrA*, *marAR*, *soxSR*, *ramAR*, *rob*, *sdiA*, *fis*, *envR*, *oqxAR*, *rarA*	*wzcwzi*
A18940/21	512	IncFIB(K) IncFII(K) IncN IncX3	$bla_{SHV-182}$ bla_{OXA-10}	bla_{KPC-2}, bla_{NDM-1}	*aac(6′)-Ib-cr*, *aac(6′)-Ib*, *aph(3′)-Ia*, *aadA2*, *aadA16*	*qnrS1*, *aac(6′)-Ib-cr*, *oqxA*, *oqxB*	*fosA*, *sul1*, *dfrA12*, *dfrA27* *catA1*	*iutA*, *mrkABCHI*	*acrR*, *marAR* *soxRS*, *ramA*, *rob*, *sdiA*, *fis*, *envR*, *oqxR*, *rarA*	*wzi*

Table 1. *Cont.*

Isolates	ST	Plasmids	Antibiotic Resistance Genes Affecting					Virulence	Efflux/Regulator	Capsule
			Other Beta-Lactams	Carba-Penems	Amino-Glycosides	Quinolones	Other			
A1746/22	35	IncC, IncR, IncFIA(HI1) IncFIB(K) IncFIB (pKPHS1) IncFIB(pQil) IncFII(K)	*bla*$_{\text{SHV-33}}$ *bla*$_{\text{OXA-10}}$ *bla*$_{\text{TEM-1B}}$ *bla*$_{\text{VEB-25}}$ *bla*$_{\text{DHA-1}}$	*bla*$_{\text{KPC-2}}$	*ant(2″)-Ia, aph(3″)-Ib aph(6)-Id, rmtB,aadA1*	*qnrB4 oqxA, oqxB*	*fosA, sul1 sul2, catA1, cmlA1, tet(A), tet(G)*	*kfuA, mrkAFHI, ybtEQTX*	*acrR, marAR, soxRS, ramA rob, sdiA, fis, envR, oqxR rarA*	*wzc, wzi*
A7213/22	11	ColRNAI IncC IncFIA(HI1) IncFIB(K)	*bla*$_{\text{OXA-1}}$ *bla*$_{\text{OXA-10}}$ *bla*$_{\text{CTX-M-15}}$ *bla*$_{\text{TEM-1B}}$ *bla*$_{\text{VEB-1}}$	*bla*$_{\text{NDM-1}}$	*aac(6′)-Ib, aac(3)-IIa, aac(6′)-Ib-cr, ant(2″)-Ia, aph(3′)-Ia, rmtB*	*aac(6′)-Ib-cr, oqxA, oqxB*	*fosA, sul2, dfrA14, tet(A)*	*mrkAC*	*soxSR, ramAR, sdiA*	*wzi*
A9974/22	11	ColRNAI IncC IncFIA(HI1) IncFIB(K) IncFII(K)	*bla*$_{\text{SHV-159}}$ *bla*$_{\text{OXA-1}}$ *bla*$_{\text{OXA-10}}$ *bla*$_{\text{CTX-M-15}}$ *bla*$_{\text{TEM-1B}}$ *bla*$_{\text{VEB-1}}$	*bla*$_{\text{NDM-1}}$	*aac(6′)-Ib-cr aac(6′)-Ib aac(3)-IIa aph(6)-Id aph(3′)-Ia aph(3″)-Ib ant(2″)-Ia aadA1, rmtB*	*aac(6′)-Ib-cr, oqxA, oqxB*	*fosA, sul2, dfrA14, tet(A), tet(G)*	*fyuA, mrk-ABCDFHIJ, ybtAEPQST*	*acrA, marAR soxR, ramAR rob, sdiA, fis, envR, oqxAB, rarA*	*wzi*
A10037/22	15	IncC IncFIA(HI1) IncFIB(K) IncFII(K)	*bla*$_{\text{SHV-100}}$ *bla*$_{\text{OXA-1}}$ *bla*$_{\text{OXA-10}}$ *bla*$_{\text{CTX-M-15}}$ *bla*$_{\text{TEM-1B}}$ *bla*$_{\text{VEB-1}}$	*bla*$_{\text{NDM-1}}$	aac(6′)-Ib-cr *aac(6′)-Ib aac(3)-IIa aph(3″)-Ib aph(6)-Id ant(2″)-Ia aadA1, rmtB*	*aac(6′)-Ib-cr, oqxA, oqxB*	*fosA, sul2 dfrA14, cmlA1, tet(A), tet(G)*	*fyuA, kfuAC, mrkABCFHJ, ybtAEPQSTU*	*acrAR, marAR soxSR, rob, sdiA fis, envR, oqxAR rarA*	*wzi*
A10143/22	11	ColRNAI IncFII(K)	*bla*$_{\text{OXA-1}}$ *bla*$_{\text{OXA-10}}$ *bla*$_{\text{CTX-M-15}}$ *bla*$_{\text{TEM-1B}}$ *bla*$_{\text{VEB-1}}$	*bla*$_{\text{NDM-1}}$	*aac(6′)-Ib-cr aac(6′)-Ib aac(3)-IIa aph(3″)-Ib aph(6)-Id ant(2″)-Ia aadA1, rmtB*	*aac(6′)-Ib-cr, oqxA, oqxB*	*fosA, sul1, sul2, dfrA12, dfrA14, cmlA1, tet(A), tet(G)*	*fyuA, mrkH, ybtEPQSX*	*marA, soxSR, ramR, sdiA, fis, envR*	*wzi*

Yersiniabactin cluster: *ybt*, *irp*, and *fyuA* genes; Aerobactin cluster: *iuc* and *iut* genes; AcrAB efflux pump: *acr*, *mar*, *sox*, *rob*, ram *sdiA*, *fis*, *and envR*, genes; OqxAB efflux pump: *oqx* and *rar* genes.

2.3. Virulence Factors and Efflux and Regulator Systems

The virulence factors detected included fimbrial genes, as 23/24 isolates possessed the *mrk* cluster (type 3 fimbriae); siderophores, as yersiniabactin genes (*ybt*, *irp*, and *fyu*A) were present in 17 isolates; and genes encoding aerobactin siderophore (*iut*) were identified in five isolates (three of them also carried yersiniabactin genes). In addition, the *kfu* gene (*Klebsiella* ferric iron uptake, which is a regulator of the iron transport system) was detected in four isolates.

Efflux and regulator system genes encoding the AcrAB efflux pump (*acr*, *mar*, *sox*, *rob*, *ram*, *sdi*A, *fis*, *and env*R) or the OqxAB efflux pump (*oqx* and *rar*) were present in all isolates (Table 1).

2.4. Genomic Comparison among CRKP Strains

Circular genome maps were generated for the four STs with more than one isolate (ST11, ST39, ST323, and ST15). It was found that isolates belonging to the same ST shared many identical regions with the reference strain. CRKP isolates C4112 and C746 of the ST11, D2452, and D6184 of the ST39, Z866 of the ST323, and B2562 of the ST15 displayed visually the highest similarity with the reference strain throughout their whole genome (Figure 2).

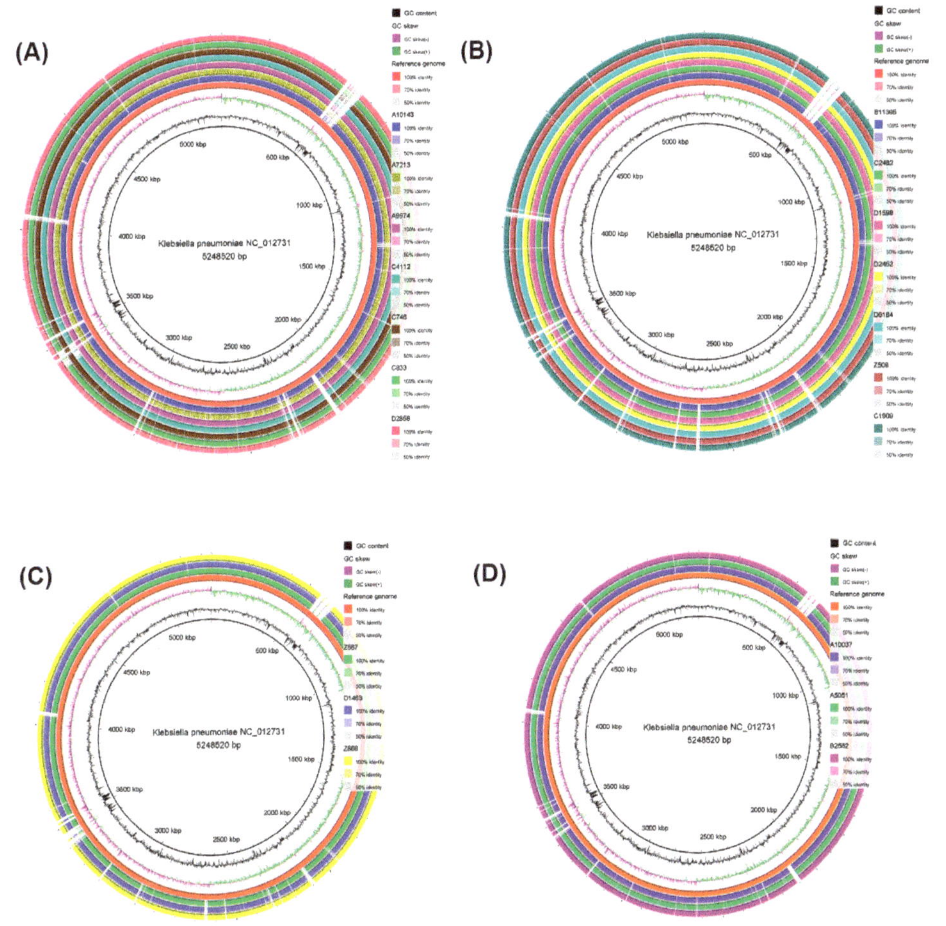

Figure 2. Circular genome map of (**A**) seven ST11 CRKP strains, (**B**) the seven ST39 CRKP strains, (**C**) the three ST323 CRKP strains, and (**D**) the three ST15 CRKP strains. The genome maps were produced through the BLAST Ring Image Generator (BRIG) software. The genomes of the seven ST11 CRKP isolates (A10143, A7213, A9974, C4112, C746, C833, and D2856), the seven ST39 CRKP isolates (B11395, C2482, D1598, D2452, D6184, Z508 and C1909), the three ST323 CRKP strains (Z557, D1463, and Z866) and the three ST15 CRKP strains (A10037, A5051, and B2562) were mapped separately to the reference genome *K. pneumoniae* NC_012731. From the inner to the external ring, the innermost ring displays the size of the genome in kbp, followed by GC content (black), GC skew (dark green and dark purple), and *K. pneumoniae* NC_012731 (red). The white vertical gaps represent sequences of the *K. pneumoniae* NC_012731 reference strain that are absent in the sequences of the ST11, ST39, ST323, and ST15 CRKP strains or present sequences <50% identical to the reference genome.

2.5. CgSNP-Based Phylogenetic Analysis

The length of the multiple core genome alignments of all CRKP isolates was 1,892,845 base pairs. The core genome length corresponds to about 36.1% of the reference genome length. Nearly 1.9 million columns in the core genome alignment (almost 98.5%) showed identical nucleotides for all isolates, meaning that 1.5% of the genome is polymorphic.

A phylogenetic tree including the cgSNP-based phylogenetic analysis of all isolates, along with the reference strain, is seen in Figure 3.

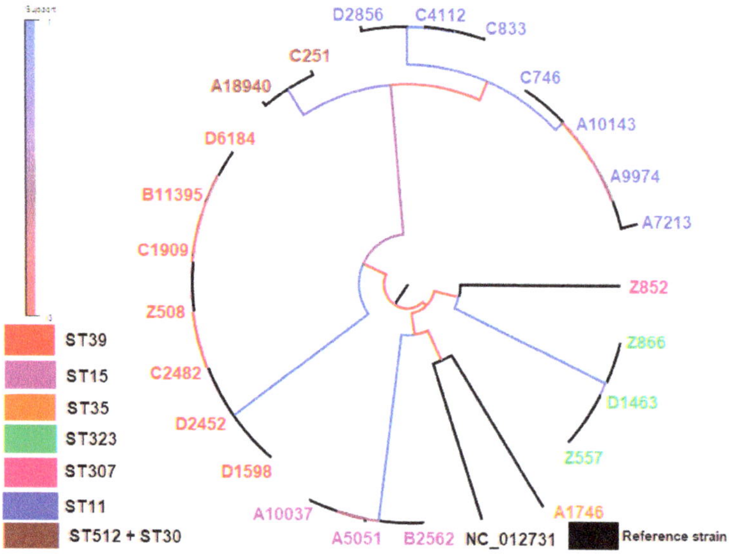

Figure 3. The tree built on the entire core genome with PhyML. The branch color indicates the fraction of supporting versus supporting plus clashing SNPs for each branch of the core tree, which supports the initial bipartition of the strains. MLST types: **ST11:** D2856, C4112, C833, C746, A10143, A9974, A7213 (blue color), **ST512:** A18940 (brown color), **ST30:** C251 (brown color), **ST307:** Z852 (pink color), **ST323:** Z866, D1463, Z557 (green color), **ST35:** A1746 (orange color), **ST15:** B2562, A5051, A10037 (purple color), **ST39:** D1598, D2452, C2482, Z508, C1909, B11395, D6184 (red color).

When performing all possible pairwise analyses among ST15 isolates, an average of 17 recombined SNPs (range 5–26) were detected; the SNP diversity along the core genome was very low, that is, 0–3 cgSNPs per kilobase. ST323 CRKP strains presented an average of six recombined SNPs (range 5–6), and the SNP diversity was extremely low, that is, 0–2 cgSNPs per kilobase. ST39 CRKP strains displayed an average of 16 recombined SNPs (range 4–43), with the SNP diversity being low, that is, 0–5 cgSNPs per kilobase. Finally, ST11 CRKP strains showed an average of 866 recombined SNPs (range 1–1652), with the SNP diversity being clearly higher, that is, 0–66 cgSNPs per kilobase.

For pairs of CRKP isolates, for instance, C4112 and C746 with relatively low diversity (7.5×10^{-4}) (Figure 4G), the impact of recombination is almost directly evident in the pattern of SNP density along the genome. While the SNP density is very low along most of the genome, that is, 0–2 SNPs per kilobase, there are a few segments where the SNP density is much higher (Figure 4A) and comparable to the typical SNP density between randomly selected highly diverged CRKP isolates, that is, 30–65 SNPs per kilobase (Figure 4C). For some pairs of isolates, for example, A18940 and D2856, with higher genetic diversity (1.5×10^{-3}) (Figure 4G), the frequency of these high-spiking recombined regions increases (Figure 4B), until most of the genome is covered by such regions when comparing highly divergent isolates (5.5×10^{-3}), for instance, A5051 and C2482 (Figure 4C,G). The distribution of SNP densities was described by a majority of clonally inherited regions with mostly up to three SNPs per kilobase and a long tail of recombined regions with up to 50 or 65 SNPs per kilobase (Figure 4D–F).

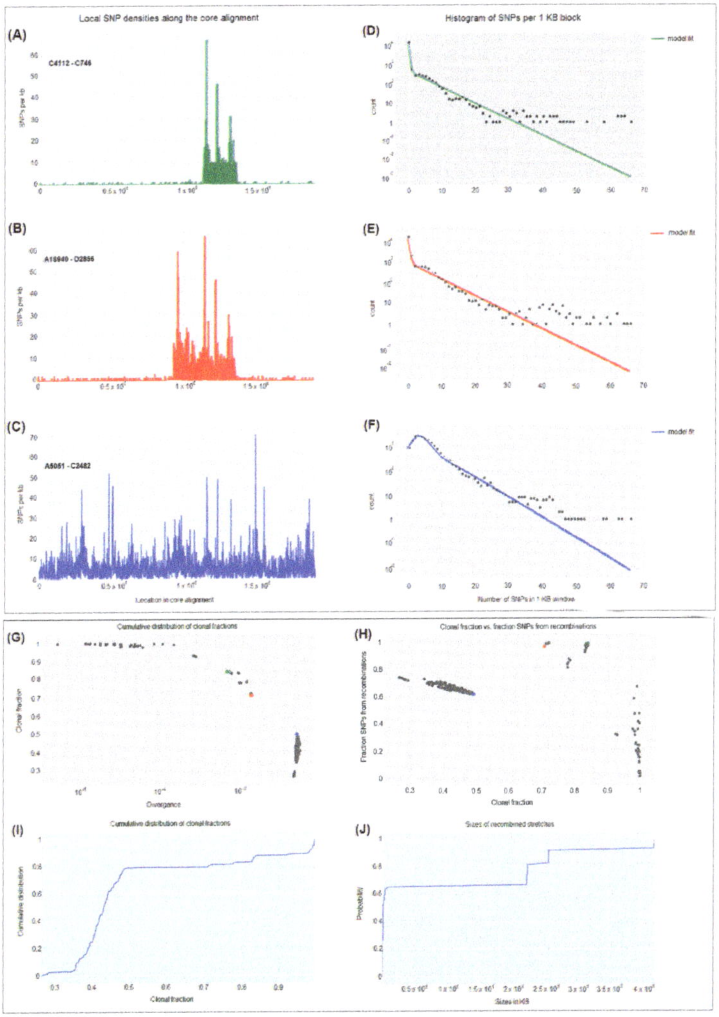

Figure 4. Pairwise analysis of recombination between carbapenem-resistant *K. pneumoniae* CRKP strains. (**A–C**) SNP densities (SNPs per kilobase) along the core genome for three pairs of CRKP strains with various nucleotide diversities. (**D–F**) Corresponding histograms for the number of SNPs per kilobase (dots) together with fits of the Poisson mixture model for C4112-C746 (green), A18940-D2856 (red), and A5051-C2482 (blue) CRKP strains. The vertical axis is on a logarithmic scale. (**G**) For each pair of CRKP strains (dots), the fraction of the genome that was inherited clonally is shown as a function of the nucleotide divergence of the pair, shown on a logarithmic scale. The three CRKP pairs that are highlighted in panels (**A–F**) are depicted as green, red, and blue dots. (**H**) Fraction of all SNPs that lie in recombined regions as a function of the clonally inherited fraction of the genome. (**I**) Cumulative distribution of the clonal fractions of CRKP pairs. (**J**) Cumulative distribution of the lengths of recombined segments for CRKP pairs that are in the mostly clonal regime.

Figure 5 highlights the distribution of the lengths of tree-compatible stretches. This distribution had a mode at $n = 2$, and stretches were normally around 7–12 consecutive SNPs and very rarely longer than 15–20 consecutive SNPs (Figure 5A). Similarly, tree-compatible segments were typically just a few hundreds of base pairs long (around 1000–1500 base pairs) and very rarely more than 2000 base pairs (Figure 5B).

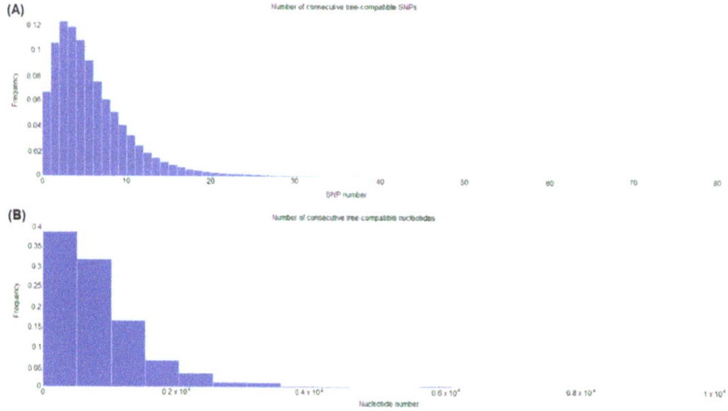

Figure 5. SNP compatibility along the core genome alignment displays the shortness of tree-compatible segments. (**A**) Probability distribution of the number of consecutive SNP columns that are consistent with a common phylogeny for the core genome alignment. (**B**) Probability distribution of the number of consecutive nucleotides consistent with a common phylogeny for the core genome alignment.

Figure 6 underscores the ratio C/S for random subsets of all 24 isolates. When comprising all 24 isolates, a ratio C/S = 0.15 was obtained for the 5% homoplasy-corrected full alignment. For small subsets of isolates, the C/S ratio displayed significant fluctuations. For instance, for subsets of $n = 7$ isolates, the C/S ratio presented a mean of 0.094 ± 0.046. However, as the number of isolates in the subset increased, the ratios converged to the value C/S = 0.15, and for large subsets of isolates, there was little variation in this ratio. Consequently, for alignments of large sets of isolates, the phylogeny should change at least approximately every seven SNPs (Figure 6). In addition, on average, each randomly chosen position on the core genome has been overwritten at least $T = 4344$ times (5% homoplasies removed).

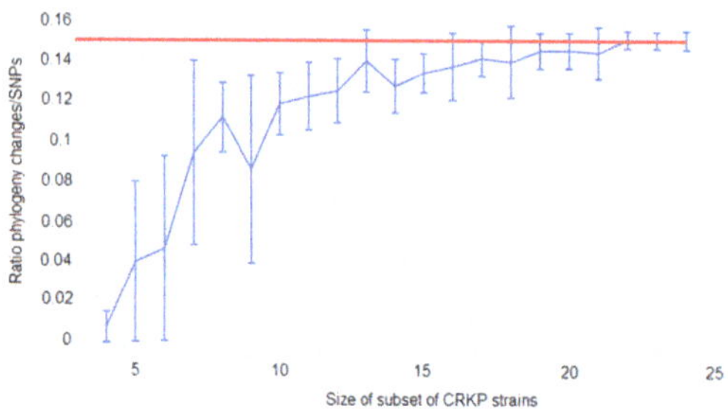

Figure 6. Ratio C/S of the minimal number of phylogeny changes C to substitutions S for random subsets of carbapenem-resistant *K. pneumoniae* (CRKP) strains using the alignment from which 5% of potentially homoplastic positions have been removed. For CRKP strain numbers ranging from $n = 4$ to $n = 24$, random subsets of n strains were collected, and the ratios C/S of phylogeny changes to SNPs in the alignment were calculated. The figure shows box-whisker plots that indicate, for each CRKP strain number n, means, and standard deviations for C/S across subsets. The horizontal red line displays C/S = 0.15.

3. Materials and Methods

3.1. Clinical Isolates-Setting

The present study included 24 CRKP isolates collected from June 2018 to July 2022 from 18 patients hospitalized in the ICU (8 females and 10 males) and six patients hospitalized in the PICU (4 females and 2 males) of a tertiary university-affiliated general hospital in Greece. The median age of the adult patients was 57.5 years (range 30–88), while the median age of the children was 6.2 years (range 0.25–17). The selection of the isolates was based on the type of carbapenemase(s) previously detected in order to include all possible combinations, as they were pre-screened as previously described [11]. Eight isolates were collected from colonization sites (rectal swabs), while 16 isolates were taken from various types of clinical samples (Table 2).

Table 2. Clinical data of carbapenem-resistant *Klebsiella pneumoniae* isolates.

Strain	Date of Isolation	Sex	Age (Years)	Ward	Biological Sample	Infection/Colonization
C2482/18	9 June 2018	F	8	PICU	Rectal swab	Colonization
C4112/18	7 September 2018	F	56	ICU	Rectal swab	Colonization
B11395/18	24 October 2018	F	1	PICU	Urine	Infection
D6184/18	19 November 2018	F	8	PICU	Bronchial aspirate	Infection
C251/19	17 January 2019	F	17	PICU	Rectal swab	Colonization
Z557/19	30 January 2019	M	57	ICU	Rectal swab	Colonization
C746/19	7 February 2019	M	73	ICU	Rectal swab	Colonization
B2562/19	4 March 2019	M	79	ICU	Urine	Infection
Z508/19	20 March 2019	F	60	ICU	Rectal swab	Colonization
D1463/19	22 March 2019	M	55	ICU	Bronchial aspirate	Infection
D1598/19	1 April 2019	M	40	ICU	Bronchial aspirate	Infection
Z852/19	12 April 2019	M	3	PICU	Rectal swab	Colonization
Z866/19	15 April 2019	M	49	ICU	Rectal swab	Colonization
D2452/19	24 May 2019	M	73	ICU	Bronchial aspirate	Infection
C833/21	31 March 2021	F	57	ICU	Wound	Infection
A5051/21	25 April 2021	F	69	ICU	Blood	Infection
D2856/21	7 July 2021	F	54	ICU	Bronchial aspirate	Infection
C1909/21	8 July 2021	M	85	ICU	Wound	Infection
A18940/21	24 December 2021	F	58	ICU	Blood	Infection
A1746/22	25 February 2022	M	0.25	PICU	Blood	Infection
A7213/22	3 May 2022	F	30	ICU	Blood	Infection
A9974/22	16 June 2022	F	67	ICU	Blood	Infection
A10037/22	16 June 2022	F	53	ICU	Blood	Infection
A10143/22	19 June 2022	M	88	ICU	Blood	Infection

ICU: Intensive Care Unit, PICU: Pediatric Intensive Care Unit.

3.2. Ethics Approval

The study was approved by the Ethics Committee of Aristotle's University Medical Faculty (no. of approval 5.160/18/12/2019).

3.3. Microbiological Methods-Antimicrobial Susceptibility Testing

CRKP isolates were identified through the automated system VITEK 2 (bioMérieux, Marcy-l'Étoile, France) using the Gram-negative identification card (GN ID). Antimicrobial susceptibility testing (AST) of isolates was performed using the AST318 and XN10 cards, while the use of minimum inhibitory concentration (MIC) test strips (Liofilchem srl, Roseto, Italy) was applied to confirm susceptibility to ceftazidime-avibactam (CAZ-AVI). Identification and susceptibility cards were interpreted according to the manufacturer's instructions. The Clinical and Laboratory Standards Institute (CLSI) susceptibility breakpoints were applied for the interpretation of results [29]. The breakpoints approved by the US Food and Drug Administration (FDA) were used to interpret the MICs of tigecycline [30].

All CRKP isolates were previously screened for carbapenemase production using a lateral flow immunochromatographic assay (LFIA) (NG-Test CARBA 5, NG Biotech, Paris, France). The results were confirmed by multiplex PCR in a single reaction following a modified protocol that was previously described [11,31].

3.4. DNA Extraction and Whole Genome Sequencing

Bacterial DNA was extracted using the DNA extraction kit (Qiagen, Hilden, Germany). The DNA concentration (sample starting concentration between 10–100 ng/μL) was measured using the Qubit double-strand DNA HS assay kit (Q32851, Life Technologies Corporation, Grand Island, NY, USA).

NGS was performed using the Ion Torrent PGM Platform (Life Technologies Corporation). All procedures regarding purification, ligation, barcoding, library preparation, emulsion PCR, and enrichment were performed following the manufacturer's instructions. PCR products were loaded on an Ion-316™ chip. The Ion PGM Hi-Q View (200) chemistry was applied using the Ion PGM Hi-Q view sequencing kit (A25592).

3.5. Assembly Assessment and Genome Annotation

The software Geneious Prime version 2021.2.1 and BLAST Ring Image Generator (BRIG) were used to produce genome assemblies and annotation data [32,33]. *K. pneumoniae* NTUH-K2044 strain (GenBank accession number NC_012731) was used as a reference sequence, which is an invasive and hypervirulent *K. pneumoniae* strain with the hypervirulent plasmid pK2044 [34].

3.6. MLST and Detection of Antimicrobial Resistance Genes and Plasmids

The sequence types (ST), plasmid types, and antimicrobial resistance genes were identified through the Center for Genomic Epidemiology website using the MLST web server and the related online databases, PlasmidFinder version 2.0 and Resfinder version 4.1, respectively [35–37]. The Comprehensive Antibiotic Resistance Database was also used, with the selection criteria for hits set to perfect (100%) and strict (>95%) identity to the reference sequence [38]. The detection of genes related to virulence capsule, efflux, and regulator systems was performed according to the protocols available from the Pasteur Institute (https://bigsdb.pasteur.fr/cgi-bin/bigsdb/bigsdb.pl?db=pubmlst_klebsiella_seqdef (accessed on 20 October 2022).

3.7. Genomic Comparison–Core Genome Single-Nucleotide Polymorphism (cgSNP)-Based Phylogenetic Analysis

The BRIG software was used to visually compare multiple genome alignments using a default minimum threshold of 50% [33]. The reference sequence alignment-based phylogeny builder version 1.12 (REALPHY; https://realphy.unibas.ch/realphy/ (accessed on 15 November 2022) was used to identify relevant SNP sites for core genome phylogenetic analysis [39]. Default input parameters were applied for cgSNP identification. The *K. pneumoniae* NTUH-K2044 sequence was used as a reference for the multiple alignments and the visual comparison of the whole genomes [34]. The produced alignment was used for the creation of an unrooted tree on the entire core genome through the approximate maximum likelihood (ML) method using PhyML [40]. Pairwise analysis between isolates was also performed using REALPHY, as previously described by Dixit et al. [41]. The calculation of SNPs between isolates was conducted through REALPHY, using a Poisson mixture model plus a negative binomial for the recombined regions. REALPHY was also used to assess to what extent mutually consistent SNPs are clustered along the alignment; the lengths of segments along the alignment that are consistent with a single phylogeny were calculated [39]. In addition, the lower bound for the ratio between the total number of phylogeny changes (C) and substitution events (S) (C/S) that occur along the core genome alignment was calculated. The same ratio was calculated for any subset of CRKP isolates by removing 5% of potentially homoplastic sites [42]. Moreover, a rough estimate for the average number of times T that a randomly chosen position in the core genome alignment has been overwritten by recombination in its history was calculated, that is, the time since the genetic ancestors of the position in the alignment diverged from a common ancestor.

4. Discussion

In the current study, the WGSs of 24 CRKP isolates were analyzed. It was found that they belonged to eight different STs, with ST11 and ST39 being the most prevalent. Ten out of 24 isolates (41.6%) carried bla_{NDM-1}, and two of them (8.3%) transferred concurrently bla_{NDM-1} and bla_{OXA-48}. Six out of 10 NDM-producers (60.0%) were assigned to ST11, which is common in NDM-positive CRKP strains and is globally distributed, especially in Asia and particularly in China, but also in Europe [43,44]. In Greece, the first outbreak of an NDM-1-producing *K. pneumoniae* ST11 clonal strain occurred during 2011–2012 [44], while the first report of a *K. pneumoniae* ST11 clinical isolate co-producing NDM-1 and OXA-48 carbapenemases was reported in 2016 [22]. NDM-1-producing ST11 CRKP strains have been isolated during the same period in other Balkan countries [45]. However, three out of 10 NDM producers (30.0%) were assigned to ST15; this has also been previously described [46].

ST39 was also the most prevalent type, encompassing seven isolates (29.1%); two of them carried bla_{KPC-2}, while one of them (14.3%) harbored bla_{KPC-33} (a KPC-2 variant). CRKP strains belonging to ST39 have been detected only in sporadic cases so far, like ST39 carrying bla_{KPC-2} [47] or bla_{VIM-1} [48]. ST39 isolates transferring KPC-33 have been recently detected in Greece [8]. The present study also included the recently reported four ST39 isolates carrying both bla_{KPC-2} and bla_{VIM-1} [21]. Recently, Kuzina et al. highlighted the concurrent isolation of bla_{NDM-1}, bla_{KPC-2}, and bla_{OXA-48} in ST39 CRKP isolates [49]. Three CRKP isolates (12.5%) were assigned to ST323; sporadic cases of ST323 CRKP have been previously reported in Greece and other European countries [50,51]. Another study from Tunisia reports the isolation of clonally related ST323 CRKP isolates [52]. It is of interest that one isolate, A1746/22, carried the VEB-25 gene (article submitted for publication); VEB-25 is a variant of VEB-1 and confers resistance to CAZ-AVI [53,54].

Twenty of the 24 isolates harbored at least two genes encoding aminoglycoside-modifying enzymes (AMEs), and six of them carried one 16S rRNA methylase gene (*arm*A, *rmt*B). In addition, 12 of these isolates carried at least one gene of the three AME classes (acetyltransferases-*aac*, phosphotransferase-*aph*, nucleotransferases-*ant*). A disagreement between phenotype and genotype has been detected in five isolates (20.8%) (detection of AME-encoding resistance genes but phenotypically sensitive to both aminoglycosides tested). A disagreement of approximately 10% between genotype and phenotype for aminoglycosides for the VITEK 2 automated system has been reported previously [55].

Fluoroquinolone resistance was observed in seven isolates (29.2%), which carried the plasmid-encoded *qnr* gene, as previously described for CRKP isolates [56]. The multidrug efflux pump oqxAB gene was detected in all strains [57], while the *aac(6′)-Ib-cr* gene (which encodes an aminoglycoside acetyltransferase that modifies not only aminoglycosides but also fluoroquinolones) has been identified in 16 isolates, as previously described [56].

For fosfomycin resistance, the gene *fos*A was detected in all isolates, but 16 of them (66.7%) were phenotypically susceptible, evidenced by an increase in the MIC, as previously described [58]. Resistance genes for cotrimoxazole (*sul* or *dfr*A) were found in all isolates except one. The Inc-type plasmids detected in all isolates and the Col-type plasmids detected in 15 isolates have been reported in other studies [21,59].

In terms of adhesive structures, almost all isolates possessed the *mrk* cluster, characteristic of type 3 fimbriae, which promotes bacterial adhesion to both abiotic and biotic surfaces [60]. Regarding siderophores, the majority of isolates carried yersiniabactin genes that are significantly associated with pathogenesis and invasive infections in humans [61], while aerobactin (detected in five isolates) is more specific for hypervirulent *K. pneumoniae* [62]. The acquisition of a chimeric plasmid including genes for the siderophores yersiniabactin and aerobactin grants cefiderocol activity against CRKP strains [63].

NGS allows for more accurate analyses regarding the clonal relatedness of strains compared to conventional typing methods [64]. Several studies have described the molecular epidemiology of CRKP isolates through NGS using a gene-by-gene comparison or SNP analysis [65,66].

In pairwise comparisons, isolates of ST15, ST323, and ST39 presented an average of 17, 6, and 16 recombined SNPs, respectively, implying the clonal relatedness of these isolates in accordance with their ST classification. Similarly, Miro et al. report the clonal relatedness of OXA-48-producing ST405 CRKP strains further studied with core genome MLST, showing between 6 and 17 cgSNPs [64]. Moreover, Onori et al. reveal the clonal relatedness of sporadic KPC-producing ST258 CRKP strains, presenting an average of 20 or 27 cgSNPs [67]. ST15 and ST39 strains display a higher number of cgSNPs compared to ST323 strains, and this is highlighted as a slightly increased diversity in the corresponding clades of the phylogenetic tree (Figure 3). However, ST15 and ST39 has been isolated over a 33- and 36-month period, respectively, while ST323 have been collected over a 2.5-month period (Table 2). This could indicate a difference in the measured paces of the molecular clock between these clades, as previously proposed [67].

On the contrary, ST11 isolates showed an average of 866 cgSNPs, indicating the polyclonality of these isolates. This indication is further strengthened by the variability in their content of carbapenemase and antimicrobial resistance genes (Table 1) and by their phylogenetic division into two distinct subclades (Figure 3). The high average of cgSNPs occurred between ST11 strains belonging to these two different subclades, resulting in SNP patterns (Figure 4A). Most probably, these high SNP density patterns result from horizontal recombination events, e.g., the transfer of a plasmid or phage, as previously proposed for other bacterial species [42]. These results underscore the higher discriminatory power of cgSNP analysis compared to MLST [64].

In addition, the current study highlighted novel ways in which these cgSNPs could be used to quantify phylogenetic structures and the role of recombination in genome evolution. It was shown that there are pairs of closely related isolates and that most of their DNA has been clonally inherited; however, most of the changes are derived from recombination events. When focusing on the core genome alignment, it has been shown that, until the current genetic state of CRKP strains, each genomic locus has been overwritten thousands of times by recombination in the history of their evolution. Sakoparnig et al. reached the same conclusions when focusing on *E. coli* isolates, and they disclosed a similar C/S ratio [42]. Moreover, an attempt was made to calculate the probability for a pair of SNPs to be consistent with a common phylogeny as a function of their genome distance [68]. Several other studies have also attempted to quantify recombination in bacteria and assess the relative rate with which different lineages have recombined, focusing on different types of bacteria [42,68]. To the best of our knowledge, our study is the first to accomplish such a task by studying CRKP strains.

One of the basic challenges among WGS-based typing methods is the definition of a cut-off value for the identification of the clonal relatedness between isolates; however, this remains doubtful, as it seems to be not only species-dependent but also population-dependent [64,67,69]. Thus, a threshold range should be evaluated along with clinical epidemiological data.

In conclusion, the current study provides insight into the genetic characterization of CRKP isolates circulating during a 4-year period in the ICUs of a Greek tertiary hospital. Phylogenetic analysis showed that some of them were epidemiologically related. Given that NGS has become more affordable, its use in hospital settings is extremely helpful, as the obtained data could guide infection control and prevention strategies.

Author Contributions: C.Z.: investigation, review, and editing. T.K.: methodology, conceptualization, formal analysis, writing the initial draft. S.P.: investigation. E.I.: provision of study materials, review, and editing. E.V.: investigation. E.R.: provision of study materials, review, and editing. A.P.: conceptualization, methodology, data curation, formal analysis, visualization, reviewing and editing, supervision, funding acquisition. All authors have read and agreed to the published version of the manuscript.

Funding: The current study was supported by the European Union's Horizon 2020 projects COMPARE (grant number 643476) and VEO (grant number 874735).

Institutional Review Board Statement: The study was approved by the Ethics Committee of Aristotle's University Medical Faculty (no. of approval 5.160/18/12/2019).

Informed Consent Statement: Not Applicable.

Data Availability Statement: The data of this study have been deposited in the European Nucleotide Archive (ENA) at EMBL-EBI under accession numbers PRJEB42192 and PRJEB46676.

Acknowledgments: We thank Surbhi Malhotra-Kumar and Minh Nguyen-Ngoc from the University of Antwerp in Belgium for their technical support.

Conflicts of Interest: The authors declare no conflict of interest.

References

1. Juan, C.H.; Fang, S.Y.; Chou, C.H.; Tsai, T.Y.; Lin, Y.T. Clinical characteristics of patients with pneumonia caused by Klebsiella pneumoniae in Taiwan and prevalence of antimicrobial-resistant and hypervirulent strains: A retrospective study. *Antimicrob. Resist. Infect. Control* **2020**, *9*, 4. [CrossRef] [PubMed]
2. Martin, R.M.; Bachman, M.A. Colonization, Infection, and the Accessory Genome of Klebsiella pneumoniae. *Front. Cell. Infect. Microbiol.* **2018**, *8*, 4. [CrossRef] [PubMed]
3. Shao, C.; Wang, W.; Liu, S.; Zhang, Z.; Jiang, M.; Zhang, F. Molecular Epidemiology and Drug Resistant Mechanism of Carbapenem-Resistant Klebsiella pneumoniae in Elderly Patients With Lower Respiratory Tract Infection. *Front. Public Health* **2021**, *9*, 669173. [CrossRef]
4. Gupta, A.; Bhatti, S.; Leytin, A.; Epelbaum, O. Novel complication of an emerging disease: Invasive Klebsiella pneumoniae liver abscess syndrome as a cause of acute respiratory distress syndrome. *Clin. Pract.* **2018**, *8*, 1021. [CrossRef] [PubMed]
5. Ferri, M.; Ranucci, E.; Romagnoli, P.; Giaccone, V. Antimicrobial resistance: A global emerging threat to public health systems. *Crit. Rev. Food Sci. Nutr.* **2017**, *57*, 2857–2876. [CrossRef] [PubMed]
6. Nelson, K.; Hemarajata, P.; Sun, D.; Rubio-Aparicio, D.; Tsivkovski, R.; Yang, S.; Sebra, R.; Kasarskis, A.; Nguyen, H.; Hanson, B.M.; et al. Resistance to Ceftazidime-Avibactam is Due to Transposition of KPC in a Porin-Deficient Strain of Klebsiella pneumoniae with Increased Efflux Activity. *Antimicrob. Agents Chemother.* **2017**, *61*, e00989-17. [CrossRef] [PubMed]
7. Wang, B.; Pan, F.; Wang, C.; Zhao, W.; Sun, Y.; Zhang, T.; Shi, Y.; Zhang, H. Molecular epidemiology of Carbapenem-resistant Klebsiella pneumoniae in a paediatric hospital in China. *Int. J. Infect. Dis.* **2020**, *93*, 311–319. [CrossRef]
8. Galani, I.; Karaiskos, I.; Angelidis, E.; Papoutsaki, V.; Galani, L.; Souli, M.; Antoniadou, A.; Giamarellou, H. Emergence of ceftazidime-avibactam resistance through distinct genomic adaptations in KPC-2-producing Klebsiella pneumoniae of sequence type 39 during treatment. *Eur. J. Clin. Microbiol. Infect. Dis.* **2021**, *40*, 219–224. [CrossRef]
9. Galani, I.; Nafplioti, K.; Adamou, P.; Karaiskos, I.; Giamarellou, H.; Souli, M. Study Collaborators Nationwide epidemiology of carbapenem resistant Klebsiella pneumoniae isolates from Greek hospitals, with regards to plazomicin and aminoglycoside resistance. *BMC Infect. Dis.* **2019**, *19*, 167.
10. Karampatakis, T.; Tsergouli, K.; Politi, L.; Diamantopoulou, G.; Iosifidis, E.; Antachopoulos, C.; Karyoti, A.; Mouloudi, E.; Tsakris, A.; Roilides, E. Molecular Epidemiology of Endemic Carbapenem-Resistant Gram-Negative Bacteria in an Intensive Care Unit. *Microb. Drug Resist.* **2019**, *25*, 712–716. [CrossRef]
11. Zarras, C.; Pappa, S.; Zarras, K.; Karampatakis, T.; Vagdatli, E.; Mouloudi, E.; Iosifidis, E.; Roilides, E.; Papa, A. Changes in molecular epidemiology of carbapenem-resistant Klebsiella pneumoniae in the intensive care units of a Greek hospital, 2018–2021. *Acta Microbiol. Immunol. Hung.* **2022**, *69*, 104–108. [CrossRef] [PubMed]
12. Karampatakis, T.; Antachopoulos, C.; Iosifidis, E.; Tsakris, A.; Roilides, E. Molecular epidemiology of carbapenem-resistant Klebsiella pneumoniae in Greece. *Future Microbiol.* **2016**, *11*, 809–823. [CrossRef] [PubMed]
13. Protonotariou, E.; Meletis, G.; Pilalas, D.; Mantzana, P.; Tychala, A.; Kotzamanidis, C.; Papadopoulou, D.; Papadopoulos, T.; Polemis, M.; Metallidis, S.; et al. Polyclonal Endemicity of Carbapenemase-Producing Klebsiella pneumoniae in ICUs of a Greek Tertiary Care Hospital. *Antibiotics* **2022**, *11*, 149. [CrossRef] [PubMed]
14. Surveillance Atlas of Infectious Diseases. 2020. Available online: https://atlas.Ecdc.Europa.Eu/public/index.Aspx (accessed on 5 December 2022).
15. Hatrongjit, R.; Kerdsin, A.; Akeda, Y.; Hamada, S. Detection of plasmid-mediated colistin-resistant and carbapenem-resistant genes by multiplex PCR. *MethodsX* **2018**, *5*, 532–536. [CrossRef]
16. Mari-Almirall, M.; Ferrando, N.; Fernandez, M.J.; Cosgaya, C.; Vines, J.; Rubio, E.; Cuscó, A.; Muñoz, L.; Pellice, M.; Vergara, A.; et al. Clonal Spread and Intra- and Inter-Species Plasmid Dissemination Associated with Klebsiella pneumoniae Carbapenemase-Producing Enterobacterales during a Hospital Outbreak in Barcelona, Spain. *Front. Microbiol.* **2021**, *12*, 781127. [CrossRef]
17. Yanat, B.; Rodriguez-Martinez, J.M.; Touati, A. Plasmid-mediated quinolone resistance in Enterobacteriaceae: A systematic review with a focus on Mediterranean countries. *Eur. J. Clin. Microbiol. Infect. Dis.* **2017**, *36*, 421–435. [CrossRef]

18. Bosch, T.; Lutgens, S.P.M.; Hermans, M.H.A.; Wever, P.C.; Schneeberger, P.M.; Renders, N.H.M.; Leenders, A.C.A.P.; Kluytmans, J.A.J.W.; Schoffelen, A.; Notermans, D.; et al. Outbreak of NDM-1-Producing Klebsiella pneumoniae in a Dutch Hospital, with Interspecies Transfer of the Resistance Plasmid and Unexpected Occurrence in Unrelated Health Care Centers. *J. Clin. Microbiol.* **2017**, *55*, 2380–2390. [CrossRef]
19. Silva, D.D.C.; Rampelotto, R.F.; Lorenzoni, V.V.; Santos, S.O.D.; Damer, J.; Horner, M.; Horner, R. Phenotypic methods for screening carbapenem-resistant Enterobacteriaceae and assessment of their antimicrobial susceptibility profile. *Rev. Soc. Bras. Med. Trop.* **2017**, *50*, 173–178. [CrossRef]
20. Tzouvelekis, L.S.; Miriagou, V.; Kotsakis, S.D.; Spyridopoulou, K.; Athanasiou, E.; Karagouni, E.; Tzelepi, E.; Daikos, G.L. KPC-producing, multidrug-resistant Klebsiella pneumoniae sequence type 258 as a typical opportunistic pathogen. *Antimicrob. Agents Chemother.* **2013**, *57*, 5144–5146. [CrossRef]
21. Karampatakis, T.; Zarras, C.; Pappa, S.; Vagdatli, E.; Iosifidis, E.; Roilides, E.; Papa, A. Emergence of ST39 carbapenem-resistant Klebsiella pneumoniae producing VIM-1 and KPC-2. *Microb. Pathog.* **2022**, *162*, 105373. [CrossRef]
22. Protonotariou, E.; Meletis, G.; Chatzopoulou, F.; Malousi, A.; Chatzidimitriou, D.; Skoura, L. Emergence of Klebsiella pneumoniae ST11 co-producing NDM-1 and OXA-48 carbapenemases in Greece. *J. Glob. Antimicrob. Resist.* **2019**, *19*, 81–82. [CrossRef]
23. Partridge, S.R.; Ginn, A.N.; Wiklendt, A.M.; Ellem, J.; Wong, J.S.; Ingram, P.; Guy, S.; Garner, S.; Iredell, J.R. Emergence of blaKPC carbapenemase genes in Australia. *Int. J. Antimicrob. Agents* **2015**, *45*, 130–136. [CrossRef] [PubMed]
24. Tang, H.J.; Ku, Y.H.; Lee, M.F.; Chuang, Y.C.; Yu, W.L. In Vitro Activity of Imipenem and Colistin against a Carbapenem-Resistant Klebsiella pneumoniae Isolate Coproducing SHV-31, CMY-2, and DHA-1. *BioMed Res. Int.* **2015**, *2015*, 568079. [CrossRef] [PubMed]
25. Kaczmarek, F.M.; Dib-Hajj, F.; Shang, W.; Gootz, T.D. High-level carbapenem resistance in a Klebsiella pneumoniae clinical isolate is due to the combination of bla(ACT-1) beta-lactamase production, porin OmpK35/36 insertional inactivation, and down-regulation of the phosphate transport porin phoe. *Antimicrob. Agents Chemother.* **2006**, *50*, 3396–3406. [CrossRef] [PubMed]
26. Paczosa, M.K.; Mecsas, J. Klebsiella pneumoniae: Going on the Offense with a Strong Defense. *Microbiol. Mol. Biol. Rev.* **2016**, *80*, 629–661. [CrossRef]
27. Brhelova, E.; Antonova, M.; Pardy, F.; Kocmanova, I.; Mayer, J.; Racil, Z.; Lengerova, M. Investigation of next-generation sequencing data of Klebsiella pneumoniae using web-based tools. *J. Med. Microbiol.* **2017**, *66*, 1673–1683. [CrossRef]
28. Enany, S.; Zakeer, S.; Diab, A.A.; Bakry, U.; Sayed, A.A. Whole genome sequencing of Klebsiella pneumoniae clinical isolates sequence type 627 isolated from Egyptian patients. *PLoS ONE* **2022**, *17*, e0265884. [CrossRef]
29. *CLSI standard M02*; Performance Standards for Antimicrobial Susceptibility Testing; 32nd Ed. Clinical and Laboratory Standards Institute: Wayne, PA, USA, 2020.
30. U.S. FDA. FDA approves new antibacterial drug Avycaz. In *FDA News Release*; U.S. FDA: Silver Spring, MD, USA, 2015.
31. Tsakris, A.; Pournaras, S.; Woodford, N.; Palepou, M.F.; Babini, G.S.; Douboyas, J.; Livermore, D.D. Outbreak of infections caused by Pseudomonas aeruginosa producing VIM-1 carbapenemase in Greece. *J. Clin. Microbiol.* **2000**, *38*, 1290–1292. [CrossRef]
32. Geneious Prime 2021.2.1. Available online: https://www.geneious.com (accessed on 1 December 2022).
33. Alikhan, N.F.; Petty, N.K.; Ben Zakour, N.L.; Beatson, S.A. BLAST Ring Image Generator (BRIG): Simple prokaryote genome comparisons. *BMC Genom.* **2011**, *12*, 402. [CrossRef]
34. Wu, K.M.; Li, L.H.; Yan, J.J.; Tsao, N.; Liao, T.L.; Tsai, H.C.; Fung, C.P.; Chen, H.J.; Liy, Y.M.; Chen, Y.T.; et al. Genome sequencing and comparative analysis of Klebsiella pneumoniae NTUH-K2044, a strain causing liver abscess and meningitis. *J. Bacteriol.* **2009**, *191*, 4492–4501. [CrossRef]
35. Larsen, M.V.; Cosentino, S.; Rasmussen, S.; Friis, C.; Hasman, H.; Marvig, R.L.; Jelsbak, L.; Sicheritz-Ponten, T.; Ussery, D.W.; Aarestrup, F.M.; et al. Multilocus sequence typing of total-genome-sequenced bacteria. *J. Clin. Microbiol.* **2012**, *50*, 1355–1361. [CrossRef] [PubMed]
36. Carattoli, A.; Zankari, E.; Garcia-Fernandez, A.; Voldby Larsen, M.; Lund, O.; Villa, L.; Møller Aarestrup, F.; Hasman, H. In silico detection and typing of plasmids using PlasmidFinder and plasmid multilocus sequence typing. *Antimicrob. Agents Chemother.* **2014**, *58*, 3895–3903. [CrossRef] [PubMed]
37. Zankari, E.; Hasman, H.; Cosentino, S.; Vestergaard, M.; Rasmussen, S.; Lund, O.; Aarestrup, F.M.; Larsen, M.V. Identification of acquired antimicrobial resistance genes. *J. Antimicrob. Chemother.* **2012**, *67*, 2640–2644. [CrossRef]
38. Jia, B.; Raphenya, A.R.; Alcock, B.; Waglechner, N.; Guo, P.; Tsang, K.K.; Lago, B.A.; Dave, B.M.; Pereira, S.; Sharma, A.N.; et al. CARD 2017, Expansion and model-centric curation of the comprehensive antibiotic resistance database. *Nucleic Acids Res.* **2017**, *45*, D566–D573. [CrossRef] [PubMed]
39. Bertels, F.; Silander, O.K.; Pachkov, M.; Rainey, P.B.; van Nimwegen, E. Automated reconstruction of whole-genome phylogenies from short-sequence reads. *Mol. Biol. Evol.* **2014**, *31*, 1077–1088. [CrossRef] [PubMed]
40. Guindon, S.; Dufayard, J.F.; Lefort, V.; Anisimova, M.; Hordijk, W.; Gascuel, O. New algorithms and methods to estimate maximum-likelihood phylogenies: Assessing the performance of PhyML 3.0. *Syst. Biol.* **2010**, *59*, 307–321. [CrossRef] [PubMed]
41. Dixit, P.D.; Pang, T.Y.; Studier, F.W.; Maslov, S. Recombinant transfer in the basic genome of Escherichia coli. *Proc. Natl. Acad. Sci. USA* **2015**, *112*, 9070–9075. [CrossRef]
42. Sakoparnig, T.; Field, C.; van Nimwegen, E. Whole genome phylogenies reflect the distributions of recombination rates for many bacterial species. *elife* **2021**, *10*, e65366. [CrossRef]

43. Zhang, R.; Liu, L.; Zhou, H.; Chan, E.W.; Li, J.; Fang, Y.; Li, Y.; Liao, K.; Chen, S. Nationwide Surveillance of Clinical Carbapenem-resistant Enterobacteriaceae (CRE) Strains in China. *eBioMedicine* **2017**, *19*, 98–106. [CrossRef]
44. Voulgari, E.; Gartzonika, C.; Vrioni, G.; Politi, L.; Priavali, E.; Levidiotou-Stefanou, S.; Tsakris, A. The Balkan region: NDM-1-producing Klebsiella pneumoniae ST11 clonal strain causing outbreaks in Greece. *J. Antimicrob. Chemother.* **2014**, *69*, 2091–2097. [CrossRef]
45. Todorova, B.; Sabtcheva, S.; Ivanov, I.N.; Lesseva, M.; Chalashkanov, T.; Ioneva, M.; Bachvarova, A.; Dobreva, E.; Kantardjiev, T. First clinical cases of NDM-1-producing Klebsiella pneumoniae from two hospitals in Bulgaria. *J. Infect. Chemother.* **2016**, *22*, 837–840. [CrossRef] [PubMed]
46. Berglund, B.; Hoang, N.T.B.; Lundberg, L.; Le, N.K.; Tarnberg, M.; Nilsson, M.; Bornefall, E.; Khu, D.T.K.; Welander, J.; Le, H.T.; et al. Clonal spread of carbapenem-resistant Klebsiella pneumoniae among patients at admission and discharge at a Vietnamese neonatal intensive care unit. *Antimicrob. Resist. Infect. Control* **2021**, *10*, 162. [CrossRef] [PubMed]
47. Liu, H.; Wilksch, J.; Li, B.; Du, J.; Cao, J.; Zhang, X.; Zhou, T. Emergence of ST39 and ST656 extensively drug-resistant Klebsiella pneumoniae isolates in Wenzhou, China. *Indian. J. Med. Microbiol.* **2017**, *35*, 145–146. [CrossRef]
48. Cabanel, N.; Rosinski-Chupin, I.; Chiarelli, A.; Botin, T.; Tato, M.; Canton, R.; Glaser, P. Evolution of VIM-1-Producing Klebsiella pneumoniae Isolates from a Hospital Outbreak Reveals the Genetic Bases of the Loss of the Urease-Positive Identification Character. *mSystems* **2021**, *6*, e0024421. [CrossRef] [PubMed]
49. Kuzina, E.S.; Kislichkina, A.A.; Sizova, A.A.; Skryabin, Y.P.; Novikova, T.S.; Ershova, O.N.; Savin, I.A.; Khokhlova, O.E.; Bogun, A.G.; Fursova, N.K. High-Molecular-Weight Plasmids Carrying Carbapenemase Genes bla(NDM-1), bla(KPC-2), and bla(OXA-48) Coexisting in Clinical Klebsiella pneumoniae Strains of ST39. *Microorganisms* **2023**, *11*, 459. [CrossRef]
50. Giakkoupi, P.; Papagiannitsis, C.C.; Miriagou, V.; Pappa, O.; Polemis, M.; Tryfinopoulou, K.; Tzouvelekis, L.S.; Vatopoulos, A.C. An update of the evolving epidemic of blaKPC-2-carrying Klebsiella pneumoniae in Greece (2009-10). *J. Antimihcrob. Chemother.* **2011**, *66*, 1510–1513. [CrossRef]
51. Piekarska, K.; Zacharczuk, K.; Wolkowicz, T.; Wolaniuk, N.; Rzeczkowska, M.; Gierczynski, R. Emergence of Enterobacteriaceae co-producing CTX-M-15, ArmA and PMQR in Poland. *Adv. Clin. Exp. Med.* **2019**, *28*, 249–254. [CrossRef]
52. Mansour, W.; Grami, R.; Ben Haj Khalifa, A.; Dahmen, S.; Chatre, P.; Haenni, M.; Aouni, M.; Madec, J.Y. Dissemination of multidrug-resistant blaCTX-M-15/IncFIIk plasmids in Klebsiella pneumoniae isolates from hospital- and community-acquired human infections in Tunisia. *Diagn. Microbiol. Infect. Dis.* **2015**, *83*, 298–304. [CrossRef]
53. Galani, I.; Karaiskos, I.; Souli, M.; Papoutsaki, V.; Galani, L.; Gkoufa, A.; Antoniadou, A.; Giamarellou, H. Outbreak of KPC-2-producing Klebsiella pneumoniae endowed with ceftazidime-avibactam resistance mediated through a VEB-1-mutant (VEB-25), Greece, September to October 2019. *Eurosurveillance* **2020**, *25*, 2000028. [CrossRef]
54. Voulgari, E.; Kotsakis, S.D.; Giannopoulou, P.; Perivolioti, E.; Tzouvelekis, L.S.; Miriagou, V. Detection in two hospitals of transferable ceftazidime-avibactam resistance in Klebsiella pneumoniae due to a novel VEB beta-lactamase variant with a Lys234Arg substitution, Greece, 2019. *Eurosurveillance* **2020**, *25*, 1900766. [CrossRef]
55. Livermore, D.M.; Struelens, M.; Amorim, J.; Baquero, F.; Bille, J.; Canton, R.; Henning, S.; Gatermann, S.; Marchese, A.; Mittermayer, H.; et al. Multicentre evaluation of the VITEK 2 Advanced Expert System for interpretive reading of antimicrobial resistance tests. *J. Antimicrob. Chemother.* **2002**, *49*, 289–300. [CrossRef] [PubMed]
56. Garcia-Fulgueiras, V.; Magallanes, C.; Reyes, V.; Cayota, C.; Galiana, A.; Vieytes, M.; Vignoli, R.; Márquez, C. In Vivo High Plasticity of Multi-Drug Resistant ST258 Klebsiella pneumoniae. *Microb. Drug Resist.* **2021**, *27*, 1126–1130. [CrossRef] [PubMed]
57. Albarri, O.; AlMatar, M.; Ocal, M.M.; Koksal, F. Overexpression of Efflux Pumps AcrAB and OqxAB Contributes to Ciprofloxacin Resistance in Clinical Isolates of *K. pneumonia*. *Curr. Protein Pept. Sci.* **2022**, *23*, 356–368. [CrossRef] [PubMed]
58. Diez-Aguilar, M.; Canton, R. New microbiological aspects of fosfomycin. *Rev. Esp. Quimioter.* **2019**, *32* (Suppl. S1), 8–18.
59. Sugita, K.; Aoki, K.; Komori, K.; Nagasawa, T.; Ishii, Y.; Iwata, S.; Tateda, K. Molecular Analysis of bla(KPC-2)-Harboring Plasmids: Tn4401a Interplasmid Transposition and Tn4401a-Carrying ColRNAI Plasmid Mobilization from Klebsiella pneumoniae to Citrobacter europaeus and Morganella morganii in a Single Patient. *mSphere* **2021**, *6*, e0085021. [CrossRef]
60. Di Martino, P.; Cafferini, N.; Joly, B.; Darfeuille-Michaud, A. Klebsiella pneumoniae type 3 pili facilitate adherence and biofilm formation on abiotic surfaces. *Res. Microbiol.* **2003**, *154*, 9–16. [CrossRef]
61. Lam, M.M.C.; Wick, R.R.; Wyres, K.L.; Gorrie, C.L.; Judd, L.M.; Jenney, A.W.J.; Brisse, S.; Holt, K.E. Genetic diversity, mobilisation and spread of the yersiniabactin-encoding mobile element ICEKp in Klebsiella pneumoniae populations. *Microb. Genom.* **2018**, *4*, e000196. [CrossRef]
62. Zhu, J.; Wang, T.; Chen, L.; Du, H. Virulence Factors in Hypervirulent Klebsiella pneumoniae. *Front. Microbiol.* **2021**, *12*, 642484. [CrossRef]
63. Di Pilato, V.; Henrici De Angelis, L.; Aiezza, N.; Baccani, I.; Niccolai, C.; Parisio, E.M.; Giordano, C.; Camarlinghi, G.; Barnini, S.; Forni, S.; et al. Resistome and virulome accretion in an NDM-1-producing ST147 sublineage of Klebsiella pneumoniae associated with an outbreak in Tuscany, Italy: A genotypic and phenotypic characterisation. *Lancet Microbe* **2022**, *3*, e224–e234. [CrossRef]
64. Miro, E.; Rossen, J.W.A.; Chlebowicz, M.A.; Harmsen, D.; Brisse, S.; Passet, V.; Navarro, F.; Friedrich, A.W.; García-Cobos, S. Core/Whole Genome Multilocus Sequence Typing and Core Genome SNP-Based Typing of OXA-48-Producing Klebsiella pneumoniae Clinical Isolates From Spain. *Front. Microbiol.* **2019**, *10*, 2961. [CrossRef]

65. Snitkin, E.S.; Zelazny, A.M.; Thomas, P.J.; Stock, F.; NISC Comparative Sequencing Program Group; Henderson, D.K.; Palmore, T.N.; Segre, J.A. Tracking a hospital outbreak of carbapenem-resistant Klebsiella pneumoniae with whole-genome sequencing. *Sci. Transl. Med.* **2012**, *4*, 148ra116. [CrossRef] [PubMed]
66. Marsh, J.W.; Krauland, M.G.; Nelson, J.S.; Schlackman, J.L.; Brooks, A.M.; Pasculle, A.W.; Shutt, K.A.; Doi, Y.; Querry, A.M.; Muto, C.A.; et al. Genomic Epidemiology of an Endoscope-Associated Outbreak of Klebsiella pneumoniae Carbapenemase (KPC)-Producing *K. pneumoniae*. *PLoS ONE* **2015**, *10*, e0144310. [CrossRef] [PubMed]
67. Onori, R.; Gaiarsa, S.; Comandatore, F.; Pongolini, S.; Brisse, S.; Colombo, A.; Cassani, G.; Marone, P.; Grossi, P.; Minoja, G.; et al. Tracking Nosocomial Klebsiella pneumoniae Infections and Outbreaks by Whole-Genome Analysis: Small-Scale Italian Scenario within a Single Hospital. *J. Clin. Microbiol.* **2015**, *53*, 2861–2868. [CrossRef]
68. Arnold, B.J.; Gutmann, M.U.; Grad, Y.H.; Sheppard, S.K.; Corander, J.; Lipsitch, M.; Hanage, W.P. Weak Epistasis May Drive Adaptation in Recombining Bacteria. *Genetics* **2018**, *208*, 1247–1260. [CrossRef] [PubMed]
69. Dallman, T.; Ashton, P.; Schafer, U.; Jironkin, A.; Painset, A.; Shaaban, S.; Hartman, H.; Myers, R.; Underwood, A.; Jenkins, C.; et al. SnapperDB: A database solution for routine sequencing analysis of bacterial isolates. *Bioinformatics* **2018**, *34*, 3028–3029. [CrossRef]

antibiotics

Article

F18:A-:B1 Plasmids Carrying $bla_{CTX\text{-}M\text{-}55}$ Are Prevalent among *Escherichia coli* Isolated from Duck–Fish Polyculture Farms

Li-Juan Zhang [1,2,3], Jin-Tao Yang [2,4], Hai-Xin Chen [2,4], Wen-Zi Liu [4], Yi-Li Ding [3], Rui-Ai Chen [1,4], Rong-Min Zhang [2,4,*] and Hong-Xia Jiang [2,4,*]

1 Zhaoqing Branch Center of Guangdong Laboratory for Lingnan Modern Agricultural Science and Technology, Zhaoqing 526000, China
2 Guangdong Key Laboratory for Veterinary Pharmaceutics Development and Safety Evaluation, College of Veterinary Medicine, South China Agricultural University, Guangzhou 510642, China
3 Life Science Department, Foshan University, Foshan 528000, China; yding@kean.edu
4 Guangdong Laboratory for Lingnan Modern Agriculture, Guangzhou 510642, China
* Correspondence: zrm@scau.edu.cn (R.-M.Z.); hxjiang@scau.edu.cn (H.-X.J.)

Abstract: We determined the prevalence and molecular characteristics of $bla_{CTX\text{-}M\text{-}55}$-positive *Escherichia coli* (*E. coli*) isolated from duck–fish polyculture farms in Guangzhou, China. A total of 914 *E. coli* strains were isolated from 2008 duck and environmental samples (water, soil and plants) collected from four duck fish polyculture farms between 2017 and 2019. Among them, 196 strains were CTX-M-1G-positive strains by PCR, and 177 (90%) $bla_{CTX\text{-}M\text{-}1G}$-producing strains were $bla_{CTX\text{-}M\text{-}55}$-positive. MIC results showed that the 177 $bla_{CTX\text{-}M\text{-}55}$-positive strains were highly resistant to ciprofloxacin, ceftiofur and florfenicol, with antibiotic resistance rates above 95%. Among the 177 strains, 37 strains carrying the F18:A-:B1 plasmid and 10 strains carrying the F33:A-:B- plasmid were selected for further study. Pulse field gel electrophoresis (PFGE) combined with S1-PFGE, Southern hybridization and whole-genome sequencing (WGS) analysis showed that both horizontal transfer and clonal spread contributed to dissemination of the $bla_{CTX\text{-}M\text{-}55}$ gene among the *E. coli*. $bla_{CTX\text{-}M\text{-}55}$ was located on different F18:A-:B1 plasmids with sizes between ~76 and ~173 kb. In addition, the presence of $bla_{CTX\text{-}M\text{-}55}$ with other resistance genes (e.g., *tetA*, *floR*, *fosA3*, bla_{TEM}, *aadA5 CmlA* and *InuF*) on the same F18:A-:B1 plasmid may result in co-selection of resistance determinants and accelerate the dissemination of $bla_{CTX\text{-}M\text{-}55}$ in *E. coli*. In summary, the F18:A-:B1 plasmid may play an important role in the transmission of $bla_{CTX\text{-}M\text{-}55}$ in *E. coli*, and the continuous monitoring of the prevalence and transmission mechanism of $bla_{CTX\text{-}M\text{-}55}$ in duck–fish polyculture farms remains important.

Keywords: duck–fish polyculture farm; *Escherichia coli*; antibiotic resistance; $bla_{CTX\text{-}M\text{-}55}$; F18:A-:B1; transmission

check for updates

Citation: Zhang, L.-J.; Yang, J.-T.; Chen, H.-X.; Liu, W.-Z.; Ding, Y.-L.; Chen, R.-A.; Zhang, R.-M.; Jiang, H.-X. F18:A-:B1 Plasmids Carrying $bla_{CTX\text{-}M\text{-}55}$ Are Prevalent among *Escherichia coli* Isolated from Duck–Fish Polyculture Farms. *Antibiotics* **2023**, *12*, 961. https://doi.org/10.3390/antibiotics12060961

Academic Editor: Theodoros Karampatakis

Received: 5 May 2023
Revised: 22 May 2023
Accepted: 22 May 2023
Published: 25 May 2023

1. Introduction

Antimicrobial resistance is a serious global public health problem associated with significant clinical, economic and social impacts. *Escherichia coli* exists as part of the commensal microbiota in the mammalian digestive tract, as a zoonotic pathogen responsible for intestinal and extraintestinal infections in both humans and animals [1,2]. Globally, the emergence of multidrug-resistant (MDR) *E. coli* producing extended-spectrum β-lactamase (ESBL) enzymes has led to empirical therapy failure, leading to high morbidity and mortality, which has raised great public concern [1,2].

Currently, CTX-M-bearing *E. coli* is the most common species related to ESBLs and more than 220 CTX-M family members have been identified. These variants were divided into five major groups (groups 1, 2, 8, 9 and 25) based on their amino acid homology. Among these groups, 1 and 9 were the most common globally [2–4]. Over the past decade, CTX-M-55, as a variant of CTX-M-15, from animal-origin *E. coli*, which was first discovered in Thailand in 2006, has spread rapidly in dozens of countries around the world, especially in China [4–6]. The dissemination of *bla*$_{CTX-M-55}$ was mainly caused by plasmid-mediated gene horizontal transfer, and epidemic self-mobilizable F33:A-:B, IncI1, IncI2 and IncHI2 plasmids played an important role in the transmission of *bla*$_{CTX-M-55}$ [6–8]. In addition to animals, *bla*$_{CTX-M-55}$ is also distributed in food products and humans, and frequently co-localized with other resistance genes, such as *fosA3*, *rmtB*, *mcr-1*, *bla*$_{TEM}$, *tet(A)* and *floR* [7–10]. The wide distribution of *bla*$_{CTX-M-55}$ and the co-transfer of *bla*$_{CTX-M-55}$ with different resistance genes worldwide represent a growing threat to public health.

As the largest producer and consumer of cultivated duck in the world, duck production plays a major role in the agricultural economy of China [11]. Duck farming in China is practiced on a large and diverse scale, and the integrated culture of fish–duck farming using untreated duck manure as fish feed, is the typical farming method throughout coastal areas of China, particularly in Guangdong Province. Previous studies have shown that fish–duck integrated farming systems have become a hotspot environment for the occurrence and proliferation of antibiotic resistance genes (ARGs) and antibiotic-resistant bacteria (ARB), which can facilitate the spread of resistance genes and have the potential to be the reservoir of novel ARGs [12–14]. Owing to the threat of *bla*$_{CTX-M-55}$ to public health and the risks of ARG transmission in fish–duck farming, monitoring the prevalence of *bla*$_{CTX-M-55}$-positive *E. coli* isolates from duck–fish polyculture farms should receive more attention.

However, CTX-M-55 has been widely reported in *E. coli* isolated from food animals, pets and humans in China [6,8,15]. Current data on the prevalence, genetic information and the transmission mechanism of CTX-M-55-producing *E. coli* from duck–fish polyculture farms are still limited. In this study, we investigated the prevalence of *bla*$_{CTX-M-55}$ and illustrated the characteristics of *bla*$_{CTX-M-55}$-bearing *E. coil* and plasmids recovered from ducks and the environment in duck–fish polyculture farms. Our findings emphasize the importance of the surveillance of *ESBL*-producing *E. coli* in duck–fish polyculture farms in China and provide knowledge for further One Health studies to control the spread of resistant bacteria from food animals to humans.

2. Results

2.1. *Identification of bla*$_{CTX-M-55}$*-Positive E. coli Isolates*

A total of 196 *bla*$_{CTX-M-1G}$-positive *E. coli* isolates were obtained from 2008 different samples, and the *bla*$_{CTX-M-1G}$-producing *E. coli*-positive rates from duck fecal samples, water, soil and foliage were 14.4% (138/957), 6.25% (31/496), 1.5% (7/476) and 1.3% (1/79), respectively (Table S1). Six different CTX-M-1G subtypes were obtained from the 196 *bla*$_{CTX-M-1G}$-positive strains, including CTX-M-55 (177 strains), CTX-M-79 (8 strains), CTX-M-123 (5 strains), CTX-M-3 (3 strains), CTX-M-64 (2 strains) and CTX-M-224 (only 1 strain). Among the 177 CTX-M-55-positive strains, 77.9% (138/177) were isolated from fecal samples, 17.5% (31/177) were isolated from water, 4.0% (7/177) were isolated from soil and only 1 (1/177, 0.6%) isolate was obtained from foliage (Table S1).

The antimicrobial susceptibility results showed that the 177 *E. coli* isolates were highly resistant to ceftiofur, ciprofloxacin and florfenicol, with resistance rates of 98.3%, 96.7% and 95.4%, respectively. The rates of resistance to colistin, ceftazidime and fosfomycin were 35.0%, 31.7% and 29.9%, respectively. However, these strains were less resistant to tigecycline, meropenem and amikacin, and the resistance rates were less than 10% (Table S2).

To further understand the molecular characteristics of *bla*$_{CTX-M-55}$-positive *E. coli*, the PCR-base replicon typing (PBRT) method was performed on the 177 strains. The results showed that 19.2% (34/177), 6.2% (11/177), 1.7% (3/177) and 1.1% (2/177) of the strains carried the IncT, IncN, IncHI1 and IncP plasmids, respectively, and 10.7% (19/177) and 5.6% (10/177) of the strains carried IncY and IncI1 plasmids, respectively. Notably, 81.4% (144/177) of the strains carried IncF plasmids. We further performed F plasmid replicon sequencing typing (RST) on these 144 IncF-type plasmid-carrying strains. The results showed that F18:A-:B1-plasmid-carrying strains accounted for the largest proportion, with 37 (37/144, 25.7%) strains, followed by F33:A-:B-plasmid-carrying strains, with 10 (10/144,

6.9%) strains carrying this type of plasmid and the remaining 97 strains carrying other types of IncF plasmids.

Based on the above results, we selected the 37 F18:A-:B1-plasmid-carrying strains and 10 F33:A-:B-plasmid-carrying strains for further study of the characteristics and transmission mechanism of plasmids carrying $bla_{CTX\text{-}M\text{-}55}$-positive *E. coli*. Among the 47 isolates, the number of strains from duck fecal samples was the largest, accounting for 81% (38/47), followed by the strains from water samples, accounting for 13% (6/47). The remaining 6.4% (3/47) of strains were isolated from soil samples (Figure 1).

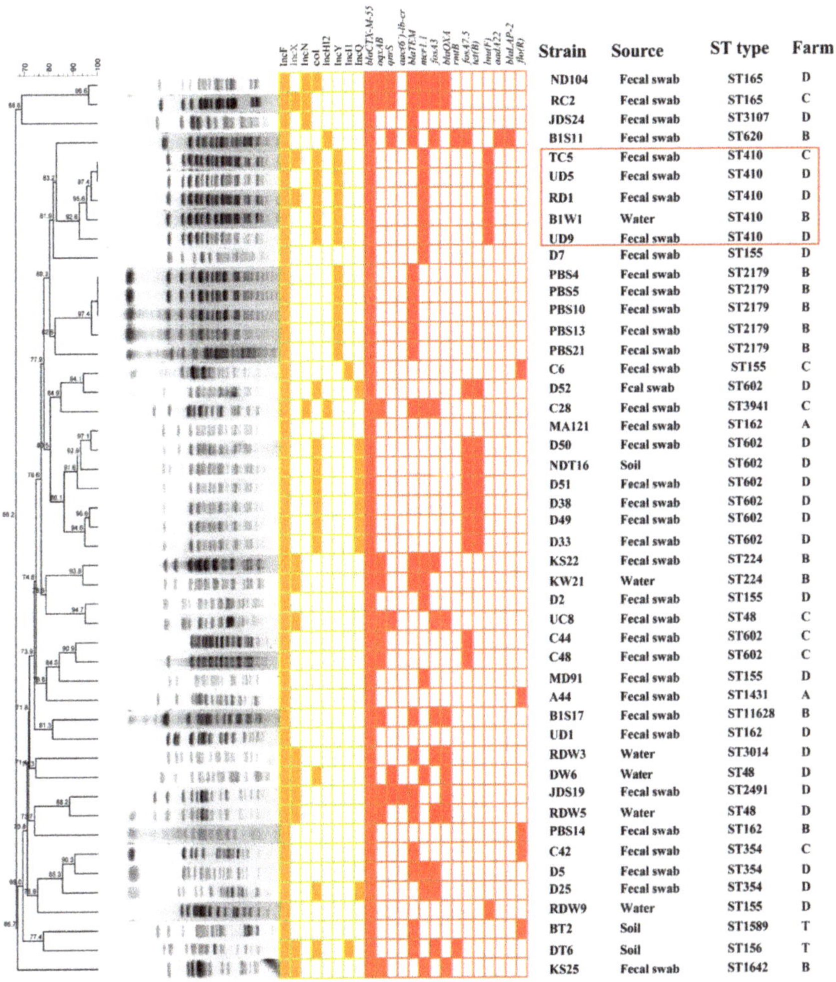

Figure 1. Pulsed-field gel electrophoresis fingerprinting patterns of *XbaI*-digested total DNA preparations from 47 strains harboring $bla_{CTX\text{-}M\text{-}55}$. In yellow is indicated the PBRT classification, in red are indicated antimicrobial resistance genes.

2.2. Molecular Characterization of $bla_{CTX\text{-}M\text{-}55}$-Positive E. coli

PFGE was successfully performed for all 47 selected strains, and the PFGE results were divided into 15 different clusters according to similarity >85%, indicating the genetically different backgrounds of strains from different sources. In addition, two clusters of strains with the same PFGE spectrum were derived from the different types of samples from farms B, C and D (TC5, UD5, RD1, B1W1, UD9 and D7), suggesting that there is clonal transmission among different farms (Figure 1).

The conjugation results indicated that 17 of the 47 strains could successfully transfer $bla_{\text{CTX-M-55}}$ from the donor strain to the recipient strain *E. coli* C600, and the transfer frequency was between 2.6×10^{-5} and 4.65×10^{-1} (Table S3). There were four strains that could transfer $bla_{\text{CTX-M-55}}$ from the donor strain to the recipient strain *E. coli* J53 with conjugative transfer frequencies between 3.89×10^{-6} and 8.35×10^{-5}. The remaining 26 strains showed unsuccessful transfer of $bla_{\text{CTX-M-55}}$ from the donor strain to the recipient strain after multiple attempts (Table S3). To further determine the location of the $bla_{\text{CTX-M-55}}$ gene in the non-conjugatively transferable strains and the genetic environment of the $bla_{\text{CTX-M-55}}$ gene, whole-genome sequencing was performed on all 47 strains. Sequence analysis showed that among the 26 non-conjugable strains, the $bla_{\text{CTX-M-55}}$ gene was located on the chromosome of 11 strains, while the $bla_{\text{CTX-M-55}}$ gene of the remaining 15 strains was located on plasmids.

2.3. Genomic Analysis of $bla_{\text{CTX-M-55}}$-Positive E. coli

Whole-genome sequencing data were generated for the 47 $bla_{\text{CTX-M-55}}$-positive *E. coli* isolates. The results of WGS demonstrated that these isolates belonged to seventeen distinct strains (STs). Among them, the most dominant ST type was ST602, with a total of nine (9/47, 19%) strains belonging to this ST type; followed by ST155, ST410 and ST2179 (five, 10.6% each); ST48, ST162 and ST354 (three, 6.4% each); and ST165 and ST224 (two, 4.3% each). Of the remaining eight ST types, only one strain belonged to each ST type.

We identified 14 ARGs that mediated resistance to 9 types of antibiotics that coexisted with $bla_{\text{CTX-M-55}}$ in these 47 *E. coli* isolates. These included genes that mediated resistance to fosfomycin, colistin, tetracycline, aminoglycosides, chloramphenicol, quinolones, macrolides, sulfonamides and lincomycin. Among these, 26% (12/47) and 13% (6/47) of the strains carried the quinolone resistance genes *oqxAB* and *qnrS*, respectively. In addition, 38% (18/47) of the strains carried the colistin resistance gene *mcr-1*, and 26% (12/47) and 21% (10/47) of the strains carried the fosfomycin resistance genes *fosA3* and *fosA7.5*, respectively. Eleven percent (5/47) of the strains carried the chloramphenicol resistance gene *floR*, and thirteen percent (6/47) of the strains carried the lincomycin resistance gene *lnuF*. Interestingly, our results showed that all strains belonging to ST602 carried the fosfomycin resistance gene *fosA7.5*, and *fosA7.5* was located on the chromosome. Additionally, all ST410 strains carried the *lnuF* gene, and the $bla_{\text{CTX-M-55}}$ gene carried by the ST410 strains in the current study could not be transferred by conjugation (Figure 1 and Table S3).

Genetic environment analysis showed that $bla_{\text{CTX-M-55}}$ was present in four genomic contexts, including types I, II, III and IV. The structure of type I was IS*Ecp1*-$bla_{\text{CTX-M-55}}$-*orf477*, which was the most prevalent, and was isolated in a total of 21 (45%) strains originating from feces, soil and water (Table S3 and Figure S1). The structure of type II was ΔIS26-ΔIS*Ecp1*-$bla_{\text{CTX-M-55}}$-*orf477*-ΔTn2, and there were 12 (26%) strains belonging to this genetic environment. There were 10 (21%) strains belonging to the type III genetic environment, and the structure of type III was ΔIS26-ΔIS*Ecp1*-$bla_{\text{CTX-M-55}}$-*orf477*-ΔTn2-ΔIS26. The type IV genetic environment was more complex, and its structure was ΔTn2-IS*Ecp1*-$bla_{\text{CTX-M-55}}$-*orf477*-ΔTn2-ΔIS26. Only three isolates belonged to this genetic environment. Notably, among the strains belonging to the type I genetic structure, the $bla_{\text{CTX-M-55}}$ gene of the ten strains was located on the plasmid, and the $bla_{\text{CTX-M-55}}$ gene of the other eleven strains was located on the chromosome. The $bla_{\text{CTX-M-55}}$ gene in the strains belonging to the type II and type III genetic structures were all located on the plasmid. Among the strains with type IV genetic structure, one strain had the $bla_{\text{CTX-M-55}}$ gene located on the chromosome, and $bla_{\text{CTX-M-55}}$ in the other two strains was located on plasmids (Table S3 and Supplementary Figure S1).

2.4. Complete Sequence Analysis of $bla_{CTX-M-55}$-Carrying F18:A-:B1 Plasmids

To better understand the characteristics of the $bla_{CTX-M-55}$-bearing F18:A-:B1 plasmid, S1-PFGE and Southern blotting were performed on the 12 conjugants successfully obtained from the F18:A-:B1 plasmid harboring $bla_{CTX-M-55}$-positive strains. The results showed that $bla_{CTX-M-55}$ from 12 strains was located on F18:A-:B1 plasmids with sizes between ~76 kb and ~173 kb (Supplementary Figure S2). To further explore the sequence features of the F18:A-:B1 plasmid, four representative strains (PBS4, B1W1, B1S11 and KW21) were subjected to long-read sequencing to obtain the complete sequence of different F18:A-:B1 plasmids. The complete sequences of the F18:A-:B1 plasmids pPBS4, pB1W1, pB1S11 and pKW21 were obtained by long-read combined with short-read sequencing, and detailed information on the four plasmids is shown in Supplementary Table S4.

Sequence analysis revealed that all four plasmids were novel IncF18:A-:B1 plasmids, three (pBS4, pB1W1 and pKW21) of which carried $bla_{CTX-M-55}$, and the remaining plasmid (pB1S11) did not carry $bla_{CTX-M-55}$. In strain B1S11, $bla_{CTX-M-55}$ was located on an IncHI2 plasmid. Linear comparison revealed that the backbone of the four F18:A-:B1 plasmids was similar to that of other typical IncF plasmids containing regions for functions of replication (*repA, repB*), conjugal transfer (*tra, trb*), partitioning (*parA, parB*) and maintenance and stability (*pemI, pemK, stbA*). Notably, in the plasmid pB1W1, ~29-kb conjugative regions containing the *traV, traR, traC, traW, traU, traN, traF, traH, traG, traD, traT, traS, trbD, trbG, trbI, trbB, trbJ* and *trbF* genes were lacking, which may explain the conjugative transfer failure of pB1W1 (Figure 2). Basic Local Alignment Search Tool (BLAST) analysis indicated that the four plasmids (pBS4, pB1W1, pB1S11 and pKW21) had high similarity to the $bla_{CTX-M-55}$-carrying plasmid pTREC8 of an *E. coli* strain isolated from wetland sediment in the United States (GenBank accession no. MN158991.1) and shared high identities of 99.9%, 100%, 99.9% and 99.8%, respectively (Figure 2).

Although the F18:A-:B1 plasmid possessed a conserved backbone region, the variable regions, especially multidrug resistance regions, were distinct. The multidrug resistance region of pKW21 was 80,273 bp, which is quite different from the resistance regions of the other three F18:A-:B1 plasmids in this study. The ~80.3 kb MDR region contained 11 ARGs ($bla_{CTX-M-55}$, *aac(3′)-IIa, aph(3′)-Ib, mph(A), aph(6′)-Id, tet(R), tet(A), floR*, bla_{TEM-1B}, *fosA3* and *aadA5*) with one or two copies, interspersed with different complete or truncated insertion sequences and transposons (IS*26*, IS4 family, ΔTn*3*, ΔTn*2*, ΔIS*Ecp1*, ΔIS*1R*, IS*5075*, IS*91* and IS*Cfr1*) (Figure 3a). BLAST analysis showed that the ~80.3 kb region shared a high identity (>99.99%) with plasmids collected from humans and pigs that did not bear $bla_{CTX-M-55}$, such as *E. coli* plasmid p14406-FII (GenBank accession no. MN823988), p13P484A-2 (GenBank accession no. CP019282) and pTEM-1-GZC065 (GenBank accession no. CP048026). Compared with p14406-FII, p13P484A-2 and pTEM-1-GZC065, the difference was that three regions of pKW21 (IS*26*-*aph(3′)-Ib*-IS*26*-*mph(A)*-IS*26*, IS*26*-*hp-hp-hp*-IS*26*-*hp-hp*-IS*26* and IS*26*-*hp-hp-hp*-ΔTn*2*-*hp-tet(R)-tet(A)-hp*-ΔTn*2*-IS*26*-*hp-hp*-IS*26*) were inverted. Notably, there was a region of pKW21 (~33 kb) containing $bla_{CTX-M-55}$ that completely mismatched with the compared plasmids (Figure 3a).

The MDR regions of pB1W1, pB1S11 and pPBS4 were ~46.0 kb, ~26.5 kb and ~38.9 kb, respectively. Even though the resistance genes in the three plasmids were varied, they all contained *floR, tet(A)* and *tet(R)* and $bla_{CTX-M-55}$, except pB1S11. In addition to the mentioned genes, the pB1S1 MDR region also contained the quinolone resistance gene *qnrS1*, which was highly similar to pGSH8M-2-1 (GenBank accession no. AP019676.1) and shared 99.8% identify with 100% coverage. pGSH8M-2-1 was recovered from the effluent of a wastewater treatment plant in Tokyo Bay, and the only difference in the MDR region between pGSH8M-2-1 and pB1S11 was that an IS*26* was inserted between two ΔTn*2* in pB1S11 (Figure 3b). Compared with pB1S11, pB1W1 and pPBS4 contained more resistance genes and carried multiple copies of IS*26* and transposons. BLAST analysis did not detect plasmids with high homology to the MDR region of pPBS4 and pB1W1, which may suggest that multiple copies of IS*26* and transposons formed the distinctive MDR regions of pPBS4 and pB1W1 through multiple complex recombination events. Notably, although pPBS4 and pB1W1 were derived from the same farm (Farm B), the genetic environments of *bla*$_{CTX-M-55}$ in pPBS4 and pB1W1 were different, indicating that the *bla*$_{CTX-M-55}$ in these two plasmids originated from different sources (Figure 3b). In addition, we also found a truncated class I integron in the plasmid pB1W1 (Figure 3b).

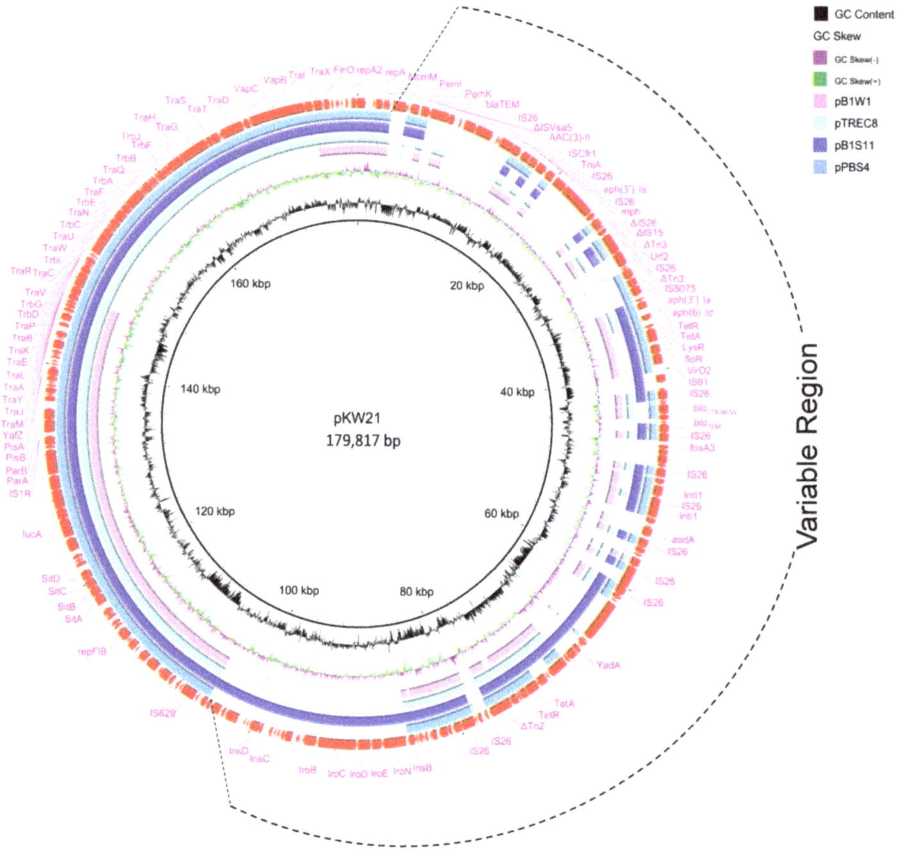

Figure 2. Circular sequence alignments of the plasmids pKW21, pB1W1, pB1S11, pPBS4 and pTREC8. Genes depicted in the outer circle belong to plasmid pKW21, which was included as a reference, and the image was generated using BRIG.

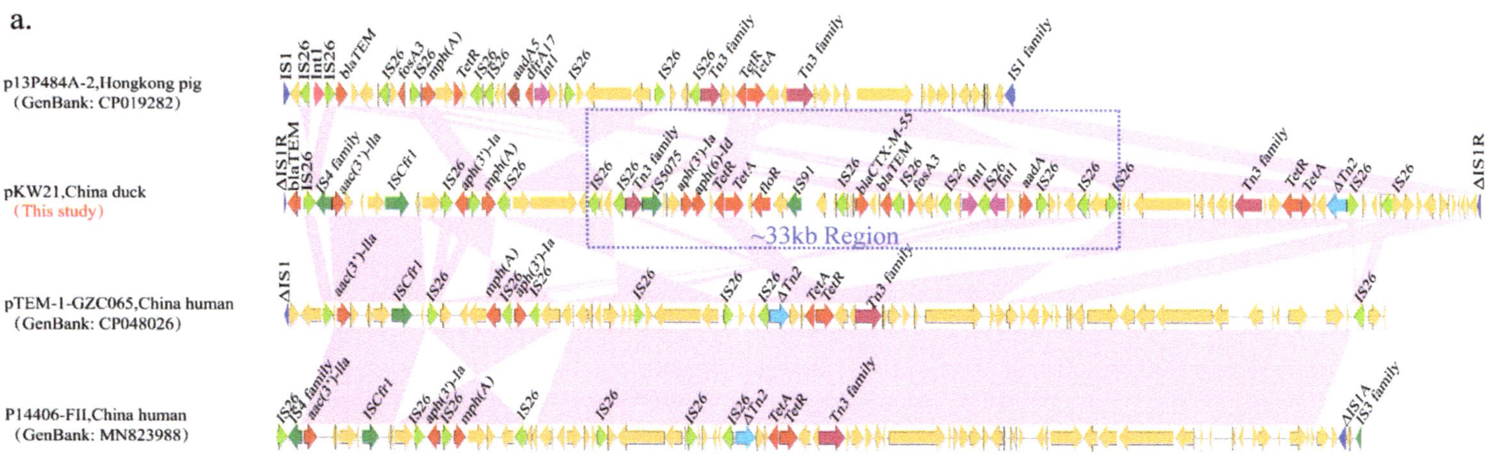

Figure 3. *Cont.*

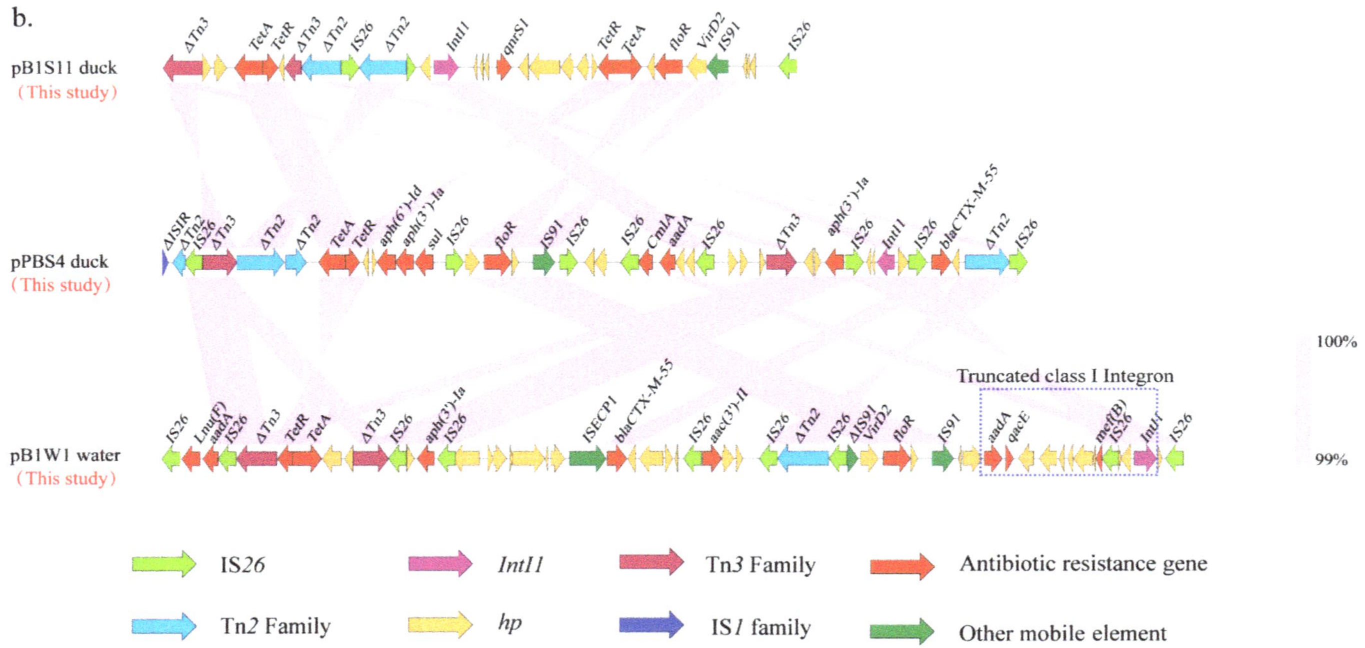

Figure 3. Comparison of the multidrug resistance regions of F18:A-:B1 plasmids identified in this study. (**a**) Comparison of pKW21 with similar plasmids p14406-FII (GenBank accession no. MN823988), p13P484A-2 (GenBank accession no. CP019282) and pTEM-1-GZC065 (GenBank accession no. CP048026). (**b**) Comparison of pB1S11, pPBS4 and pB1W1 with the similar plasmid pGSH8M-2-1 (GenBank accession no. AP019676.1). Genes are denoted by arrows. Genes, mobile elements and other features are colored based on functional classification. Shaded regions denote shared DNA homology (>95% nucleotide identity).

3. Materials and Methods

3.1. Sampling Information, Bacterial Isolation and Identification

From March 2017 to August 2019, a total of 2008 nonduplicate samples including 496 water samples, 476 soil samples, 957 duck fecal samples and 79 foliage samples were collected from four duck–fish polyculture farms in the Panyu District of Guangzhou, China (Figure 1; see also Table S1 in the Supplemental Material). All samples were screened for cefotaxime-resistant *E. coli* by a selective isolation procedure. In brief, each sample was suspended in 10 mL of buffered peptone water (BPW; BD Difco, Sparks, MD, USA) and incubated at 37 °C for 24 h. Then, subsequent selective cultivation on MacConkey (MC; BD Difco) agar supplemented with 1 mg/L cefotaxime (CTX) was performed. For each sample, only one red colony was selected and identified as *E. coli* by MALDI-TOF MS AximaTM (Shimadzu-Biotech Corp., Kyoto, Japan) and 16S rRNA sequencing by Sanger sequencing. In all cefotaxime-resistant *E.coli*, $bla_{\text{CTX-M-1G}}$ was detected by PCR using previously reported primers [16] and sequencing (Supplementary Table S1).

3.2. Antimicrobial Susceptibility Testing

Antibiotic susceptibility testing was performed by the agar dilution method and interpreted according to the Clinical and Laboratory Standards Institute guidelines (CLSI M100-S29) for the following antimicrobials: amikacin, meropenem, cefotaxime, ceftazidime, ceftiofur, florfenicol, ciprofloxacin and fosfomycin [17]. Susceptibility to colistin and tigecycline was assessed by broth microdilution as recommended by the European Committee on Antimicrobial Susceptibility Testing (EUCAST Version 9.0) [18]. *E. coli* ATCC 25922 was used as the quality control strain.

3.3. Molecular Typing

The incompatibility (Inc) groups of all $bla_{\text{CTX-M-55}}$-producing *E. coli* were assigned by PCR-base replicon typing (PBRT) [19]. To better characterize IncFII plasmid, replicon sequencing typing (RST) was performed, according to protocols described previously [20]. Based on the results of PBRT analysis, a total of 47 strains (37 strains harboring the F18:A-:B1 plasmid and 10 strains harboring the F33:A-:B- plasmid) were selected for further study to explore the transmission characteristics and molecular mechanism of $bla_{\text{CTX-M-55}}$ in these strains.

The genetic typing of the 47 selected $bla_{\text{CTX-M-55}}$-producing *E. coli* isolates was performed by digestion with restriction endonuclease *XbaI* and pulsed-field gel electrophoresis (PFGE) according to our previous study [21]. The band patterns were analyzed with BioNumerics software version 5.10 (Applied Maths, Austin, TX, USA).

3.4. Conjugation Assay and Southern Blotting

To investigate the transferability of the resistance genes, a conjugation assay was performed for all selected $bla_{\text{CTX-M-55}}$-positive *E.coli* isolates with streptomycin-resistant *E. coli* C600 or sodium-azide-resistant *E. coli* J53 as the recipient strain. For *E. coli* C600, donor strains and *E. coli* C600 were mixed and applied to a 0.22 μm filter in Luria-Bertani (LB) plates for 16~18 h. The mixed culture was then diluted and spread on selective MacConkey agar plates containing both 1 mg/L of cefotaxime and 2 g/L of streptomycin to recover transconjugants. For *E. coli* J53, donor strains and *E. coli* J53 were mixed and applied to a 0.22 μm filter in LB plates for 16–18 h. The mixed culture was then diluted and spread on selective MacConkey agar plates supplemented with 0.5 mg/L cefotaxime and 0.2 g/L sodium azide. Transconjugants were confirmed by PCR. S1-PFGE and Southern blotting were performed to determine plasmid size according to previous study, and the *Salmonella enterica* serotype, Braenderup H9812, was used as the standard size marker [22].

3.5. DNA Extraction and Whole-Genome Sequencing

Total DNA was extracted from 47 $bla_{\text{CTX-M-55}}$-producing *E. coli* isolates using a Genomic DNA Purification Kit (TIANGEN, Beijing, China) according to the manufacturer's instructions. WGS was performed with the Illumina HiSeq 2500 System (Novogene Guangzhou, China) using the paired-end 2 × 150-bp sequencing protocol. The draft genome was assembled using the tools available at EnteroBase (https://enterobase.warwick.ac.uk/species/ecoli, accessed on 28 November 2020) with default parameters. All genome assemblies of the 47 sequenced *E. coli* isolates were deposited in GenBank and are registered with BioProject number PRJNA934699. Then, the sequence types, replicon types and antibiotic resistance genes of all the sequenced isolates were identified by the Center for Genomic Epidemiology (http://www.genomicepidemiology.org/, accessed on 20 March 2021). Four representative strains (PBS4, B1W1, B1S11 and KW21) were further selected for whole-genome sequencing on the PacBio RS II sequencing platform (Biochip Company, Tianjin, China) to obtain the complete sequence of the F18:A-:B1 plasmid. Sequences of those strains were assembled using HGAP version 4.0. to analyze the genetic features. BRIG (Vision 0.95) and Easyfig (Vision 2.2.5) software were used for comparative analysis with other plasmid sequences published in NCBI. The complete sequence of the plasmid was annotated and analyzed using RAST (https://rast.nmpdr.org/rast.cgi, accessed on 12 March 2021), ISfinder (https://www-is.biotoul.fr/, accessed on 20 March 2021), Resfinder (https://cge.cbs.dtu.dk//services/ResFinder/, accessed on 22 March 2021) and BLAST (https://blast.ncbi.nlm.nih.gov/Blast.cgi, accessed on 23 March 2021).

3.6. Nucleotide Sequence Accession Numbers

The complete sequences of the plasmids (pPBS4, pB1W1, pB1S11 and pKW21) have been deposited in GenBank under accession numbers CP117716, CP117722, CP117718 and CP117673.

4. Discussion

Duck–fish polyculture is a common circular farming model in the Pearl River Delta in southern China, specifically in Guangdong Province. In this model, duck manure is discharged directly without treatment, and a large number of ARGs or residual agents can directly contaminate fish ponds, promoting the transmission of ARGs between ducks and fish [14,23,24]. Previous studies have shown that fish–duck integrated farming systems have become a hotspot environment for the occurrence and proliferation of ARGs and ARB, and both ARGs and pathogen-related ARB have been detected in the water and sediment of this culture system [12–14]. Moreover, fish–duck integrated aquaculture farms have significantly higher levels of antibiotic resistance compared to monoculture fish farms, suggesting a higher risk of transmission of ARGs and mobile genetic elements (MGEs) to humans or the environment [13]. The Pearl River Delta water system is intricate, which provides a unique opportunity to develop freshwater aquaculture, but there is also the risk of contaminating river or sea areas on a large scale by ARG and ARB dissemination via aquatic water or sediment [24,25]. Based on the "one health" concept, considering the common duck–fish freshwater aquaculture system in the Pearl River Delta in southern China, greater attention should be given to the transfer risk of ARGs in integrated duck–fish farming to promote the healthy development of Chinese aquaculture and the environment.

In this study, the antibiotic susceptibility of 177 $bla_{\text{CTX-M-55}}$-positive strains was tested with 10 antibiotics commonly used in both veterinary and human medicine. The results showed that the $bla_{\text{CTX-M-55}}$-positive strains were highly resistant strains, and almost all strains were multidrug-resistant. In addition to resistance to cephalosporins, they were also highly resistant to ciprofloxacin and florfenicol (with resistance rates at 96.7 and 95.4%, respectively). These strains also had a high resistance rate to ceftazidime (31.7%), colistin (35.0%) and fosfomycin (29.9%), and the resistance rate to tigecycline, meropenem and amikacin was less than 10%. Two recent studies also showed that *E. coli* strains isolated from ducks and the environment were not only resistant to cephalosporins but also resistant to chloramphenicols, aminoglycosides, quinolones and tetracyclines. Our results were similar to those reported in their study. Notably, the rate of detection of strains carrying the *bla*$_{CTX-M-55}$ gene was higher than that of strains without this gene for both the antibiotic resistance spectra and ARGs [10,26].

Previous studies have shown that *bla*$_{CTX-M-55}$ is mainly transmitted by the horizontal transfer of epidemical plasmids, and IncI1 and F33:A-:B- plasmids were the most important types of plasmids that mediated the spread of *bla*$_{CTX-M-55}$ in human and animal *E. coli* in China [6,15]. In the present study, we selected 37 *bla*$_{CTX-M-55}$-positive strains carrying the F18:A-:B1 plasmid and 10 *bla*$_{CTX-M-55}$-positive strains carrying the F33:A-:B- plasmid to explore the transmission mechanism and molecular characteristics of *bla*$_{CTX-M-55}$ among these strains. The PFGE results showed that the 47 strains could be divided into 15 different clusters, and the genetic backgrounds were relatively different. However, strains from different farms also had the same PFGE profile. CTX-M-positive *E. coli* isolates from ducks in Korea also had significant differences in PFGE profiles, but the same PFGE profiles were found in different livestock farms and slaughterhouses, which was consistent with the results of our study [27].

Multilocus sequence typing results indicated that ST602, ST155, ST410 and ST2179 were the prevalent ST types in our study, and the internationally prevalent ST10 and ST131 clones were not detected. ST602 has not been widely reported in previous studies, but a recent epidemiological surveillance of ESBL *E. coli* from human and food-chain-derived samples from England, Wales and Scotland found that the CTX-M-1G-positive ST602 strain was widely present in chicken samples [28]. Notably, all ST602 strains carried the *bla*$_{CTX-M-55}$ gene chromosomally, and co-carried the chromosomal fosfomycin resistance gene *fosA7.5*. *fosA7.5* is a new member of the fosfomycin resistance gene *fosA7* gene family recently reported in *E. coli* isolates in Canadian hospitals. Its distribution is limited to *E. coli*, and it can be located on both plasmids and chromosomes [29]. Because fosfomycin was effective as a first-line therapy for urinary tract infections, the emergence of *fosA7.5* and *bla*$_{CTX-M-55}$ co-carrying ST602 clone needs more attention.

ST410 *E. coli* is an emerging multidrug-resistant pathogen. Two major sublineages are currently circulating in Europe and North America, one is a fluoroquinolone- and extended-spectrum cephalosporin-resistant clade that emerged in the 1980s, and the other is a carbapenem-resistant clone that emerged in 2003. ST410 has been considered a "high-risk" clone similar to ST131 owing to its high transmissibility, its capacity to cause recurrent infections and its ability to persist in the gut [30,31]. This clone has also been found to harbor *mcr-1* in isolates recovered from food and human samples worldwide, and *tet(X)*-carrying ST410 *E. coli* in China and South Asia has been recently reported [10,31–33]. Given the potential of ST 410 *E. coli* to acquire resistance to last-resort antimicrobials, this clone should arouse regional and global concern.

The genetic contexts of *bla*$_{CTX-M-55}$ were divided into four types in the current study. In the genetic context of type III (ΔIS26-ΔIS*Ecp1*-*bla*$_{CTX-M-55}$-*orf477*-ΔTn2-ΔIS*26*) and type IV (ΔTn2-IS*Ecp1*-*bla*$_{CTX-M-55}$-*orf477*-ΔTn2-ΔIS26), ΔTn2 and ΔIS26 were located both downstream and upstream of ΔIS*Ecp1*/IS*Ecp1*-*bla*$_{CTX-M-55}$-*orf477*'s structure. This structure was found not only in *E. coli* but also in *Klebsiella*, *Vibrio parahaemolyticus* and *Salmonella*. This structure, possible formed by a copy of IS*26* and an *ISEcp1*-mediated transposon carrying *bla*$_{CTX-M-55}$ and *orf477*, was inserted into a *tnpA* gene; this finding stresses the need for further assessment of the mobility of IS26 or its variant [34].

Plasmids play a key role in the horizontal transfer of the ESBL gene among *E. coli* strains. Our results suggest that the F18:A-:B1 plasmid may play an important role in the transmission of the *bla*$_{CTX-M-55}$ gene in *E. coli* isolated from duck farms and the environment. The F18:A-:B1 plasmid was first reported in avian pathogenic *E. coli* in 2010. It is characterized by the lack of an iron uptake gene (*eitABCD*), hemagglutinin and a survival gene. The *bla*$_{CTX-M-55}$-bearing F18:A-:B1 plasmid was first reported in *E. coli* isolated from patients with urinary tract infections in the United States in 2016, and carried the *mcr* gene [20,33]. Subsequently, the F18:A-:B1-plasmid-carrying *bla*$_{CTX-M-55}$ was reported in *E. coli* isolated from human clinical samples in China [35]. Recently, a study in Southeast Asia reported that the highly pathogenic clone *E. coli* ST410 isolated from the environment and humans carried the $bla_{CTX-M-55}$-bearing F18:A-:B1 plasmid with a high prevalence [30]. In addition, studies from Tunisia and China also reported the high prevalence of the F18:A-:B1 subtype in animal-derived *E. coli* strains, which also contained *fosA3, oqxAB*, $bla_{CTX-M-14}$ and other resistance genes [36–38]. These studies emphasize the possibility of horizontal transfer of the F18:A-:B1 plasmid between humans and animals.

However, to the best of our knowledge, only one recent study reported $bla_{CTX-M-55}$, *floR* and *fosA3* carrying the F18:A-:B1 plasmid obtained from ducks [10]. Our study is the first to report the high prevalence of the $bla_{CTX-M-55}$-carrying F18:A-:B1 plasmid in *E. coli* isolated from duck–fish polyculture farms. Complete plasmid sequence analysis showed that $bla_{CTX-M-55}$ was colocalized with *tetA, floR, fosA3*, bla_{TEM}, *aadA5*, *aph(3′)-Ib*, *aph(6′)-Id, CmlA, InuF* and other ARGs on the F18:A-:B1 plasmid. These ARGs, linked by different kinds of insertion sequences and transposable sequences and multiple copies of IS*26*, constitute the complex multidrug resistance region of the F18:A-:B1 plasmid. IS*26* is commonly associated with ARGs in multidrug-resistant Gram-negative and Gram-positive species, and is most widespread in Gram-negative bacteria. Clusters of ARGs can be generated by directly oriented IS*26* interspersed in multiple resistant pathogens [34,39]. As the best-studied member of the IS*26* family, IS*26* is known to form cointegrates using conservative transposition, homologous recombination and replicative transposition [39,40]. Multiple copies of IS*26* with different orientations located on the F18:A-:B1 plasmid are very likely to form cointegrates to promote the transmission of ARGs.

5. Conclusions

We reported the high prevalence of $bla_{CTX-M-55}$-carrying *E. coli* isolated from duck–fish polyculture farms. Both horizontal transfer and clonal spread contributed to the dissemination of the $bla_{CTX-M-55}$ gene among *E. coli* strains isolated from ducks and their environment, and the F18:A-:B1 plasmid may play an important role in the spread of $bla_{CTX-M-55}$. Coexistence of $bla_{CTX-M-55}$ and other resistance genes (eg., *tetA, floR, fosA3*, bla_{TEM}, *aadA5 CmlA, InuF*) on the same F18: A-: B1 plasmid may result in the co-selection of these resistance determinants and accelerate the dissemination of $bla_{CTX-M-55}$ in *E. coli*. In addition, our study is the first to report the emergence of a *fosA7.5* and $bla_{CTX-M-55}$ co-carrying ST602 clone in *E. coli* isolated from ducks and soil. These findings emphasize the importance of the ongoing surveillance of $bla_{CTX-M-55}$-positive *E. coli* in duck–fish polyculture farms in China.

Supplementary Materials: The following supporting information can be downloaded at: https://www.mdpi.com/article/10.3390/antibiotics12060961/s1, Table S1: Sample and isolate information from different sources; Table S2: The MIC results of 177 $bla_{CTX-M-55}$-positive *E. coli* strains; Table S3: Bacterial information and antimicrobial resistance profiles; Table S4: Characteristics of sequenced strains and obtained F18:A-:B1 plasmids and Figure S1: Comparison of four types of $bla_{CTX-M-55}$ genomic contexts in 47 strains. Figure S2: Pulsed field gels of S1 digested genomic DNA and Southern blot in-gel hybridization with $bla_{CTX-M-55}$ probe. M: H9812, 1-12: $bla_{CTX-M-55}$-positive transconjugants MA121, D5, KW21, PBS4, KS22, PBS10, BT2, PBS5, PBS13, PBS21, UD1 and A44.

Author Contributions: H.-X.J., R.-M.Z. and L.-J.Z. conceived the research. J.-T.Y., H.-X.C. and L.-J.Z. carried out the experiments. H.-X.J., R.-M.Z., W.-Z.L. and L.-J.Z. collected and analyzed the data. L.-J.Z. drafted the manuscript, and H.-X.J., R.-M.Z., R.-A.C. and Y.-L.D. revised the manuscript. All authors have read and agreed to the published version of the manuscript.

Funding: This research was supported by grants from the National Natural Science Foundation of China (NSFC) (31972734), the Local Innovative and Research Teams Project of Guangdong Pearl River Talents Program (2019BT02N054) and the independent project of Zhaoqing Branch Center of Guangdong Laboratory for Lingnan Modern Agricultural Science and Technology (P20211154-0101-09).

Institutional Review Board Statement: Not applicable.

Informed Consent Statement: Not applicable.

Data Availability Statement: The whole genome sequence generated in this study is available from the National Center for Biotechnology Information (Accession No. PRJNA934699).

Conflicts of Interest: The authors declare no conflict of interest.

References

1. Laurent, P.; Jean-Yves, M.; Agnese, L.; Anne-Kathrin, S.; Nicolas, K.; Patrice, N.; Stefan, S. Antimicrobial Resistance in *Escherichia coli*. *Microbiol. Spectr.* **2018**, *6*, 0026.
2. Peirano, G.; Pitout, J. Extended-Spectrum β-Lactamase-Producing Enterobacteriaceae: Update on Molecular Epidemiology and Treatment Options. *Drugs* **2019**, *79*, 1529–1541. [CrossRef] [PubMed]
3. Akya, A.; Ahmadi, M.; Khodamoradi, S.; Rezaei, M.R.; Karani, N.; Elahi, A.; Lorestani, R.C.; Rezaei, M. Prevalence of bla_{CTX-M}, $bla_{CTX-M-2}$, $bla_{CTX-M-8}$, $bla_{CTX-M-25}$ and $bla_{CTX-M-3}$ Genes in *Escherichia coli* Isolated from Urinary Tract Infection in Kermanshah City, Iran. *J. Clin. Diagn. Res.* **2019**, *13*, 4–7. [CrossRef]
4. Bevan, E.R.; Jones, A.M.; Hawkey, P.M. Global epidemiology of CTX-M β-lactamases: Temporal and geographical shifts in genotype. *J. Antimicrob. Chemother.* **2017**, *72*, 2145–2155. [CrossRef] [PubMed]
5. Kiratisin, P.; Apisarnthanarak, A.; Saifon, P.; Laesripa, C.; Kitphati, R.; Mundy, L.M. The emergence of a novel ceftazidime-resistant CTX-M extended-spectrum β-lactamase, CTX-M-55, in both community-onset and hospital-acquired infections in Thailand. *Diagn. Microbiol. Infect. Dis.* **2007**, *58*, 349–355. [CrossRef] [PubMed]
6. Huang, Y.; Zeng, L.; Doi, Y.; Lv, L.; Liu, J.-H. Extended-spectrum β-lactamase-producing *Escherichia coli*. *Lancet Infect. Dis.* **2020**, *20*, 404–405. [CrossRef] [PubMed]
7. Yang, X.; Liu, W.; Liu, Y.; Wang, J.; Lv, L.; Chen, X.; He, D.; Yang, T.; Hou, J.; Tan, Y. F33:A-:B-, IncHI2/ST3, and IncI1/ST71 plasmids drive the dissemination of *fosA3* and $bla_{CTX-M-55/-14/-65}$ in *Escherichia coli* from chickens in China. *Front. Microbiol.* **2014**, *5*, 688. [CrossRef]
8. Wang, J.; Zeng, Z.L.; Huang, X.Y.; Ma, Z.B.; Guo, Z.W. Evolution and Comparative Genomics of F33:A-:B- Plasmids Carrying $bla_{CTX-M-55}$ or $bla_{CTX-M-65}$ in *Escherichia coli* and *Klebsiella pneumoniae* Isolated from Animals, Food Products, and Humans in China. *mSphere* **2018**, *3*, e00137-18. [CrossRef]
9. Agnese, L.; Estelle, S.; Jean-Yves, M.; Marisa, H. Emergence of $bla_{CTX-M-55}$ associated with *fosA*, *rmtB* and *mcr* gene variants in *Escherichia coli* from various animal species in France. *J. Antimicrob. Chemother.* **2018**, *73*, 867–872.
10. Liu, F.; Tian, A.; Wang, J.; Zhu, Y.; Xie, Z.; Zhang, R.; Jiang, S. Occurrence and molecular epidemiology of fosA3-bearing *Escherichia coli* from ducks in Shandong province of China. *Poult. Sci.* **2022**, *101*, 101620. [CrossRef]
11. Wang, M.G.; Yu, Y.; Wang, D.; Yang, R.S.; Liao, X.P. The Emergence and Molecular Characteristics of New Delhi Metallo β-Lactamase-Producing *Escherichia coli* From Ducks in Guangdong, China. *Front. Microbiol.* **2021**, *12*, 677633. [CrossRef] [PubMed]
12. Wang, X.R.; Lian, X.L.; Su, T.T.; Long, T.F.; Sun, J. Duck wastes as a potential reservoir of novel antibiotic resistance genes. *Sci. Total. Environ.* **2021**, *771*, 144828. [CrossRef]
13. Xu, C.; Lv, Z.; Shen, Y.; Liu, D.; Shen, J. Metagenomic insights into differences in environmental resistome profiles between integrated and monoculture aquaculture farms in China. *Environ. Int.* **2020**, *144*, 106005. [CrossRef]
14. Zhou, M.; Xu, Y.; Ouyang, P.; Ling, J.; Zheng, L. Spread of resistance genes from duck manure to fish intestine in simulated fish-duck pond and the promotion of cefotaxime and As. *Sci. Total. Environ.* **2020**, *731*, 138693. [CrossRef] [PubMed]
15. Lv, L.; Partridge, S.R.; He, L.; Zeng, Z.; He, D.; Ye, J.; Liu, J.H. Genetic characterization of IncI2 plasmids carrying $bla_{CTX-M-55}$ spreading in both pets and food animals in China. *Antimicrob. Agents Chemother.* **2013**, *57*, 2824–2827. [CrossRef]
16. Carattoli, A.; Garcia-Fernandez, A.; Varesi, P.; Fortini, D.; Gerardi, S. Molecular Epidemiology of *Escherichia coli* Producing Extended-Spectrum -Lactamases Isolated in Rome, Italy. *J. Clin. Microbiol.* **2008**, *46*, 103–108. [CrossRef]
17. *M100*; Performance Standards for Antimicrobial Susceptibility Testing, 29th Edition. Clinical and Laboratory Standards Institute: Wayne, PA, USA, 2019.
18. The European Committee on Antimicrobial Susceptibility Testing. Breakpoint Tables for Interpretation of MICs and Zone Diameters, version 9.0. 2019. Available online: http://www.eucast.org (accessed on 1 January 2020).
19. Carattoli, A.; Bertini, A.; Villa, L.; Falbo, V.; Hopkins, K.L.; Threlfall, E.J. Identification of plasmids by PCR-based replicon typing. *J. Microbiol. Methods* **2005**, *63*, 219–228. [CrossRef]
20. Laura, V.; Aurora, G.F.; Daniela, F.; Alessandra, C. Replicon sequence typing of IncF plasmids carrying virulence and resistance determinants. *J. Antimicrob. Chemother.* **2010**, *65*, 2518–2529.
21. Jiang, H.X.; Song, L.; Liu, J.; Zhang, X.H.; Ren, Y.N.; Zhang, W.H.; Zhang, J.Y.; Liu, Y.H.; Webber, M.A.; Ogbolu, D.O. Multiple transmissible genes encoding fluoroquinolone and third-generation cephalosporin resistance co-located in non-typhoidal *Salmonella* isolated from food-producing animals in China. *Int. J. Antimicrob. Agents* **2014**, *43*, 242–247. [CrossRef]
22. Zhang, L.J.; Gu, X.X.; Zhang, J.; Yang, L.; Lu, Y.W.; Fang, L.X.; Jiang, H.X. Characterization of a *fosA3* Carrying IncC-IncN Plasmid From a Multidrug-Resistant ST17 *Salmonella* Indiana Isolate. *Front. Microbiol.* **2020**, *11*, 1582. [CrossRef]
23. Lv, L.C.; Lu, Y.Y.; Gao, X.; He, W.Y.; Gao, M.Y.; Mo, K.B.; Liu, J.H. Characterization of NDM-5-producing *Enterobacteriaceae* isolates from retail grass carp (*Ctenopharyngodon idella*) and evidence of bla_{NDM-5}-bearing IncHI2 plasmid transfer between ducks and fish. *Zool. Res.* **2022**, *43*, 255–264. [CrossRef] [PubMed]

24. Zhou, Q.; Wang, M.; Zhong, X.; Liu, P.; Xie, X.; Wangxiao, J.; Sun, Y. Dissemination of resistance genes in duck/fish polyculture ponds in Guangdong Province: Correlations between Cu and Zn and antibiotic resistance genes. *Env. Sci. Pollut. Res. Int.* **2019**, *26*, 8182–8193. [CrossRef] [PubMed]
25. Huang, L.; Xu, Y.B.; Xu, J.X.; Ling, J.Y.; Chen, J.L.; Zhou, J.L.; Zheng, L.; Du, Q.P. Antibiotic resistance genes (ARGs) in duck and fish production ponds with integrated or non-integrated mode. *Chemosphere* **2017**, *168*, 1107–1114. [CrossRef] [PubMed]
26. Zhang, Y.; Peng, S.; Xu, J.; Li, Y.; Pu, L.; Han, X.; Feng, Y. Genetic context diversity of plasmid-borne $bla_{CTX\text{-}M\text{-}55}$ in *Escherichia coli* isolated from waterfowl. *J. Glob. Antimicrob. Resist.* **2022**, *28*, 185–194.
27. Na, S.H.; Dong, C.M.; Choi, M.J.; Oh, S.J.; Lim, S.K. Antimicrobial Resistance and Molecular Characterization of Extended-Spectrum β-Lactamase-Producing *Escherichia coli* Isolated from Ducks in South Korea. *Foodborne Pathog. Dis.* **2019**, *16*, 799–806. [CrossRef] [PubMed]
28. Day, M.J.; Hopkins, K.L.; Wareham, D.W.; Toleman, M.A.; Elviss, N.; Randall, L.; Teale, C.; Cleary, P.; Wiuff, C.; Doumith, M.; et al. Extended-spectrum β-lactamase-producing *Escherichia coli* in human-derived and foodchain-derived samples from England, Wales, and Scotland: An epidemiological surveillance and typing study. *Lancet Infect. Dis.* **2019**, *19*, 1325–1335. [CrossRef] [PubMed]
29. Milner, K.A.; Bay, D.C.; Alexander, D.; Walkty, A.; Zhanel, G.G. Identification and Characterization of a Novel *FosA7* Member from Fosfomycin-Resistant *Escherichia coli* Clinical Isolates from Canadian Hospitals. *Antimicrob. Agents Chemother.* **2020**, *65*, e00865-20. [CrossRef]
30. Nadimpalli, M.L.; de Lauzanne, A.; Phe, T.; Borand, L.; Jacobs, J.; Fabre, L.; Naas, T.; Le Hello, S.; Stegger, M. *Escherichia coli* ST410 among humans and the environment in Southeast Asia. *Int. J. Antimicrob. Agents* **2019**, *54*, 228–232.
31. Roer, L.; Overballe-Petersen, S.; Hansen, A.F.; Schønning, K.; Wang, C.M.; Hinkle, M.; Whitman, T.; Lesho, E.; Schaecher, K.E. *Escherichia coli* Sequence Type 410 Is Causing New International High-Risk Clones. *Msphere* **2018**, *3*, e00337-18. [CrossRef]
32. Li, R.; Mohsin, M.; Lu, X.; Abdullah, S.; Munir, A.; Wang, Z. Emergence of Plasmid-Mediated Resistance Genes *tet(X)* and *mcr-1* in *Escherichia coli* Clinical Isolates from Pakistan. *mSphere* **2021**, *6*, e0069521. [CrossRef]
33. McGann, P.; Snesrud, E.; Maybank, R.; Corey, B.; Ana, C.; Clifford, R. *Escherichia coli* Harboring *mcr-1* and $bla_{CTX\text{-}M}$ on a Novel IncF Plasmid: First Report of mcr-1 in the United States. *Antimicrob. Agents Chemother.* **2016**, *60*, 4420–4421. [CrossRef]
34. Zheng, Z.; Li, R.; Ye, L.; Chan, W.C.; Xia, X.; Chen, S. Genetic Characterization of $bla_{CTX\text{–}M\text{–}55}$-Bearing Plasmids Harbored by Food-Borne Cephalosporin-Resistant Vibrio parahaemolyticus Strains in China. *Front. Microbiol.* **2019**, *10*, 1338. [CrossRef]
35. Xia, L.; Liu, Y.; Xia, S.; Kudinha, T.; Xiao, S.N.; Zhong, N.S.; Ren, G.S.; Zhuo, C. Prevalence of ST1193 clone and IncI1/ST16 plasmid in *E. coli* isolates carrying $bla_{CTX\text{-}M\text{-}55}$ gene from urinary tract infections patients in China. *Sci. Rep.* **2017**, *7*, 44866. [CrossRef]
36. He, W.Y.; Zhang, X.X.; Gao, G.L.; Gao, M.Y.; Zhong, F.G.; Lv, L.C.; Cai, Z.P.; Si, X.F.; Yang, J.; Liu, J.H. Clonal spread of *Escherichia coli* O101:H9-ST10 and O101:H9-ST167 strains carrying *fosA3* and $bla_{CTX\text{-}M\text{-}14}$ among diarrheal calves in a Chinese farm, with Australian Chroicocephalus as the possible origin of *E. coli* O101:H9-ST10. *Zool. Res.* **2021**, *42*, 8. [CrossRef] [PubMed]
37. BenSallem, R.; BenSlama, K.; Rojo-Bezares, B.; Porres-Osante, N.; Jouini, A.; Klibi, N.; Boudabous, A.; Sáenz, Y.; Torres, C. IncI1 plasmids carrying $bla_{CTX\text{-}M\text{-}1}$ or $bla_{CMY\text{-}2}$ genes in *Escherichia coli* from healthy humans and animals in Tunisia. *Microb.* **2014**, *20*, 495–500.
38. Jiang, W.; Men, S.; Kong, L.; Ma, S.; Yang, Y.; Wang, Y.; Yuan, Q.; Cheng, G.; Zou, W.; Wang, H. Prevalence of Plasmid-Mediated Fosfomycin Resistance Gene *fosA3* Among CTX-M-Producing *Escherichia coli* Isolates from Chickens in China. *Foodborne Pathog. Dis.* **2017**, *14*, 210–218. [CrossRef] [PubMed]
39. Harmer, C.J.; Hall, R.M. Targeted Conservative Cointegrate Formation Mediated by IS*26* Family Members Requires Sequence Identity at the Reacting End. *mSphere* **2021**, *6*, e01321-20. [CrossRef] [PubMed]
40. Partridge, S.R.; Kwong, S.M.; Neville, F.; Jensen, S.O. Mobile Genetic Elements Associated with Antimicrobial Resistance. *Clin. Microbiol. Rev.* **2018**, *31*, e00088-17. [CrossRef]

antibiotics

MDPI

Article

Cloning and Molecular Characterization of the *phlD* Gene Involved in the Biosynthesis of "Phloroglucinol", a Compound with Antibiotic Properties from Plant Growth Promoting Bacteria *Pseudomonas* spp.

Payal Gupta [1,†], Prasanta K. Dash [1,*,†], Tenkabailu Dharmanna Sanjay [1], Sharat Kumar Pradhan [2,3], Rohini Sreevathsa [1] and Rhitu Rai [1,*]

1 ICAR-National Institute for Plant Biotechnology, Pusa Campus, New Delhi 110012, India
2 ICAR-National Rice Research Institute, Cuttack 753006, India
3 Indian Council of Agriculture Research, Krishi Bhawan, New Delhi 110001, India
* Correspondence: prasanta01@yahoo.com (P.K.D.); rhitunrcpb@yahoo.com (R.R.); Tel.: +91-1125841787 (P.K.D.); Fax: +91-1125843984 (P.K.D.)
† These authors contributed equally to this work and share first authorship.

Abstract: *phlD* is a novel kind of polyketide synthase involved in the biosynthesis of non-volatile metabolite phloroglucinol by iteratively condensing and cyclizing three molecules of malonyl-CoA as substrate. Phloroglucinol or 2,4-diacetylphloroglucinol (DAPG) is an ecologically important rhizospheric antibiotic produced by pseudomonads; it exhibits broad spectrum anti-bacterial and anti-fungal properties, leading to disease suppression in the rhizosphere. Additionally, DAPG triggers systemic resistance in plants, stimulates root exudation, as well as induces phyto-enhancing activities in other rhizobacteria. Here, we report the cloning and analysis of the *phlD* gene from soil-borne gram-negative bacteria—*Pseudomonas*. The full-length *phlD* gene (from 1078 nucleotides) was successfully cloned and the structural details of the PHLD protein were analyzed in-depth via a three-dimensional topology and a refined three-dimensional model for the PHLD protein was predicted. Additionally, the stereochemical properties of the PHLD protein were analyzed by the Ramachandran plot, based on which, 94.3% of residues fell in the favored region and 5.7% in the allowed region. The generated model was validated by secondary structure prediction using PDBsum. The present study aimed to clone and characterize the DAPG-producing *phlD* gene to be deployed in the development of broad-spectrum biopesticides for the biocontrol of rhizospheric pathogens.

Keywords: phloroglucinol; polyketide; DAPG; *phlD*; Ramachandran plot; PMDB

Citation: Gupta, P.; Dash, P.K.; Sanjay, T.D.; Pradhan, S.K.; Sreevathsa, R.; Rai, R. Cloning and Molecular Characterization of the *phlD* Gene Involved in the Biosynthesis of "Phloroglucinol", a Compound with Antibiotic Properties from Plant Growth Promoting Bacteria *Pseudomonas* spp. *Antibiotics* **2023**, *12*, 260. https://doi.org/10.3390/antibiotics12020260

Academic Editors: Theodoros Karampatakis and Grzegorz Wegrzyn

Received: 30 October 2022
Revised: 21 December 2022
Accepted: 23 December 2022
Published: 28 January 2023

1. Introduction

Plant growth-promoting rhizobacteria (PGPR) are bacteria that colonize some or all parts of the rhizosphere environment and have the capability to promote plant growth [1–3] either directly by antibiosis or indirectly by quorum sensing. PGPR produce non-volatile metabolites that can directly stimulate plant growth, inhibit plant pathogens, and/or induce host–defense mechanisms against pathogens [4,5]. *Pseudomonas fluorescens* is an important group of PGPR that suppress root and seedling diseases by producing non-volatile secondary metabolite phloroglucinol. Genetic methods [6–8] and direct isolation from the soils of diseased plants [9–11] have shown the importance of DAPG and its derivatives as biocontrol activity agents. These compounds act as antibiotics, antimicrobials or antifungals, signaling molecules, and pathogenicity factors. Several antibacterial and antifungal compounds from plants have been characterized [12] and their mechanisms of action have been delineated [13]; biosynthesis and genetic regulation of DAPG in the *Pseudomonas* spp. have been the focus of active research.

Most of the genes required for the biosynthesis of DAPG have been cloned and characterized from different strains of *Pseudomonas*. These genes are recognized as conserved in 2,4-DAPG, producing pseudomonads that have been isolated from various soil samples collected from around the world [14–16]. All of the genes are arranged as an operon on the *phl* locus, spanning a genomic fragment of ~6.5 kb, comprising six genes, viz. *phl*A, *phl*B, *phl*C, *phl*D, *phl*E, and *phl*F. While *phl*A, *phl*B, *phl*C, and *phl*D are transcribed as an operon; they are flanked on either side by *phl*E and *phl*F. *phl*E codes for an efflux protein and *phl*F encodes a repressor protein (Figure 1).

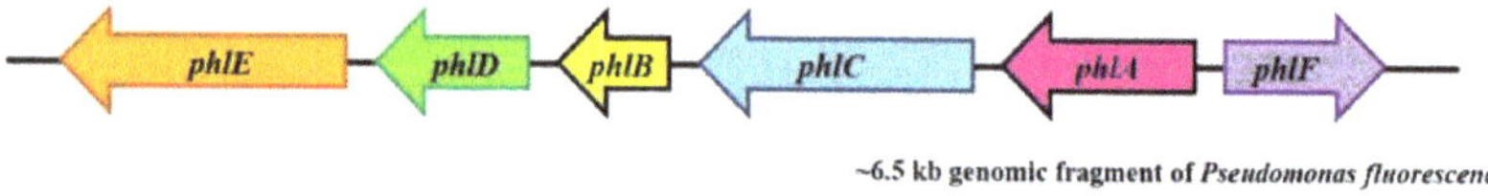

Figure 1. Schematic representation of a ~6.5 kb genomic fragment of *Pseudomonas* harboring the genes responsible for the biosynthesis of 2,4-DAPG by *phl* operon. *phl* operon comprises four genes *phl*A, *phl*B, *phl*C, and *phl*D. The operon is flanked on either side by *phl*E and *phl*F genes that are separately transcribed and coded for the putative efflux and regulatory (repressor) proteins, respectively. They are not required for phloroglucinol production.

The DAPG gene cluster is self-sufficient for the biosynthesis and regulation of 2,4-diacetylphloroglucinol. Amongst all genes, *phl*D is the key gene responsible for the production of (MAPG), while, *phl*A, *phl*B, and *phl*C are necessary to convert MAPG to 2,4-DAPG. Products of these genes resemble neither type I nor type II PKS enzyme systems. Rather, *Phl*D shows similarity to plant chalcone synthases, indicating that phloroglucinol synthesis is mediated by a novel kind of PKS [17–19].

Apart from changes in gene expression, the production of 2,4-DAPG in many strains of fluorescent *Pseudomonas* spp. is stimulated by physical factors, such as a concentration of glucose [20] or concentrations of sucrose/ethanol [19,21]. Moreover, zinc sulfate and ammonium molybdate have been reported to favor 2,4-DAPG production in some strains, whereas inorganic phosphate in general has an inhibitory effect [20].

*phl*D gene, responsible for the bio-synthesis of MAPG, shows similarity to novel type III polyketide synthase (PKS) [22]. *phl*D iteratively condenses three molecules of malonyl-CoA that subsequently cyclize [23] to form mono-acetyl phloroglucinol (MAPG). However, it is delineated that *Phl*D catalyzes the condensation of three molecules of malonyl-CoA into 3,5-diketoheptanedioate [24] and this polyketide intermediate through decarboxylation/cyclization, forms phloroglucinol [25].

Our laboratory has isolated and cloned and characterized *phl*A, *phl*B, and *phl*C; the downstream genes [26–28] of *phl* operon, and the current study focuses on the cloning and characterization of the *phl*D gene, an upstream/first committed step of the *phl* operon of *Pseudomonas* and its in-depth characterization to obtain deep insight into the PHLD function. This will help in fine-tuning (upregulating/downregulating) the bio-synthesis of 2,4-DAPG in response to potent fungal and bacterial pathogens for improving the biocontrols of plant pathogens.

2. Materials and Methods

2.1. Genomic DNA Isolation and PCR Amplification of the phlD Gene

Genomic DNA of the *Pseudomonas* spp. strain RS9 (KP057506) [29] was isolated as described earlier [26]. The *phl*D gene from *Pseudomonas* spp. was amplified by the polymerase chain reaction-based strategy. The forward and reverse primers were designed using the PRIMER 3 tool, viz, *phl*D (FP): CCG<u>ACTAGT</u>AGGACTTGTCATGTCTACTCTTTG and *phl*D (RP): GGA<u>AAGCTT</u>CGTGCAATGTGTTGGTCTGTCA were designed using the nucleotide sequence of *Pseudomonas fluorescens* (U41818) available in the EMBL database. Restriction sites for the enzymes *Spe*I and *Hind*III were incorporated at the 5′ ends of forward and reverse primers (underlined sequences), respectively. The PCR reaction mixture consisted of 10 pmol of each primer, 50 ng of template DNA, 50 mM KCl, 10 mM Tris-HCl (pH 9.0), 0.1% Triton, 2.5 mM $MgCl_2$, 0.2 mM of each dNTP, and a 1.25 unit of Phusion *Taq* DNA polymerase in 100 µL of volume. The thermal cycling was performed with an initial denaturation cycle of 3 min at 98 °C, followed by 30 cycles of (i) denaturation at 98 °C for 20 s; (ii) annealing for 30 s at 55 °C; (iii) extension for 30 s at 72 °C, as well as one cycle of the final extension for 7 min at 72 °C.

2.2. Cloning of the phlD Gene in pBluescript (SK+) Vector

For cloning of the PCR amplified *phl*D gene into the pBluescript (SK+) vector, the PCR product (insert) and pBluescript (SK+) vector DNA were double digested with *Spe*I and *Hind*III. The reaction mixture was incubated at 37 °C for 3 hours. Restricted DNA was gel-purified using a Zymo clean gel DNA recovery kit. Purified 50 ng of linearized pBluescript vector and 100 ng of a double-digested PCR product were ligated using T_4 DNA ligase. The ligated mixture was incubated at 4 °C overnight for ligation and used for transformation into *E. coli*. The transformed colonies (white in color) obtained after overnight incubation at 37 °C were picked and streaked onto fresh LA-carbenicillin (100 µg/mL) plates.

2.3. Confirmation of Cloning and Sequencing

Recombinant colonies were confirmed by restriction digestion with *Spe*I+*Hind*III enzymes. The restricted DNA samples were analyzed on 1.2% agarose gel. The complete nucleotide sequence was determined by the Sanger di-deoxy sequencing. M13F and M13R primers were used for sequencing. *phl*D gene-specific primers were also used for confirming the sequence. The final sequence was determined from both strands and a comparison of *phl*D nucleic acid and amino acid sequences with already existing sequences was performed. The deduced amino acid sequence of PHLD from *Pseudomonas* RS-9 was compared with type III PKS from gram-positive bacteria and other plants by a multiple sequence alignment using MAFFT version 7.271 program [30] with the L-INS-I strategy and output in Phylip format. A similarity score for each nucleotide of the aligned sequences was calculated by ESPRIPT 3.0 [31] (https://espript.ibcp.fr/ESPript/ESPript/, accessed on 20 February 2022) with default parameters. Conserved domain annotation analysis was performed using InterProScan [32].

2.4. Phylogenetic Analysis

For estimation of the phylogenetic relationship between PHLD from various Pseudomonas strains and type III PKS from gram-positive bacteria and plants, the amino acid sequences were retrieved from the NCBI database. A multiple sequence alignment for the respective amino acid sequences was performed by Clustal Omega [33] and an un-rooted tree was constructed in MEGA10 [34] using the maximum likelihood (ML) method. Tree topology was searched using the nearest neighbor interchanges (NNIs) algorithm [35]. The LG+G+I substitution model was employed. The gamma shape parameter was estimated directly from the data and the analysis was performed using 1000 bootstrap replicates. The proportion of invariable sites was fixed. The tree was obtained in the Newick format.

2.5. Structure Prediction

The model of the PHLD protein was predicted using the I-TASSER server (http://zhanglab.ccmb.med.umich.edu/I-TASSER/, accessed on 25 February 2022) [36]. I-TASSER (Iterative Threading ASSEmbly Refinement) is a hierarchical approach to protein structures and function prediction. Structural templates were first identified from PDB by the multiple threading approach, LOMETS; full-length atomic models were then constructed by iterative template fragment assembly simulations. The generated model was refined using ModRefiner (http://zhanglab.ccmb.med.umich.edu/ModRefiner/, accessed on 3 March 2022). ModRefiner is an algorithm for high-resolution protein structure refinement. Both side-chain and backbone atoms were completely flexible during structure refinement simulations. ModRefiner allowed the assignment of a second structure that was used as a reference to which the refinement simulations were driven. The ModRefiner was used to draw the initial starting model of PHLD closer to its native state.

2.6. Ramachandran Plot Analysis

The stereochemical properties of the PHLD protein were assessed by the Ramachandran plot analysis using RAMPAGE [37]. This allowed visualization of energetically allowed regions for backbone dihedral angles ψ against φ of amino acid residues in the PHLD protein structure. The residues in the disallowed region were further refined by using Modloop (https://modbase.compbio.ucsf.edu/modloop/, accessed on 5 March 2022). Modloop relies on MODELLER, which predicts the loop conformations of PHLD by the satisfaction of spatial restraints, without relying on a database of known protein structures [38].

2.7. Validation and Visualization of Modeled Structure

The validation of the modeled structure was performed using PDBsum [39] and PROCHECK [40]. Structure visualization was performed using PyMOL. The predicted model of the protein was submitted to the Protein Model Database [41] (http://srv00.recas.ba.infn.it/PMDB/main.php, accessed on 10 March 2022).

3. Results

PCR amplification of the *phl*D gene from the genomic DNA of the *Pseudomonas* spp. strain RS9 (KP057506) [29] resulted in a fragment of 1 kb (Figure 2). The amplified PCR product and pBluescript control vector were then restricted with *Spe*I and *Hind*III restriction enzymes. This resulted in a 1 kb fragment PCR product with sticky ends (insert) and 3 kb of linearized control vector pBluescript (SK+) with sticky ends for *Spe*I and *Hind*III (Figure 3). The purified double-digested PCR product (insert) was ligated into the linearized pBluescript vector.

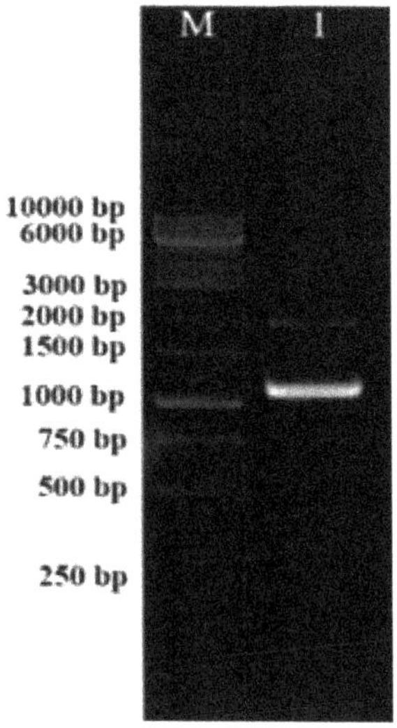

Figure 2. PCR amplification of the *phl*D gene from genomic DNA of *Pseudomonas fluorescens*. Lane M: 1 kb DNA ladder; Lane 1: ~1078 kb amplicon of the *phl*D gene.

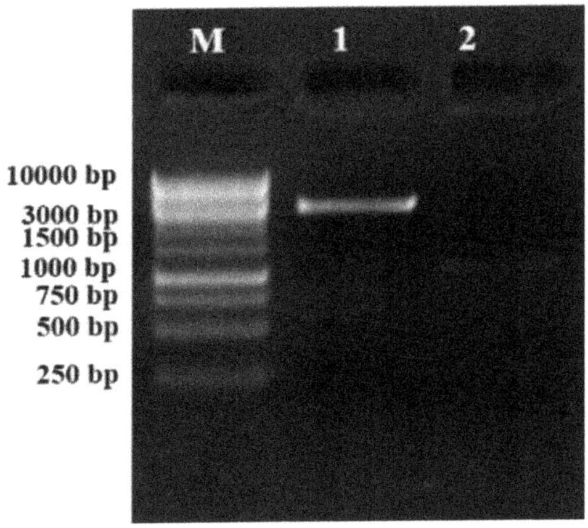

Figure 3. Restriction profile of the double-digested PCR product and control pBluescript vector. Lane M: 1 kb DNA ladder. Lane 1: ~3 kb fragment of pBS digested with *Spe*I+*Hind*III. Lane 2: ~1.078 kb fragment of purified PCR product. The PCR product was digested with *Spe*I+*Hind*III.

The ligated mixture was transformed into *E. coli* DH-5α competent cells and five randomly picked white colonies were used for the plasmid isolation. The presence of the *phl*D gene was confirmed by restriction digestion with *Spe*I+*Hind*III enzymes that released the expected fragment of ~1 kb (Figure 4).

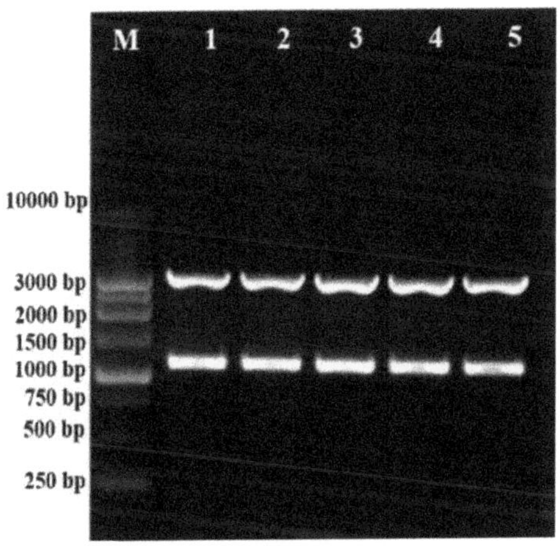

Figure 4. Screening and confirmation of positive putative clones by restriction digestion, respectively. Lanes 1–5: restriction digestion with *Spe*I+*Hind*III released a fragment of ~1.078 kb of the *phlD* gene and ~3 kb fragment of pBluescript (SK+) vector backbone.

Sanger sequencing of the cloned *phl*D was carried to check its identity and the results revealed that it consisted of 1078 nucleotides with an open reading frame of 1050 bp. Based on the blast results, the *phl*D gene was found to be of full-length coding for 349 amino acids. The cloned *phl*D gene showed considerable homology with the other known genes, indicating a common descent. The deduced amino acid sequence of 349 amino acids (~38.3 KDa) showed significant similarity with the homologs of PHLD (Figures 5 and 6).

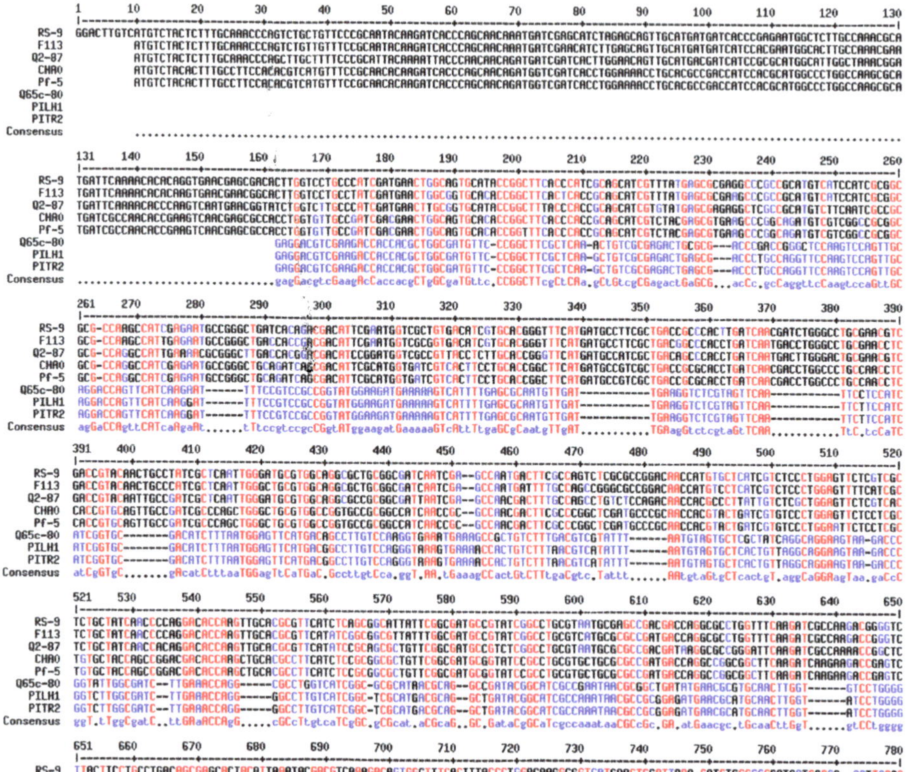

Figure 5. Pairwise sequence alignment of the cloned *phl*D gene (RS-9) with the reported DAPG producing *phl*D gene from *Pseudomonas* strains from the database (http://multalin.toulouse.inra.fr/multalin/, accessed on 15 March 2022). Our cloned *phl*D gene from the RS-9 strain showed homology with already reported *phl*D genes from *Pseudomonas* strain F113, Q2-87, CHAO, Pf-5, Q65c-80, PILH1, and PITR2.

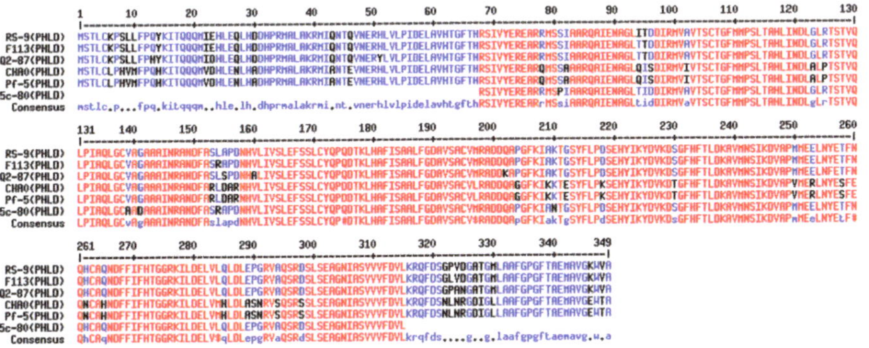

Figure 6. Pairwise sequence alignment of *Pseudomonas* spp. RS-9 PHLD protein with the PHLD protein from five *Pseudomonas* strains (http://multalin.toulouse.inra.fr/multalin/, accessed on 15 March 2022). The deduced amino acid sequence from the cloned *phl*D gene *Pseudomonas* strain RS-9 showed homology with the already reported PHLD protein from *Pseudomonas* strain F113, Q2-87, CHAO, and Pf-5.

The deduced amino acid sequence from *Pseudomonas* RS-9 was compared with type III PKSs from gram-positive bacteria and CHS/STS from plants. The functional roles of key amino acid residues found in type III PKSs/CHS/STS were found in PHLD proteins and other bacterial-type III PKSs (Figure 7), such as plant C169 (cysteine-169), responsible for the catalytic activities of plant CHSs, S158 in plants (serine-158), and Q166 in plants (glutamine-166), and were conserved in PHLD proteins. C135 (cysteine-135) and C195 (cysteine-195) played roles in substrate specificity and K180 (lysine-180), which are important for the enzymatic structure and function in plant-type III PKSs, and are replaced in the bacterial counterparts. Threonine, serine, and asparagine, respectively, replaced these amino acids in the PHLD sequences (Figure 7).

Figure 7. Alignment of predicted amino acid sequences for PHLD from various strains of *Pseudomonas* and type III polyketide synthase (PKS) from gram-positive bacteria and plant chalcone synthases/Stilbene synthase (CHS/STS). Conserved residues are indicated by boxes. (1) Cysteine (C135) from plant CHS implicated in substrate specificity and corresponding to threonine (T104) in PHLD. (2) Serine (S158) subunit contact site corresponding to S127 in PHLD. (3) Glutamine (Q166) residue conserved in most plant CHSs and corresponding to Q166 in the PKS of *Streptomyces griseus* and Q135 in PHLD proteins. (4) The glycine cysteine (GC) box corresponds to the conserved region with its catalytic cysteine residue. (5) Lysine (K180) residue, which corresponds to asparagine (N149) in PHLD, conserved strictly in the plant. (6) C195 involved in the product specificity in the plant CHS.

Analyses of the conserved domains and annotations were performed using InterProScan. The results revealed the presence of two InterPro domains viz. the chalcone/stilbene_synthases_N-terminal domain (37-200 a.a., IPR001099) and the chalcone/stilbene_synthases_C-terminal domain (213-344 a.a., IPR012328). These two domains belong to the Polyketide_synthase_type III InterPro family (IPR011141). A maximum likelihood (ML) tree was constructed to compare the phylogenetic relation of PHLD from *Pseudomonas* to Type III PKSs from bacteria and plants (PKS from other gram-positive bacteria and CHS/STS from plants). Upon comparison, eight PHLD sequences clustered into a separate group along with PKS from *Streptomyces griseus* (PKS). PKS from gram-positive bacteria and CHS/STS from plants also clustered in a separate group in the ML tree (Figure 8).

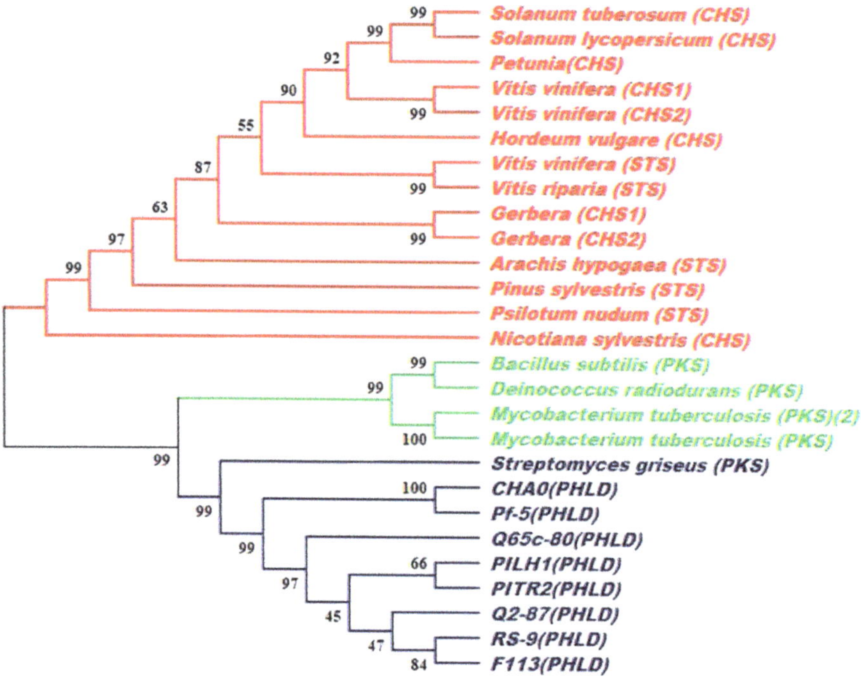

Figure 8. Diversity analysis between predicted amino acid sequences of PHLD from eight *Pseudomonas* strains and type III polyketide synthase from gram-positive bacteria (PKS) and plants (CHS/STS). The maximum likelihood (ML) tree was generated by the neighbor-joining method implemented in MEGA10.

The three-dimensional model of the PHLD protein using the deduced amino acid sequence of 349 amino acids was generated using the I-TASSER server (Figure 9) for the detailed structural analysis. The homology modeling of the initial structure was based on the template crystal structure of *Mycobacterium tuberculosis* polyketide synthase 11 (PKS11) (PDB entry 4JAP) [42]. The PHLD model had a C-score of 1.61 and a TM score of 0.94 ± 0.05.

This initial model was refined using ModRefiner and Ramachandran plot analysis and the results of both analyses revealed that only 89.5% of residues were in the favored region, while 1% of residues were in the outlier region. The initial model was iteratively re-refined using Modloop until 0% of residues fell in the outlier region, 94.3% of residues fell in the favored region, and 5.7% in the allowed region (Figure 10). The predicted model was submitted to the Protein Model Database (PMDB) and was assigned the identifier PM0080923.

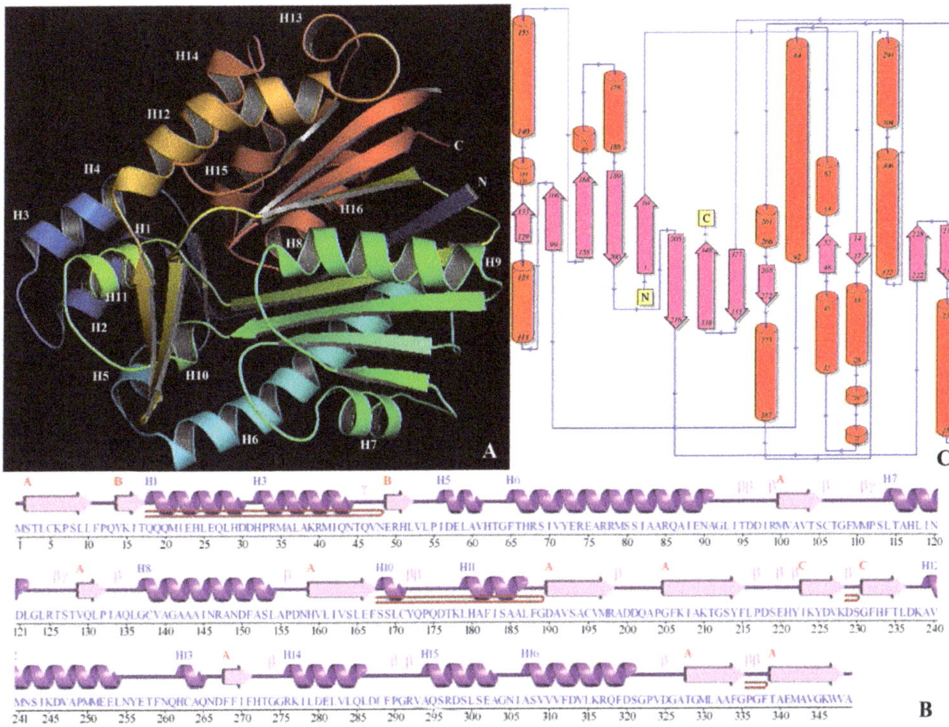

Figure 9. Structure of PHLD from *Pseudomonas* spp. (**A**) Stereo-ribbon diagram of the PHLD monomer (chain *A*) color-coded from the N-terminus (blue) to the C-terminus (red). Helices (H1–H16). (**B**) Diagram showing the secondary-structure elements of PHLC superimposed on its primary sequence. The labeling of the secondary-structure elements is in accordance with *PDBsum* (http://www.ebi.ac.uk/pdbsum): α-helices are labeled H1–H16, the β-strands are labeled β, β-turns and γ-turns are designated by their respective Greek letters (β, γ) and red loops indicate β-hairpins. (**C**) Topology of the PHLD protein showing the orientation of α-helices and β-strands.

The three-dimensional model generated by homology modeling was in accordance with the secondary structure predicted by PDBsum (Figure 9B,C). The PHLD secondary structure was dominated by the presence of α-helices (40.7%) followed by β-strands (26.1%) and 3-10 helices (2.6%). The PHLD structure revealed the presence of 3 beta sheets and 16 alpha-helices. β-sheet A contained 9 mixed β-strands with topology -3X -1X 2X 1 2X 3X -1X -1, β-sheet B and C contained 2 anti-parallel β-strands with topology 1 [43]. Protein also contained 16 α-helices viz; α1 {Gln18-Gln27 (10 residues)}, α2 {Leu28-Asp30 (3 residues)}, α3{His32-Arg34 (3 residues)}, α4{Met35-Asn44 (10 residues)}, α5{Ile56-Val61 (6 residues)}, α6{Phe65-Ala91 (27 residues)}, α7{Leu114-Leu122 (9 residues)}, α8{Gly137-Val139 (3 residues)}, α9 {Ala140-Leu154 (15 residues)}, α10 {Ser168-Cys171 (4 residues)}, α11{Leu179-Leu187 (9 residues)}, α12{Ala239-Leu254 (16 residues)}, α13{His262-Gln265 (4 residues)}, α14{Arg276-Leu286 (11 residues)}, α15{Ala294-Ala303 (10 residues)} and α16{Ala307-Ser321 (15 residues)}. The 4 β-hairpins of class 29:31, 20:22, 2:2 I, and 3:3 were identified [44]. The 14 helix–helix interactions, 12 H-H types (between α1 and α4, α1, and α5, α1 and α11, α4 and α11, α6 and α7, α9, and α16, α11 and α15, α12 and α13, α12 and α14, α13 and α14, α14 and α15, and α15 and α16), and 2 H-G type (α5 and α10 and α6 and α10) were identified.

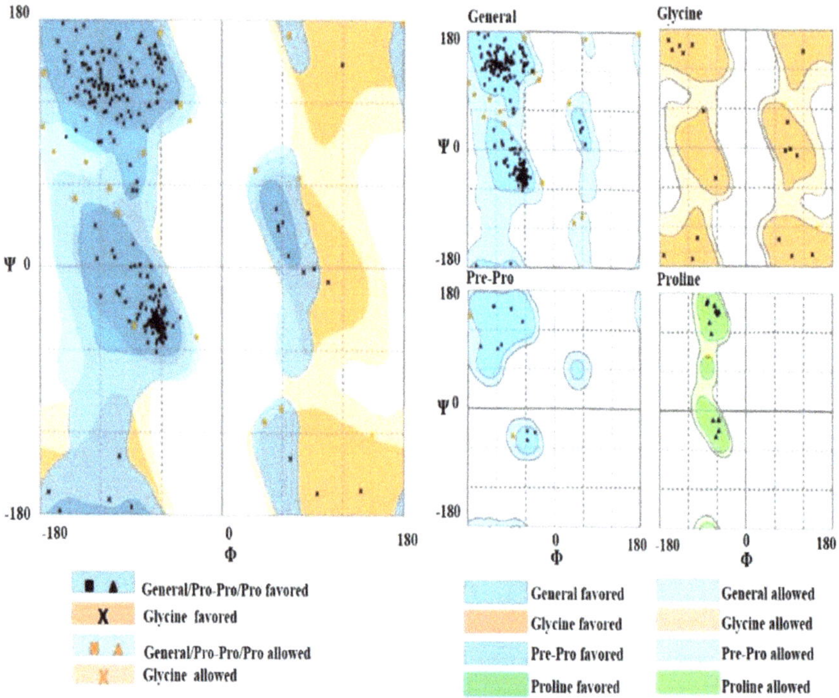

Figure 10. Stereochemical structure stability was analyzed using the Ramachandran plot analysis using RAMPAGE; 94.3% of residues fall in the favored region and 5.7% in the allowed region.

Two β-α-β motifs with 23 loops and 11 helices, and 56 loops and 42 helices participations were identified along with four β-bulges, one each of an anti-parallel classic type and anti-parallel wide type, and 2 anti-parallel G1 types were also identified. There are 21 β-turns in total belonging to 5 classes: I {(Thr95-Ile98), (Ser105-Gly108), (Tyr172-Gln175) and (Ser219-Tyr222)}; II{(Val324-Ala327) and (Gly335-Phe338)}; II′(Ile133-Leu136); IV{(Ile94-Asp97), (Ile98-Val101), (Met110-Ser113), (Arg125-Thr128), (Ala155-Asn158), (Cys171-Pro174), (Arg198-Asp201), (Leu216-Ser219), (Lys228-Gly231), (Thr273-Arg276), (Glu289-Arg292), (Pro336-Thr339) and (Gly291-Ala294)}; and VIII(His221-Lys224) [45]. Four γ turns of inverse types: (Thr45-Val47), (Met111-Ser113), (Thr126-Thr128), and (Thr273-Gly275) were also recognized.

4. Discussion

Special attention has been given to the antibiotic-producing fluorescent species of *Pseudomonas* due to their antibacterial [10,46], antifungal [47–50], and antiviral [51] abilities to control a wide variety of plant diseases. Advances in molecular techniques have also improved our potential to study the DAPG-producing antibacterial strains for their mechanisms of pathogen suppression and growth promotion. Breakthroughs in genomics [52–55] and transgenic [56,57] research to impart biotic/abiotic tolerance [58–60] or engineer traits in crops [61,62] have also driven the research in the field of biocontrol using DAPG producing strains. Genetic engineering approaches have been employed for the high-level production of phloroglucinol. Since the genetic background and metabolism of *Pseudomonas* have not been elucidated completely and the host does not respond well to genetic manipulation, the heterologous expression of *phl*D in *E. coli* is a great approach for increasing the accumulation of phloroglucinol in cultures [63–65].

DAPG is known to have antifungal properties and is produced by tandem activities of six genes viz. *phl*A, *phl*B, *phl*C, *phl*D, *phl*E, and *phl*F. These genes are organized as an operon onto a single nucleotide fragment of size ~ 6.5 kb. Among these six genes, *phl*D alone is important for the synthesis of MAPG. Although *phl*A, *phl*B, and *phl*C are also required for the synthesis of MAPG, *phl*D is the most essential. It has been proved that MAPG is synthesized only in the presence of *phl*D and in its absence, the cells converted exogenous MAPG to 2,4-DAPG but were unable to produce either compound themselves. This attribute makes *phl*D an important and useful marker of the genetic diversity and population structure among the 2,4-DAPG producers [14]. Thus, probes and primers specific for *phl*D have been used in combination with colony hybridization and polymerase chain reaction (PCR) to quantify the population sizes of 2,4-DAPG producers in the rhizosphere [11,66,67].

*phl*D shows a remarkable similarity to CHS/STS enzymes from plants. This is surprising because most of the microbial antibiotic enzymes are known to be synthesized via type I or type II PKSs [45,68–71]. Structural similarities between *phl*D and CHS/STS enzymes point to the common evolutionary descent and similarities in the roles they play during plant defense strongly support the instances of gene exchange between plants and bacteria [46,72]. The absence of the acyl carrier protein gene from the *phl* operon further confirms the similarity with the CHS/STS gene family. *Pseudomonas* spp. strain RS-9 was used for the isolation and cloning of the full-length *phl*D gene. The primers were designed using the *Pseudomonas fluoresens* (U41818) *phl*D gene sequence as a reference and *Pseudomonas* spp. strain RS-9 as the template. To amplify the full-length *phl* gene, we first standardized the PCR conditions. A gradient PCR was set in a temperature range of 50 °C to 60 °C to optimize the Tm for the reaction. At a lower Tm, multiple bands were obtained and at a very high Tm, faint amplification was obtained. The optimum amplification of ~1 kb for the *phl*D gene was obtained at 55 °C. This amplicon was restricted, purified, and ligated to the pBluescript vector and transformed into *E. coli* cells.

The cloned *phl*D gene through our investigation was confirmed by Sanger sequencing and it consisted of 1078 nucleotides with an open reading frame from 10 to 1059. The longest ORF of the *phl*D gene was found to be 1050 bp. Based on the blast results, the cloned *phl*D gene was found to be full-length, coding for 349 amino acids. This is consistent with the other reports [73] (Figure 5) on the length of amino acid coding *phl* genes. The cloned *phl*D gene showed 93% homology with *phl*D genes from different *Pseudomonas* strains, such as *Pseudomonas* sp. Q12-87, *Pseudomonas* sp. K96.27, *Pseudomonas* sp. PITR2, *Pseudomonas* sp. Q37-87, *Pseudomonas* sp. 12. The deduced amino acid sequence of 349 amino acids (~38.3 KDa) showed 97% similarity with the homologs of PHLD [73]. The *phl*D gene is of utmost importance to the DAPG gene cluster as MAPG synthesis does not occur without it.

The deduced amino acid sequence of the cloned *phl*D gene consisting of 349 amino acids was aligned pairwise with the *Pseudomonas fluorescens* (U41818) PHLD protein. The alignment revealed that there were few mutations in the protein sequence of the cloned gene. These mutations were authentic as we amplified and cloned the gene using high-fidelity Phusion *Taq* polymerase. Since this *Taq* polymerase has 3′ of proofreading activity, the chances of mis-amplifying or incorporating wrong bases are meager. These mutations need further characterization by site-directed mutagenesis.

A comparison of the deduced amino acid sequence from *Pseudomonas* RS-9, type III PKS from plants (CHS/STS), and other gram-positive bacteria indicated that PHLD and plant CHSs displayed common features. Comparison of the active site region indicated replacements of C135, C195, and K180 with threonine, serine, and asparagine that might have influenced their substrate specificities. Lysine and asparagine codons differ only at the third nucleotide position, and a single transversion can yield an asparagine instead of a lysine. The cluster analysis clearly distinguished between PhlD and plant CHS/STS by clustering them into separate groups [74–76]. Type III PKS from gram-positive bacteria clustered between CHS/STS from plants and PHLD from eight *Pseudomonas* strains. PKS closest to PHLD was the *Streptomyces gresius* PKS based on cluster analysis. The possibility that type III PKSs from fluorescent pseudomonads, gram-positive bacteria, and higher plants arose independently and may represent convergent evolution of the key enzymes involved in the biosynthesis of secondary metabolites as speculated earlier [46,73], corroborating out results of the cluster analysis.

The three-dimensional structure of the PHLD protein predicted using the I-TASSER server was based on the template crystal structure of *Mycobacterium tuberculosis* polyketide synthase 11 (PKS11) (PDB entry 4JAP) [42]. The PHLD model had an excellent C-score of 1.61 indicating a good quality model. C-score ranged from −5 to 2 and this higher value indicates the high quality of the model. Similarly, a TM score >0.5 indicated a model of correct topology, and a TM score <0.17 meant a random similarity. A TM score of 0.94 ± 0.05 for PHLD indicates the precision of the predicted topology. Moreover, >90% residues (94.3% residues) in the favored region of the Ramachandran plot reaffirmed the stereochemical stability of the generated refined molecule.

5. Conclusions

Pseudomonas strains have been used as potent biocontrol agents for controlling plant diseases because of the production of metabolites with antibiotic properties [77]. Amongst them, fluorescent pseudomonads are suitable for application as biocontrol agents and are best-characterized by biocontrol PGPR [78]. The biocontrol property of *Pseudomonas* is attributed to the synthesis of phloroglucinol—a secondary metabolite with antibiotic properties—produced by genes encoded by the *phl* operon. The *phl*D gene present in the *phl* operon is singularly involved in the synthesis of MAPG, which is processed into phloroglucinol. Though *Phl*D exhibits condensing activity on malonyl CoA to produce phloroglucinol; the substrate specificity of the enzyme is not limited to malonyl CoA compared to other type III PKS enzymes. It also catalyzes "C4–C12 aliphatic acyl-CoAs and phenylacetyl-CoA" as substrates to form tri- to heptaketide pyrones [25]. The same is evidenced by the homology modelling of *Phl*D that reveals the presence of a buried tunnel that protrudes out of the active site to accommodate the binding of acyl-CoAs. Structural details revealed from our findings can be used for targeted mutagenesis and rational designs to successfully alter the substrate specificity of *Phl*D to produce derivatized products with higher potency for antibiosis. Since, *phl*D is the first committed step of DAPG biosynthesis, targeting substrate specificity of *Phl*D would be a prudent way to enhance the biocontrol activities of *Pseudomonas* spp. that otherwise are present as long-lasting indigenous communities in several agro-ecosystems to augment the capability to protect the plant root system from numerous soil-borne plant diseases. Our results of cloning and structural delineation of *phl*D will provide novel strategies for combinatorial biosynthesis of natural but pharmaceutically important metabolites with enhanced antibacterial and biocontrol effects.

Author Contributions: R.R., R.S., S.K.P. and P.K.D. conceptualized the work and designed the experiments. P.G. and P.K.D. carried out the research. R.R., P.K.D., T.D.S., R.S. and S.K.P. analyzed the data. P.G. wrote the manuscript (first draft) with input from all authors. R.R. and P.K.D. edited the manuscript. All authors have read and agreed to the published version of the manuscript.

Funding: This research received no external funding.

Institutional Review Board Statement: This is NIPB publication number NIPB/2022/Res-35.

Informed Consent Statement: Not applicable.

Data Availability Statement: Not applicable.

Acknowledgments: The present study was carried out by DBT, GoI, grant to R.R. and as an in-house RPP project to R.R. and P.K.D. (RPP—Sanction no. 7–5/2010-IC-IV) by the Indian Council of Agricultural Research (ICAR), New Delhi, India. Financial support from ICAR-NPFGGM (RR, PKD) and a scholarship to P.G. is acknowledged. This is an NIPB publication (number NIPB/2022/Res-35).

Conflicts of Interest: The authors declare no conflict of interest.

References

1. Kloepper, J.; Lifshitz, R.; Schroth, M. *Pseudomonas Inoculants to Benefit Plant Protection;* Institute for Scientific Information: Philadelphia, PA, USA, 1988.
2. Troppens, D.; Moynihan, J.; Barret, M.; O'Gara, F.; Morrissey, J. *Molecular Microbial Ecology of the Rhizosphere;* John Wiley and Sons, Ltd.: Hoboken, NJ, USA, 2013.
3. Mandryk-Litvinkovich, M.N.; Muratova, A.A.; Nosonova, T.L.; Evdokimova, O.V.; Valentovich, L.N.; Titok, M.A.; Kolomiets, E.I. Molecular Genetic Analysis of Determinants Defining Synthesis of 2,4-Diacetylphloroglucinol by Pseudomonas Brassicacearum BIM B-446 Bacteria. *Appl. Biochem. Microbiol.* **2017**, *53*, 31–39. [CrossRef]
4. Thomashow, L.; Weller, D. Current Concepts in the Use of Introduced Bacteria for Biological Disease Control: Mechanisms and Antifungal Metabolites. In *Plant-Microbe Interactions;* Springer US: Boston, MA, USA, 1996; pp. 187–235.
5. van Loon, L.C.; Bakker, P.A.H.M.; Pieterse, C.M.J. Systemic Resistance Induced By Rhizosphere Bacteria. *Annu. Rev. Phytopathol.* **1998**, *36*, 453–483. [CrossRef]
6. Almario, J.; Moënne-Loccoz, Y.; Muller, D. Monitoring of the Relation between 2,4-Diacetylphloroglucinol-Producing Pseudomonas and Thielaviopsis Basicola Populations by Real-Time PCR in Tobacco Black Root-Rot Suppressive and Conducive Soils. *Soil Biol. Biochem.* **2013**, *57*, 144–155. [CrossRef]
7. Haichar, F.; Fochesato, S.; Achouak, W. Host Plant Specific Control of 2,4-Diacetylphloroglucinol Production in the Rhizosphere. *Agronomy* **2013**, *3*, 621–631. [CrossRef]
8. Zhou, T.-T.; Li, C.-Y.; Chen, D.; Wu, K.; Shen, Q.-R.; Shen, B. PhlF− Mutant of Pseudomonas Fluorescens J2 Improved 2,4-DAPG Biosynthesis and Biocontrol Efficacy against Tomato Bacterial Wilt. *Biol. Control* **2014**, *78*, 1–8. [CrossRef]
9. Bonsall, R.F.; Weller, D.M.; Thomashow, L.S. Quantification of 2,4-Diacetylphloroglucinol Produced by Fluorescent Pseudomonas Spp. In Vitro and in the Rhizosphere of Wheat. *Appl. Environ. Microbiol.* **1997**, *63*, 951–955. [CrossRef]
10. Duffy, B.K.; Défago, G. Zinc Improves Biocontrol of Fusarium Crown and Root Rot of Tomato by *Pseudomonas fluorescens* and Represses the Production of Pathogen Metabolites Inhibitory to Bacterial Antibiotic Biosynthesis. *Phytopathology* **1997**, *87*, 1250–1257. [CrossRef]
11. Raaijmakers, J.M.; Weller, D.M. Natural Plant Protection by 2,4-Diacetylphloroglucinol-Producing *Pseudomonas* Spp. in Take-All Decline Soils. *Mol. Plant-Microbe Interact.* **1998**, *11*, 144–152. [CrossRef]
12. Sulaiman, M.; Jannat, K.; Nissapatorn, V.; Rahmatullah, M.; Paul, A.K.; de Lourdes Pereira, M.; Rajagopal, M.; Suleiman, M.; Butler, M.S.; bin Break, M.K.; et al. Antibacterial and Antifungal Alkaloids from Asian Angiosperms: Distribution, Mechanisms of Action, Structure-Activity, and Clinical Potentials. *Antibiotics* **2022**, *11*, 1146. [CrossRef]
13. Hansanant, N.; Smith, L. Occidiofungin: Actin Binding as a Novel Mechanism of Action in an Antifungal Agent. *Antibiotics* **2022**, *11*, 1143. [CrossRef]
14. Delany, I.; Sheehan, M.M.; Fenton, A.; Bardin, S.; Aarons, S.; O'Gara, F. Regulation of Production of the Antifungal Metabolite 2,4-Diacetylphloroglucinol in Pseudomonas Fluorescens F113: Genetic Analysis of PhlF as a Transcriptional Repressor The GenBank Accession Number for the Sequence Reported in This Paper Is AF129856. *Microbiology* **2000**, *146*, 537–546. [CrossRef] [PubMed]
15. Keel, C. Suppression of Root Diseases by *Pseudomonas fluorescens* CHA0: Importance of the Bacterial Secondary Metabolite 2,4-Diacetylphloroglucinol. *Mol. Plant-Microbe Interact.* **1992**, *5*, 4. [CrossRef]
16. Mavrodi, O.V.; McSpadden Gardener, B.B.; Mavrodi, D.V.; Bonsall, R.F.; Weller, D.M.; Thomashow, L.S. Genetic Diversity of *PhlD* from 2,4-Diacetylphloroglucinol-Producing Fluorescent *Pseudomonas* spp. *Phytopathology* **2001**, *91*, 35–43. [CrossRef] [PubMed]
17. Voisard, C.; Bull, C.; Keel, C.; Laville, J.; Maurhofer, M.; Schnider, U.; Dfago, G.; Haas, D. Biocontrol of Root Diseases ByPseudomonas Fluorescens CHA0: Current Concepts and Experimental Approaches. In *Molecular Ecology of Rhizosphere Microorganisms;* Wiley-VCH Verlag GmbH: Weinheim, Germany, 1994; pp. 67–89.
18. Whistler, C.A.; Corbell, N.A.; Sarniguet, A.; Ream, W.; Loper, J.E. The Two-Component Regulators GacS and GacA Influence Accumulation of the Stationary-Phase Sigma Factor ς^{S} and the Stress Response in *Pseudomonas Fluorescens* Pf-5. *J. Bacteriol.* **1998**, *180*, 6635–6641. [CrossRef] [PubMed]
19. Yuan, Z.; Cang, S.; Matsufuji, M.; Nakata, K.; Nagamatsu, Y.; Yoshimoto, A. High Production of Pyoluteorin and 2,4-Diacetylphloroglucinol by Pseudomonas Fluorescens S272 Grown on Ethanol as a Sole Carbon Source. *J. Ferment. Bioeng.* **1998**, *86*, 559–563. [CrossRef]
20. Duffy, B.K.; Défago, G. Environmental Factors Modulating Antibiotic and Siderophore Biosynthesis by *Pseudomonas Fluorescens* Biocontrol Strains. *Appl. Environ. Microbiol.* **1999**, *65*, 2429–2438. [CrossRef]
21. Shanahan, P.; O'Sullivan, D.J.; Simpson, P.; Glennon, J.D.; O'Gara, F. Isolation of 2,4-Diacetylphloroglucinol from a Fluorescent Pseudomonad and Investigation of Physiological Parameters Influencing Its Production. *Appl. Environ. Microbiol.* **1992**, *58*, 353–358. [CrossRef]
22. Biessy, A.; Filion, M. Phloroglucinol Derivatives in Plant-Beneficial Pseudomonas spp.: Biosynthesis, Regulation, and Functions. *Metabolites* **2021**, *11*, 182. [CrossRef]
23. Shimizu, Y.; Ogata, H.; Goto, S. Type III Polyketide Synthases: Functional Classification and Phylogenomics. *ChemBioChem* **2017**, *18*, 50–65. [CrossRef]
24. Achkar, J.; Xian, M.; Zhao, H.; Frost, J.W. Biosynthesis of Phloroglucinol. *J. Am. Chem. Soc.* **2005**, *127*, 5332–5333. [CrossRef]

25. Zha, W.; Rubin-Pitel, S.B.; Zhao, H. Characterization of the Substrate Specificity of PhlD, a Type III Polyketide Synthase from Pseudomonas Fluorescens. *J. Biol. Chem.* **2006**, *281*, 32036–32047. [CrossRef] [PubMed]
26. Gupta, P.; Rai, R.; Dash, P.K. Isolation, Cloning and Characterization of PhlA Gene from an Indigenous Pseudomonas Strain from Indian Soil. *Int. J. Trop. Agric.* **2015**, *33*, 3195–3200.
27. Dash, P.; Gupta, P.; Panwar, B.S.; Rai, R. Isolation, Cloning and Characterization of PhlB Gene from an Indian Strain of Gram Negative Soil Bacteria Pseudomonas Fluorescens. *Indian J. Exp. Biol.* **2020**, *58*, 412–419.
28. Gupta, P.; Karthik, K.; Sreevathsa, R.; Rai, R.; Dash, P.K. Cloning and Characterization of Phloroglucinol Biosynthetic Gene PhlC from An Indian Strain of Pseudomonas Fluorescens. *Indian J. Exp. Biol.* **2022**, *60*, 607–614. [CrossRef]
29. Rai, R.; Srinivasamurthy, R.; Dash, P.K.; Gupta, P. Isolation, Characterization and Evaluation of the Biocontrol Potential of Pseudomonas Protegens RS-9 against Ralstonia Solanacearum in Tomato. *Indian J. Exp. Biol.* **2017**, *55*, 595–603.
30. Katoh, K.; Standley, D.M. MAFFT Multiple Sequence Alignment Software Version 7: Improvements in Performance and Usability. *Mol. Biol. Evol.* **2013**, *30*, 772–780. [CrossRef]
31. Robert, X.; Gouet, P. Deciphering Key Features in Protein Structures with the New ENDscript Server. *Nucleic Acids Res.* **2014**, *42*, W320–W324. [CrossRef]
32. Jones, P.; Binns, D.; Chang, H.-Y.; Fraser, M.; Li, W.; McAnulla, C.; McWilliam, H.; Maslen, J.; Mitchell, A.; Nuka, G.; et al. InterProScan 5: Genome-Scale Protein Function Classification. *Bioinformatics* **2014**, *30*, 1236–1240. [CrossRef]
33. Sievers, F.; Wilm, A.; Dineen, D.; Gibson, T.J.; Karplus, K.; Li, W.; Lopez, R.; McWilliam, H.; Remmert, M.; Söding, J.; et al. Fast, Scalable Generation of High-Quality Protein Multiple Sequence Alignments Using Clustal Omega. *Mol. Syst. Biol.* **2011**, *7*, 539. [CrossRef]
34. Kumar, S.; Stecher, G.; Li, M.; Knyaz, C.; Tamura, K. MEGA X: Molecular Evolutionary Genetics Analysis across Computing Platforms. *Mol. Biol. Evol.* **2018**, *35*, 1547–1549. [CrossRef]
35. Guindon, S.; Dufayard, J.F.; Lefort, V.; Anisimova, M.; Hordijk, W.; Gascuel, O. New Algorithms and Methods to Estimate Maximum-Likelihood Phylogenies: Assessing the Performance of PhyML 3.0. *Syst. Biol.* **2010**, *59*, 307–321. [CrossRef] [PubMed]
36. Zhang, Y. I-TASSER: Fully Automated Protein Structure Prediction in CASP8. *Proteins Struct. Funct. Bioinform.* **2009**, *77*, 100–113. [CrossRef] [PubMed]
37. Lovell, S.C.; Davis, I.W.; Arendall, W.B.; de Bakker, P.I.W.; Word, J.M.; Prisant, M.G.; Richardson, J.S.; Richardson, D.C. Structure Validation by Cα Geometry: ϕ,ψ and Cβ Deviation. *Proteins Struct. Funct. Bioinform.* **2003**, *50*, 437–450. [CrossRef] [PubMed]
38. Fiser, A.; Sali, A. ModLoop: Automated Modeling of Loops in Protein Structures. *Bioinformatics* **2003**, *19*, 2500–2501. [CrossRef] [PubMed]
39. Laskowski, R.A. PDBsum More: New Summaries and Analyses of the Known 3D Structures of Proteins and Nucleic Acids. *Nucleic Acids Res.* **2004**, *33*, D266–D268. [CrossRef] [PubMed]
40. Laskowski, R.A.; MacArthur, M.W.; Moss, D.S.; Thornton, J.M. PROCHECK: A Program to Check the Stereochemical Quality of Protein Structures. *J. Appl. Crystallogr.* **1993**, *26*, 283–291. [CrossRef]
41. Castrignano, T. The PMDB Protein Model Database. *Nucleic Acids Res.* **2006**, *34*, D306–D309. [CrossRef]
42. Gokulan, K.; O'Leary, S.E.; Russell, W.K.; Russell, D.H.; Lalgondar, M.; Begley, T.P.; Ioerger, T.R.; Sacchettini, J.C. Crystal Structure of Mycobacterium Tuberculosis Polyketide Synthase 11 (PKS11) Reveals Intermediates in the Synthesis of Methyl-Branched Alkylpyrones. *J. Biol. Chem.* **2013**, *288*, 16484–16494. [CrossRef]
43. Richardson, J.S. The Anatomy and Taxonomy of Protein Structure. *Adv. Protein Chem.* **1981**, *34*, 167–339.
44. Sibanda, B.L.; Blundell, T.L.; Thornton, J.M. Conformation of β-Hairpins in Protein Structures. *J. Mol. Biol.* **1989**, *206*, 759–777. [CrossRef]
45. Hutchinson, C.R.; Fujii, I. POLYKETIDE SYNTHASE GENE MANIPULATION: A Structure-Function Approach in Engineering Novel Antibiotics. *Annu. Rev. Microbiol.* **1995**, *49*, 201–238. [CrossRef] [PubMed]
46. Cook, R.J.; Thomashow, L.S.; Weller, D.M.; Fujimoto, D.; Mazzola, M.; Bangera, G.; Kim, D.S. Molecular Mechanisms of Defense by Rhizobacteria against Root Disease. *Proc. Natl. Acad. Sci. USA* **1995**, *92*, 4197–4201. [CrossRef] [PubMed]
47. Fenton, A.M.; Stephens, P.M.; Crowley, J.; O'Callaghan, M.; O'Gara, F. Exploitation of Gene(s) Involved in 2,4-Diacetylphloroglucinol Biosynthesis to Confer a New Biocontrol Capability to a Pseudomonas Strain. *Appl. Environ. Microbiol.* **1992**, *58*, 3873–3878. [CrossRef] [PubMed]
48. Sharifi-Tehrani, A.; Zala, M.; Natsch, A.; Moënne-Loccoz, Y.; Défago, G. Biocontrol of Soil-Borne Fungal Plant Diseases by 2,4-Diacetylphloroglucinol-Producing Fluorescent Pseudomonads with Different Restriction Profiles of Amplified 16S RDNA. *Eur. J. Plant Pathol.* **1998**, *104*, 631–643. [CrossRef]
49. Stutz, E.W. Naturally Occurring Fluorescent Pseudomonads Involved in Suppression of Black Root Rot of Tobacco. *Phytopathology* **1986**, *76*, 181. [CrossRef]
50. Tamietti, G.; Ferraris, L.; Matta, A.; Abbattista Gentile, I. Physiological Responses of Tomato Plants Grown in Fusarium Suppressive Soil. *J. Phytopathol.* **1993**, *138*, 66–76. [CrossRef]
51. Suresh, P.; Rekha, M.; Gomathinayagam, S.; Ramamoorthy, V.; Sharma, M.P.; Sakthivel, P.; Sekar, K.; Valan Arasu, M.; Shanmugaiah, V. Characterization and Assessment of 2, 4-Diacetylphloroglucinol (DAPG)-Producing Pseudomonas Fluorescens VSMKU3054 for the Management of Tomato Bacterial Wilt Caused by Ralstonia Solanacearum. *Microorganisms* **2022**, *10*, 1508. [CrossRef]

52. Dash, P.K.; Cao, Y.; Jailani, A.K.; Gupta, P.; Venglat, P.; Xiang, D.; Rai, R.; Sharma, R.; Thirunavukkarasu, N.; Abdin, M.Z.; et al. Genome-Wide Analysis of Drought Induced Gene Expression Changes in Flax (*Linum usitatissimum*). *GM Crops Food* **2014**, *5*, 106–119. [CrossRef] [PubMed]
53. Wang, Z.; Hobson, N.; Galindo, L.; Zhu, S.; Shi, D.; McDill, J.; Yang, L.; Hawkins, S.; Neutelings, G.; Datla, R.; et al. The Genome of Flax (*Linum Usitatissimum*) Assembled *de Novo* from Short Shotgun Sequence Reads. *Plant J.* **2012**, *72*, 461–473. [CrossRef]
54. Dash, P.K.; Rai, R.; Mahato, A.K.; Gaikwad, K.; Singh, N.K. Transcriptome Landscape at Different Developmental Stages of a Drought Tolerant Cultivar of Flax (Linum Usitatissimum). *Front. Chem.* **2017**, *5*, 82. [CrossRef] [PubMed]
55. Dash, P.K.; Gupta, P.; Pradhan, S.K.; Shasany, A.K.; Rai, R. Analysis of Homologous Regions of Small RNAs MIR397 and MIR408 Reveals the Conservation of Microsynteny among Rice Crop-Wild Relatives. *Cells* **2022**, *11*, 3461. [CrossRef] [PubMed]
56. Shivakumara, T.N.; Sreevathsa, R.; Dash, P.K.; Sheshshayee, M.S.; Papolu, P.K.; Rao, U.; Tuteja, N.; UdayaKumar, M. Overexpression of Pea DNA Helicase 45 (PDH45) Imparts Tolerance to Multiple Abiotic Stresses in Chili (*Capsicum annuum* L.). *Sci. Rep.* **2017**, *7*, 2760. [CrossRef] [PubMed]
57. Kesiraju, K.; Tyagi, S.; Mukherjee, S.; Rai, R.; Singh, N.K.; Sreevathsa, R.; Dash, P.K. An Apical Meristem-Targeted in Planta Transformation Method for the Development of Transgenics in Flax (Linum Usitatissimum): Optimization and Validation. *Front. Plant Sci.* **2021**, *11*, 562056. [CrossRef]
58. Gupta, P.; Dash, P.K. Precise Method of in Situ Drought Stress Induction in Flax (*Linum usitatissimum*) for RNA Isolation towards down-Stream Analysis. *Ann. Agric. Res.* **2015**, *36*, 10–17.
59. Bastia, R.; Pandit, E.; Sanghamitra, P.; Barik, S.R.; Nayak, D.K.; Sahoo, A.; Moharana, A.; Meher, J.; Dash, P.K.; Raj, R.; et al. Association Mapping for Quantitative Trait Loci Controlling Superoxide Dismutase, Flavonoids, Anthocyanins, Carotenoids, γ-Oryzanol and Antioxidant Activity in Rice. *Agronomy* **2022**, *12*, 3036. [CrossRef]
60. Pradhan, K.C.; Pandit, E.; Mohanty, S.P.; Moharana, A.; Sanghamitra, P.; Meher, J.; Jena, B.K.; Dash, P.K.; Behera, L.; Mohapatra, P.M.; et al. Development of Broad Spectrum and Durable Bacterial Blight Resistant Variety through Pyramiding of Four Resistance Genes in Rice. *Agronomy* **2022**, *12*, 1903. [CrossRef]
61. Sedeek, K.E.M.; Mahas, A.; Mahfouz, M. Plant Genome Engineering for Targeted Improvement of Crop Traits. *Front. Plant Sci.* **2019**, *10*, 114. [CrossRef]
62. Pradhan, S.K.; Pandit, E.; Pawar, S.; Bharati, B.; Chatopadhyay, K.; Singh, S.; Dash, P.; Reddy, J.N. Association Mapping Reveals Multiple QTLs for Grain Protein Content in Rice Useful for Biofortification. *Mol. Genet. Genom.* **2019**, *294*, 963–983. [CrossRef]
63. Yang, F.; Cao, Y. Biosynthesis of Phloroglucinol Compounds in Microorganisms—Review. *Appl. Microbiol. Biotechnol.* **2012**, *93*, 487–495. [CrossRef]
64. Rao, G.; Lee, J.-K.; Zhao, H. Directed Evolution of Phloroglucinol Synthase PhlD with Increased Stability for Phloroglucinol Production. *Appl. Microbiol. Biotechnol.* **2013**, *97*, 5861–5867. [CrossRef]
65. Yu, S.; Guo, L.; Zhao, L.; Chen, Z.; Huo, Y. Metabolic Engineering of E. Coli for Producing Phloroglucinol from Acetate. *Appl. Microbiol. Biotechnol.* **2020**, *104*, 7787–7799. [CrossRef] [PubMed]
66. Raaijmakers, J.M.; Weller, D.M.; Thomashow, L.S. Frequency of Antibiotic-Producing Pseudomonas Spp. in Natural Environments. *Appl. Environ. Microbiol.* **1997**, *63*, 881–887. [CrossRef] [PubMed]
67. Picard, C.; di Cello, F.; Ventura, M.; Fani, R.; Guckert, A. Frequency and Biodiversity of 2,4-Diacetylphloroglucinol-Producing Bacteria Isolated from the Maize Rhizosphere at Different Stages of Plant Growth. *Appl. Environ. Microbiol.* **2000**, *66*, 948–955. [CrossRef] [PubMed]
68. Donadio, S.; Staver, M.; McAlpine, J.; Swanson, S.; Katz, L. Modular Organization of Genes Required for Complex Polyketide Biosynthesis. *Science (1979)* **1991**, *252*, 675–679. [CrossRef]
69. Katz, L.; Donadio, S. POLYKETIDE SYNTHESIS: Prospects for Hybrid Antibiotics. *Annu. Rev. Microbiol.* **1993**, *47*, 875–912. [CrossRef] [PubMed]
70. Nowak-Thompson, B.; Gould, S.J.; Loper, J.E. Identification and Sequence Analysis of the Genes Encoding a Polyketide Synthase Required for Pyoluteorin Biosynthesis in Pseudomonas Fluorescens Pf-5. *Gene* **1997**, *204*, 17–24. [CrossRef]
71. McDaniel, R.; Ebert-Khosla, S.; Hopwood, D.A.; Khosla, C. Engineered Biosynthesis of Novel Polyketides. *Science (1979)* **1993**, *262*, 1546–1550. [CrossRef]
72. Bangera, M.G.; Weller, D.M.; Thomshaw, L.S. Genetic Analysis of the 2,4-Diacetylphloroglucinol Biosynthetic Locus from Pseudomonas Fluorescens Q2-87. In *Advances in Molecular Genetics of Plant-Microbe Interactions*; Daniels, M., Downie, J., Osbourn, A., Eds.; Kluwer Academic Publishers: Dordrecht, The Netherlands, 1994.
73. Bangera, M.G.; Thomashow, L.S. Identification and Characterization of a Gene Cluster for Synthesis of the Polyketide Antibiotic 2,4-Diacetylphloroglucinol from *Pseudomonas fluorescens* Q2-87. *J. Bacteriol.* **1999**, *181*, 3155–3163. [CrossRef]
74. Ramette, A.; Moënne-Loccoz, Y.; Défago, G. Polymorphism of the Polyketide Synthase Gene *PhlD* in Biocontrol Fluorescent Pseudomonads Producing 2,4-Diacetylphloroglucinol and Comparison of PhlD with Plant Polyketide Synthases. *Mol. Plant-Microbe Interact.* **2001**, *14*, 639–652. [CrossRef]
75. Tropf, S.; Lanz, T.; Rensing, S.A.; Schrder, J.; Schrder, G. Evidence That Stilbene Synthases Have Developed from Chalcone Synthases Several Times in the Course of Evolution. *J. Mol. Evol.* **1994**, *38*, 610–618. [CrossRef]
76. de Luca, D.; Lauritano, C. In Silico Identification of Type III PKS Chalcone and Stilbene Synthase Homologs in Marine Photosynthetic Organisms. *Biology* **2020**, *9*, 110. [CrossRef] [PubMed]

77. Höfte, M. *The Use of Pseudomonas spp. as Bacterial Biocontrol Agents to Control Plant Disease*; Burleigh Dodds: Cambridge, UK, 2021.
78. Panpatte, D.G.; Jhala, Y.K.; Shelat, H.N.; Vyas, R.V. Pseudomonas Fluorescens: A Promising Biocontrol Agent and PGPR for Sustainable Agriculture. In *Microbial Inoculants in Sustainable Agricultural Productivity*; Springer India: New Delhi, India, 2016; pp. 257–270.

antibiotics

Article

Klebsiella pneumonia in Sudan: Multidrug Resistance, Polyclonal Dissemination, and Virulence

Einas A. Osman [1], Maho Yokoyama [2], Hisham N. Altayb [3], Daire Cantillon [4], Julia Wille [5,6], Harald Seifert [5,6], Paul G. Higgins [5,6,7] and Leena Al-Hassan [2,*]

1 Bioscience Research Institute, Ibn Sina University, Khartoum 11111, Sudan
2 Department of Global Health and Infection, Brighton & Sussex Medical School, Brighton BN1 9PX, UK
3 Department of Biochemistry, Faculty of Sciences, King Abdulaziz University, Jeddah 21589, Saudi Arabia
4 Department of Tropical Disease Biology, Liverpool School of Tropical Medicine, Liverpool L3 5QA, UK
5 Institute for Medical Microbiology, Immunology and Hygiene, Faculty of Medicine and University Hospital Cologne, University of Cologne, 50935 Cologne, Germany
6 German Center for Infection Research (DZIF), Partner Site Bonn-Cologne, 50935 Cologne, Germany
7 Center for Molecular Medicine Cologne, Faculty of Medicine and University Hospital Cologne, University of Cologne, 50931 Cologne, Germany
* Correspondence: l.al-hassan@bsms.ac.uk; Tel.: +44-(0)-1278877817

Abstract: The emergence and global expansion of hyper-virulent and multidrug resistant (MDR) *Klebsiella pneumoniae* is an increasing healthcare threat worldwide. The epidemiology of MDR *K. pneumoniae* is under-characterized in many parts of the world, particularly Africa. In this study, *K. pneumoniae* isolates from hospitals in Khartoum, Sudan, have been whole-genome sequenced to investigate their molecular epidemiology, virulence, and resistome profiles. Eighty-six *K. pneumoniae* were recovered from patients in five hospitals in Khartoum between 2016 and 2020. Antimicrobial susceptibility was performed by disk-diffusion and broth microdilution. All isolates underwent whole genome sequencing using Illumina MiSeq; cgMLST was determined using Ridom SeqSphere+, and 7-loci MLST virulence genes and resistomes were identified. MDR was observed at 80%, with 35 isolates (41%) confirmed carbapenem-resistant. Thirty-seven sequence types were identified, and 14 transmission clusters (TC). Five of these TCs involved more than one hospital. *Ybt9* was the most common virulence gene detected, in addition to some isolates harbouring *iuc* and *rmp1*. There is a diverse population of *K. pneumoniae* in Khartoum hospitals, harbouring multiple resistance genes, including genes coding for ESBLs, carbapenemases, and aminoglycoside-modifying enzymes, across multiple ST's. The majority of isolates were singletons and transmissions were rare.

Keywords: *Klebsiella pneumoniae*; multidrug resistance; transmission; Sudan

Citation: Osman, E.A.; Yokoyama, M.; Altayb, H.N.; Cantillon, D.; Wille, J.; Seifert, H.; Higgins, P.G.; Al-Hassan, L. *Klebsiella pneumonia* in Sudan: Multidrug Resistance, Polyclonal Dissemination, and Virulence. *Antibiotics* **2023**, *12*, 233. https://doi.org/10.3390/antibiotics12020233

Academic Editor: Norbert Solymosi

Received: 26 December 2022
Revised: 13 January 2023
Accepted: 18 January 2023
Published: 21 January 2023

1. Introduction

Klebsiella pneumoniae is an important global pathogen causing a variety of infections in community and healthcare-associated infections, such as pneumonia, urinary tract infections (UTI), and bloodstream infections. It poses a serious threat to human health and is one of the six highly virulent and antibiotic resistant bacterial pathogens: *Enterococcus faecium*, *Staphylococcus aureus*, *Klebsiella pneumoniae*, *Acinetobacter baumannii*, *Pseudomonas aeruginosa*, and *Enterobacter* spp. (ESKAPE), requiring urgent global attention [1]. Multidrug resistance and carbapenem resistance in *K. pneumoniae* (MDR-Kp and CR-Kp) is of particular concern as treatment options are limited.

The population structure of *K. pneumoniae* appears to be diverse yet highly structured, with distinct clonal groups (CG) subdivided into MDR- and hypervirulent (Hv)-clones [2]. A subset of these clones contribute to global diseases and outbreaks and are referred to as 'global problem clones', involving the transfer and spread of antibiotic resistance genes and endemic plasmids in highly disseminating successful clones worldwide [3].

Antimicrobial resistance (AMR) is a global health issue, with low- and middle-income countries (LMICs) carrying the largest burden of infection and AMR, both in the community and healthcare settings [4,5]. Global efforts such as the World Health Assembly's Global Action Plan on AMR (GAP-AMR), which was followed by the adoption of National Action Plans (NAP)s by member countries, and the World Health Organisation (WHO) Global Antimicrobial Resistance and Use Surveillance System (GLASS) are all strategies to contain AMR [6]. However, lack of systematic surveillance and scarcity of data on the burden of infectious diseases and AMR is largely due to limited infrastructure and financial support.

Sudan is a lower middle-income country in Africa, with a population of approximately 44 million (in 2021) covering an area of 1.8 million km^2. Median age is 19.7, and life expectancy at birth is 64.1 years [7]. Sudan's health care system is fragile, hospital-centric, and fragmented in both the public and private sectors, and facing several challenges. The WHO data show that health-associated out-of-pocket expenditure paid by households are ~74%, which cover curative care, medicines, and medical consumables [7]. There is disparity in the distribution of health care personnel between the public and private sectors and between urban and rural areas. Moreover, the high turnover and migration of health professionals continue to threaten the capacity to respond to the increased demand for health services [7]. An analysis of the AMR situation by the Sudanese NAP revealed high rates of resistance, and an urgent need to address the situation by improving awareness, surveillance, hygiene, and infection control, and the optimization of antimicrobial medicines (https://www.who.int/publications/m/item/sudan-national-action-plan-on-antimicrobial-resistance; last accessed on 7 January 2023). However, the implementation of antimicrobial stewardship and improving the access to antimicrobials is difficult due to the lack of supervisory systems to ensure rational prescribing, combined with poor accessibility of essential antibiotics in rural areas.

MDR *K. pneumoniae* is on the WHO global priority pathogen list [8], but despite global efforts there are significant gaps in robust epidemiological data, particularly in LMICs. Recognising the significance of MDR- and CR-Kp globally and the need for accurate epidemiological typing of the organism, it is important to gain insight into circulating clones in order to contextualise data on national and international lineages, and to inform locally-relevant infection prevention and control based on local data. In Sudan, *K. pneumoniae* is highly prevalent in hospitals and has been reported in numerous studies as the most common Gram-negative organism identified [9–11]. *K. pneumoniae* was the second most frequent organism contributing to urinary tract infection (UTI) in diabetic patients [12]. In a study on the presence of pathogenic bacteria isolated from bank notes in Khartoum, *K. pneumoniae* was the most prevalent organism identified [13]. Despite this prevalence, few studies have included molecular methods for isolate identification, confirmation, or resistance surveillance. We have previously conducted a study on the local epidemiology of CR-Kp in Khartoum by using MLST [14]. In this study we used whole genome sequencing (WGS) to generate more robust data on the molecular epidemiology, transmission, resistome, and virulence profiles of MDR *K. pneumoniae* from multiple hospitals in Khartoum, Sudan.

2. Results

A total of 86 isolates were included in this study from five different hospitals in the Sudanese capital of Khartoum: 2 isolates from 2016, 24 from 2018–2019, and 60 from March–September 2020 (detailed sampling procedure in the Methods). Eighty-four *K. pneumoniae* isolates were confirmed by MALDI-TOF and Kleborate, while 2 isolates (LH_F129 and LH_F72) were identified as *Klebsiella quasipneumoniae* subsp. *quasipneumoniae* and *Klebsiella variicola* subsp. *variicola*, respectively. The organisms came from a variety of samples: blood (n = 25, 29.4%), pus (n = 2, 2.4%), sputum (n = 9, 10.6%), urine (n = 37, 43.5%), and wound swabs (n = 12, 14.1%) (Figure 1; Table 1).

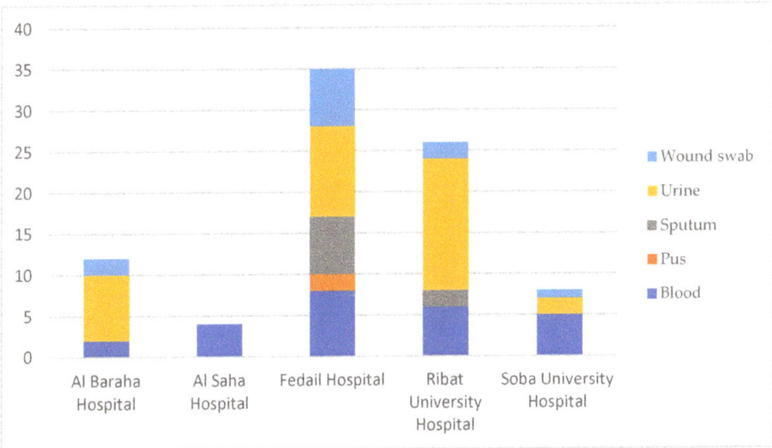

Figure 1. Details of isolate sources and hospitals.

Table 1. Isolate details: date and type of sample, and hospital.

Sample Name	Collection Date	Hospital	Isolation Source	Additional Information
HishK1	2016	RUH	Urine	
HishK2	2016	RUH	Blood	NICU
HishK10	2018	SUH	Blood	NICU
HishK3	2018	SUH	Urine	
HishK4	2018	SUH	Urine	
HishK5	2018	SUH	Blood	NICU
HishK7	2018	SUH	Blood	NICU
HishK8	2018	SUH	Blood	NICU
HishK9	2018	SUH	Wound swab	
K10	2018	ABH	Wound swab	
HishK11	2019	ASH	Blood	
HishK12	2019	ASH	Blood	
HishK13	2019	ASH	Blood	
HishK14	2019	ASH	Blood	
K11	2019	ABH	Urine	
K12	2019	ABH	Urine	
K13	2019	ABH	Wound swab	
K15	2019	ABH	Urine	
K2	2019	ABH	Urine	
K3	2019	ABH	Blood	NICU
K4	2019	ABH	Urine	
K5	2019	ABH	Urine	
K6	2019	ABH	Blood	
K8	2019	ABH	Urine	
K9	2019	ABH	Urine	
LH_R146	08 February 2020	RUH	Urine	
LH_S25	23 February 2020	SUH	Blood	
LH_R100	26 February 2020	RUH	Sputum	ICU
LH_R92	28 February 2020	RUH	Urine	
LH_R107	01 March 2020	RUH	Urine	NICU
LH_R120	03 March 2020	RUH	Urine	
LH_R154	11 March 2020	RUH	Urine	Outpatient

Table 1. *Cont.*

Sample Name	Collection Date	Hospital	Isolation Source	Additional Information
LH_R167	18 March 2020	RUH	Blood	NICU
LH_R182	19 March 2020	RUH	Blood	Inpatient
LH_R195	22 March 2020	RUH	Blood	ICU
LH_R164	23 March 2020	RUH	Sputum	Inpatient
LH_R174	23 March 2020	RUH	Wound swab	Inpatient
LH_R162	23 March 2020	RUH	Urine	Outpatient
LH_R208	26 March 2020	RUH	Urine	Outpatient
LH_R223	28 March 2020	RUH	Wound swab	Outpatient
LH_R219	28 March 2020	RUH	Urine	Outpatient
LH_R275	12 July 2020	RUH	Urine	Inpatient
LH_R289	18 July 2020	RUH	Blood	Outpatient
LH_R290	27 July 2020	RUH	Urine	Urology unit
LH_R314	30 July 2020	RUH	Blood	
LH_R313	06 August 2020	RUH	Urine	
LH_R323	08 August 2020	RUH	Urine	
LH_R344	12 August 2020	RUH	Urine	
LH_F2	16 August 2020	FH	Urine	
LH_F15	16 August 2020	FH	Sputum	
LH_F18	17 August 2020	FH	Sputum	
LH_F25	17 August 2020	FH	Blood	
LH_F50_1	18 August 2020	FH	Urine	
LH_F35	18 August 2020	FH	Urine	
LH_F68	20 August 2020	FH	Wound swab	
LH_F64	20 August 2020	FH	Wound swab	
LH_F66	20 August 2020	FH	Urine	
LH_F82	21 August 2020	FH	Pus	
LH_F86	21 August 2020	FH	Urine	
LH_F97	22 August 2020	FH	Urine	
LH_F101	22 August 2020	FH	Urine	
LH_R384	22 August 2020	RUH	Urine	
LH_F104	23 August 2020	FH	Urine	
LH_F102	23 August 2020	FH	Blood	
LH_F122	25 August 2020	FH	Blood	
LH_F126	25 August 2020	FH	Wound swab	
LH_F139	26 August 2020	FH	Wound swab	
LH_F134	26 August 2020	FH	Blood	
LH_F137	27 August 2020	FH	Sputum	
LH_R387	27 August 2020	RUH	Urine	
LH_F143	27 August 2020	FH	Blood	
LH_F146	28 August 2020	FH	Urine	
LH_F149	29 August 2020	FH	Pus	
LH_F164	01 September 2020	FH	Wound swab	
LH_F158	01 September 2020	FH	Blood	
LH_F159	01 September 2020	FH	Blood	
LH_F169	02 September 2020	FH	Blood	
LH_F174	02 September 2020	FH	Wound swab	
LH_F175	03 September 2020	FH	Wound swab	
LH_F176	03 September 2020	FH	Sputum	
LH_F190	05 September 2020	FH	Sputum	
LH_F192	05 September 2020	FH	Sputum	
LH_F281	20 September 2020	FH	Urine	

RUH: Ribat University Hospital, SUH: Soba University Hospital, ASH: Al-Saha Hospital, FH: Fedail Hospital, ABH: Al-Baraha Hospital, NICU: neonatal intensive care unit.

Most isolates came from Fedail Hospital, followed by Ribat University Hospital, both of which are two of the largest hospitals in Khartoum. Urine and blood samples were most common across all isolates.

Sixty-eight isolates (80%) were MDR (resistant to >3 antibiotic classes), 35 of which (51%) were carbapenem resistant (imipenem andor meropenem MIC $\geq$ 4 mgL). The remaining 16 isolates were susceptible to all tested antibiotics. All isolates were colistin susceptible (MIC < 2 mgL). Multiple resistance mechanisms were identified (Table 2 and Supplementary Table S1), contributing to the observed MDR phenotypes. Aminoglycoside resistance was present in 49 isolates (57%), mediated by multiple aminoglycoside-modifying enzymes. β-lactam resistance was detected in 58 isolates and was mediated by acquired β-lactamases: OXA-1 (*n* = 19), OXA-9 (*n* = 4, two of which were co-harbouring OXA-1 and TEM-1), CMY (*n* = 3, one of which was co-harbouring OXA-1), DHA-1 (*n* = 3, two of which were co-harbouring TEM-1), and SCO-1 (*n* = 3, one of which was co-harbouring TEM-1); TEM-1 was the most prevalent β-lactamase, identified in 40 isolates. The acquired extended-spectrum β-lactamase (ESBL) CTX-M-15 was present in 64 isolates, and CTX-M-14 was present in four isolates, three of which co-harboured CTX-M-15. Seventeen isolates did not harbour any ESBLs. SHV-variants were present in all *K. pneumoniae* isolates.

Table 2. Resistome and Virulence details of STs.

ST	ESBL *bla*	Carb R *bla*	Intrinsic β-Lactamase *bla*	Virulence
101 (6)	*CTX-M-15* (6)	*NDM-1* (1); *ompK* substitution (1)	*SHV-1*	*ybt9* (6)
11 (1)	*CTX-M-15*	*NDM-4*	*SHV-11*	*ybt9*
1198 (1)	-	-	*SHV-11*	
13 (1)	*CTX-M-15*	-	*SHV-1*	
1147 (1)	-	-	*SHV-27*	
147 (9)	*CTX-M-15* (9)	*NDM-1* (5); *NDM-5* (1)	*SHV-11*	*ybt9* (6)
ST15 (6)	*CTX-M-15* (5)	*OmpK35* substitution (1)	*SHV-28*	*ybt9* (1)
ST17 (4)	*CTX-M-15* (2)	*NDM-1*	*SHV-11*	
ST20 (7)	*CTX-M-15* (7)	*NDM-1* (5)	*SHV-187*	
ST218-3LV	-	-	*SHV-93*	
ST219 (2)	*CTX-M-15* (2)	-	*SHV-1*	
ST231 (2)	*CTX-M-15* (1)	*OXA-232* (1)	*SHV-1*	*ybt14* (1); *iuc* (1)
ST237 (2)	-	-	*SHV-11*	
ST24-1LV	-	-	*SHV-11*	
ST2459 (1)	*CTX-M-15*	-	*SHV-1*	
ST2674 (1)	*CTX-M-15*	*NDM-1*	*SHV-11*	
ST2735 (1)	-	-	*SHV-11*	
ST29-1LV (1)	*CTX-M-15*	-	-	
ST292 (1)	*CTX-M-15*	-	*SHV-11*	
ST307 (6)	*CTX-M-15* (6)	*NDM-1* (4)	*SHV-28*	*ybt10* (3)
ST3161 (1)	-	-	*SHV-11*	
ST3430 (1)	*CTX-M-15*	-	*SHV-77*	
ST38 (1)	*CTX-M-15*	*NDM-1*	*SHV-11*	
ST383 (5)	*CTX-M-14* (4);*CTX-M-15* (4)	*NDM-5* (4); *OXA-48* (4)	*SHV-1*	*iuc1* (4); *rmp1* (2)
ST39 (1)	-	-	*SHV-1*	*ybt4*

Table 2. *Cont.*

ST	ESBL *bla*	Carb R *bla*	Intrinsic β-Lactamase *bla*	Virulence
ST437 (7)	*CTX-M-15* (7)	*NDM-1* (1); *NDM-5* (6)	*SHV-11*	*ybt9* (7)
ST45 (1)	*CTX-M-15*	*OmpK36* variant	*SHV-1*	*ybt10*
ST469 (1)	*CTX-M-15*	-	*SHV-11*	
ST474 (1)	*CTX-M-15*	-	*SHV-11*	
ST501 (1)	*CTX-M-15*	-	*SHV-11*	
ST514 (1)	*CTX-M-15*	-	*SHV-63*	
ST530 (2)	*CTX-M-15* (2)	*NDM-1* (2)	-	*ybt10* (2)
ST664 (2)	*CTM-X-15* (2)	*NDM-1* (2)	*SHV-11*	
ST882 (2)	-	-	-	*ybt* (2)
ST901 (1)	*CTX-M-15*	-	*SHV-1*	

ST: sequence type; ESBL: extended-spectrum β lactamase; Carb: carbapanemase; OmpK: outer-membrane porin substitutions contributing to carbapenem resistance; ybt: Yersiniabectin; Numbers in brackets indicate the number of isolates. Table S1 contains details of each isolate.

Carbapenem resistance was mediated by NDM-1 (n = 23), NDM-4 (n = 1), and NDM-5 (n = 11, three of which co-harboured OXA-48); one isolate harboured OXA-48 on its own, and one isolate harboured OXA-232 (OXA-48-like enzyme). Three isolates displayed phenotypic carbapenem resistance, but no acquired carbapenemases were detected, and the observed carbapenem resistance was associated with a combination of an ESBL and modifications in the OmpK35OmpK36 (Table 2 and Table S1). LH_F164 possessed a truncated $bla_{\text{NDM-1}}$ gene but also harboured OXA-1, CTX-M-15, and a nucleotide substitution (G690A) in the OmpK35 encoding gene, leading to a premature stop codon. LH_R313, on the other hand, lacked an *ompK35* gene, and carbapenem resistance was mediated by a modification of OmpK36 containing an LSP insertion in the amino acid sequence at position 184.

A total of 37 different sequence types (STs) were identified in the study, highlighting the epidemiological diversity of the isolates. We observed 14 small transmission clusters (TC) by cgMLST, with isolates differing by 0–5 alleles (Figure 2). TC-1 (ST17) comprises three isolates collected from Al-Saha Hospital (ASH) in 2019; however, they differed in their respective resistomes. Metadata relating to the location of patients or date of isolation are not available for these isolates. LH_HishK11 and LH_HishK13 are both carbapenem-susceptible, while LH_HishK12 is MDR and harbours multiple aminoglycoside-modifying enzymes (*aac(3)-IIa-like;aac(6')-Ib-cr-like;aadA5*) and β-lactamases: $bla_{\text{CTX-M-15}}$, $bla_{\text{OXA-1}}$, and $bla_{\text{NDM-1}}$. All other TCs comprise isolates with identical resistomes (Figure 2 and Table S1).

The most frequently isolated ST in the study was ST147 (n = 9), followed by ST20 (n = 7), ST437 (n = 7), ST101 (n = 6), ST15 (n = 6), ST307 (n = 6), ST383 (n = 5), and ST17 (n = 4), all of which are high-risk global clones (GC), defined as epidemic high-risk clones over-represented globally [2], with the remaining 29 singleton STs. Isolates within the same ST exhibited some similarities at the genomic level, but as seen in Table 2 and Table S1, some had different resistance phenotypes and genes. For example, ST147 isolates came from multiple hospitals, with isolates from FH all collected within a 1-month timeframe (July–August 2020), but they differed by $\geq$21 alleles and transmission was ruled out. Moreover, differences were seen in their virulence and resistome profiles, with LH_F134 harbouring no recognised virulence factors, while LH_F149, LH_F18, and LH_F97 all harboured *ybt9* on a ICEKp3 structure. LH_F134, LH_F149, and LH_F18 were carbapenem resistant, mediated by $bla_{\text{NDM-1}}$; however, LH_F97 was carbapenem susceptible. Similar data are observed in ST-147 isolates from RUH which had identical resistomes, except R344, which harboured $bla_{\text{NDM-5}}$ and not $bla_{\text{NDM-1}}$ and differed by >50 allelic differences per the cgMLST and was therefore not part of a single transmission cluster. The only TC in this group were two identical isolates from Al Baraha Hospital, which also had identical resistomes.

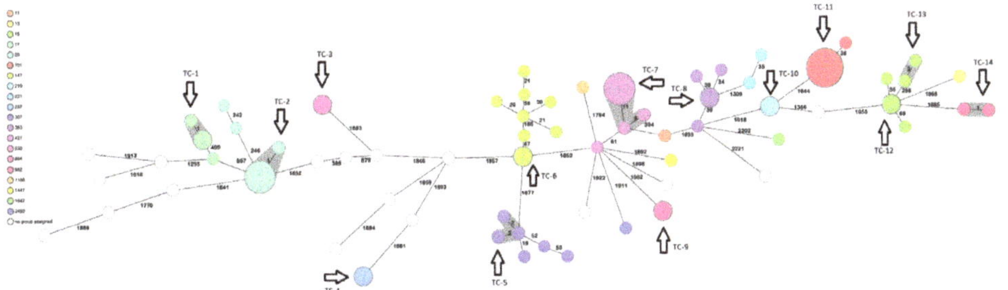

Figure 2. Ridom SeqSphere+ minimum spanning tree (MST) for 84 samples based on 2358 columns, pairwise ignoring missing values, logarithmic scale, *K. pneumoniae* MLST Pasteur (7). Cluster distance threshold: 15. Isolates are grouped by colour, indicating the different STs. Thirty-seven different STs were identified, in addition to 14 transmission clusters, represented by shaded nodes and arrows. Numbers between the nodes indicate the number of allelic differences.

ST20 was also found in multiple hospitals (SUH, ASH, ABH, and RUH), and comprised one TC from two hospitals (TC-2; Figure S1), but there were differences in their resistome. All ST437 were MDR and contained the hypervirulence gene *ybt9* on an ICEKp3 structure. This ST comprised four identical isolates forming a TC (TC-2) with two other isolates, and a singleton. Key resistome differences were observed in aminoglycoside-modifying enzymes and carbapenemases: LH_K4 (collected in 2018 from ABH) had *aac(3)-IId-like;aac(6')-Ib'.v1;aadA;aph(3')-VI* and bla_{NDM-1}, whereas the remaining isolates (collected from FH in 2020) all had *rmtB* and bla_{NDM-5}. The imipenem MIC was also notably less for LH_K4, harbouring bla_{NDM-1} at 64 mgL vs. >128 mgL for the remaining NDM-5 isolates.

ST101 was only present in isolates in 20182019 (SUH, BH; Supplementary Figure S1) and comprised five identical isolates (TC-11) and a singleton. All ST-101 isolates were MDR, but only two were carbapenem-resistant: LH_HishK5 and LH_HishK7, mediated by NDM-1 (LH_HishK5) and modification in *ompK35* (G947A, leading to a premature stop codon at position 316), combined with bla_{OXA-1} and $bla_{CTX-M-15}$ (LH_HishK7).

All ST15 isolates were MDR (and comprised two, TC-12 and TC-13, and two singletons), but were carbapenem-susceptible, except for LH_F164, which was carbapenem-resistant (imipenem MIC 16 mgL), associated with a G690A substitution in the *ompK35* gene (leading to a premature stop codon) combined with bla_{OXA-1}. LH_F164 was the only strain which was also positive for *ybt9* (on ICEK3p) within this ST. Isolates within ST307 came from two different hospitals (RUH and FH), with 56 isolates from RUH. Three of these isolates form a transmission cluster. Interestingly, one isolate, collected 4 months later in July 2020 (LH_R275) from RUH appears to have lost the MDR and CR pheno- and geno-types.

The ST-383 isolates were all MDR, CR, and were comprised of three singletons from Fedail Hospital, and one TC from Soba University Hospital (TC-8). Of these, 35 (LH_HishK8, LH_HishK9, and LH_F190) isolates co-harboured multiple resistance genes: $bla_{CTX-M-14}$ and $bla_{CTX-M-15}$, and bla_{NDM-5} and bla_{OXA-48}. The remaining two isolates (LH-F15 and LH-F35) harboured either $bla_{CTX-M-14}$ and bla_{OXA-48} or $bla_{CTX-M-15}$ and bla_{NDM-5}, respectively. Of the three ST17 isolates, only one was MDRCR and harboured bla_{NDM-1}.

Of the remaining 29 singleton STs, 11 of them were MDR and CR, whereas the remaining 18 were susceptible to all tested antibiotics.

Thirty-six isolates harboured one or more virulence genes, namely *ybt9*, *ybt10*, *ybt4*, *ybt14*, *iuc*, *rmp1*, *KpVP-1*, and *rmpA2* (Table 2 and Table S2). Yersiniabactin *ybt9* located on ICEKp3 was the most common structure detected ($n = 21$), found in ST147, ST101, ST15, and ST437 (all of which are GCs), as well as ST11, and was present in multiple isolates and hospitals in our collection. Seven isolates harboured *ybt10* on ICEKp4, associated with ST530, ST307 (GC), ST3161, and ST45. Aerobactin *iuc1* was associated with ST383 ($n = 4$, of which two co-harboured *rmp1*), and 1 ST231 co-harboured *ybt14* and *iuc*.

Forty different K-loci were identified, as detailed in Supplementary Table S2, along with seven variants of O-loci, most common of which were O1 (n = 30) and O2 (n = 21). The O-locus variant could not be identified in 10 isolates.

3. Discussion

The aim of this study was to conduct a snapshot analysis of the molecular epidemiology of the *K. pneumoniae* population from hospitals in Khartoum, Sudan. We have included isolates from five hospitals through several sampling timeframes over the years. *K. pneumoniae* was implicated in multiple infections, including blood, skin and soft tissue (SST), and urine. The majority of isolates (81%) were MDR, of which more than half were also carbapenem resistant, mediated mainly by NDM-1-, NDM-5-, and OXA-48-like enzymes. Interestingly, KPC was absent from the isolate collection, despite its global prevalence in CR-Kp, particularly in ST258. Carbapenem resistance was also mediated by the presence of an ESBL and modifications in OmpK35 and OmpK36 in three isolates (Table 2). Modifications in outer-membrane porins which restricts antibiotic entry is an important carbapenem resistance mechanism [15]. *OmpK35* is usually truncated due to a mutation encoding a frame shift that results in a premature stop codon [16,17]. *OmpK36* is more heterogenous, rarely truncated, and resistance mutations are common leading either to abundance of OmpK36 in the outer membrane or constriction of the pore size [15,18]. Carbapenem MIC for LH_R290 showed intermediate resistance (IMI 4 mgL) without an acquired carbapenamase; however, the isolate harbours SCO-1, a plasmid-encoded ESBL able to hydrolyse not only penicillins but also, to a lesser degree, cephalosporins and carbapenems. Since its discovery in 2007, the bla_{SCO-1} gene (GenBank accession no. EF063111) has been identified in *Acinetobacter baumannii, Escherichia coli, Serratia marcescens, Klebsiella aerogenes, Salmonella enterica*, and only four *K. pneumoniae* isolates [19].

The epidemiological analysis of *K. pneumoniae* in Sudan revealed a large diversity of 37 different STs, with 13 transmission clusters (Figure 1). Five of these clusters have been collected from different hospitals, thereby indicating intra-hospital circulation of clones. Global problem high-risk MDR clones (GC), identified by Wyres et al., 2020, have been found in this study; however, we noted that not all were MDR. We have identified multiple losses of resistance within these GCs over time. For example, isolates in ST307 at RUH collected in FebruaryMarch 2020 were carbapenem-resistant; however, isolate LH_R275 collected in July 2020 was neither MDR nor CR. Similarly, isolates within ST147: LH_F134, LH_F149, LH_F18, were CR harbouring bla_{NDM-1}; however, LH_F97 (isolated from the same hospital) is carbapenem-susceptible, but had acquired the *ybt9* virulence gene on ICEKp3. ST437 isolates appear to have lost bla_{NDM-1} (from 2018) and acquired bla_{NDM-5} in the isolates collected in 2020. However, we did not specifically look for plasmids in this study, and further detailed investigations of possible plasmid acquisition and loss events must be conducted to confirm the observed results. The differences in the resistomevirulomes of isolates within the same ST and transmission cluster is important to note for outbreak and epidemiological studies. The data provided by genomic analysis are an important tool where molecular epidemiology is combined with patient clinical data and the resistomes to obtain accurate information on transmission of resistant and virulent pathogens.

Yersinibectin (*ybt9* and *ybt10*) were the most identified virulence genes in the study and were mostly associated with GCs: ST101, ST147, and ST307, in addition to their occurrence in singleton non-GCs, such as ST39, ST437, and ST882 (Table 2). *ybt* is usually present in 30–40% of *K. pneumoniae* human HAI isolates and up to 13% of community-acquired isolates [20]. Other virulence genes also identified in the study included aerobactin, *iuc1*, and hypermucoidy-associated *rmpA* genes. We have noted AMR-hypervirulence convergence events in 24 isolates, which harboured a virulence gene and were MDR: ST530, ST11, ST147, ST101, ST231, ST307, ST15, ST383, ST437, and ST45. It is important to note that we have not confirmed the virulence phenotypes of the isolates, and more research into the clinical significance of hypervirulence is required.

ST383 isolates were particularly unique in their resistomes and virulomes. The isolates co-harboured multiple ESBLs and carbapenemases simultaneously, CTX-M-14 and CTX-M-15, in addition to NDM-5 and OXA-48 in the same isolates and *iuc-1, rmp1* on KpVP-1, and *rmpA2*. ST383 is a prevalent MDR clone in China, Australia, the UK, and Germany [21]. In a study from Egypt, ST383 was identified in a single isolate simultaneously encoding CTX-M-14 (on an IncLM plasmid) and CTX-M-15 (encoded on a hybrid IncHI1BIncFIB plasmid) in addition to both NDM-5 and OXA-48 [22]. Results by PlasmidFinder show the same plasmids are present in our ST383 isolates (incFIBIncHI1B and incL).This clearly indicates the endemicity of this particular clone in the region, as the data presented in the Egyptian isolate is identical to that in our ST383 cluster.

In our previous study on *K. pneumonia* isolated in Khartoum, 2015–2016, we performed MLST on 117 isolates. All were MDR, and 42.8% were CR-Kp [14]. A similar diversity was observed in the current study, with 52 different STs, with the most common being ST383 (n = 8), ST101 (n = 5), and ST48 (n = 2), in addition to assignment of 15 novel STs. NDM was the most common carbapenemase, in addition to VIM, which has not been detected in the current study. A study conducted in Sudan by Adam et al. from 2015–2016 [23] revealed that metallo-β-lactamases (VIM and IMP) were prevalent in *K. pneumoniae* (1520 isolates), and found in combination with NDM in 520 isolates, in contrast to our study where VIM and IMP were not detected and NDM was the most prevalent MBL. Comparing the data from these studies indicates the maintenance of ST101, ST383, ST219, and ST437 lineages in Khartoum hospitals over the years, and the presence of novel lineages, as reported in this study.

In comparison to the present study, we also investigated *A. baumannii* from several different hospitals in Khartoum during the same time period and found that the *A. baumannii* isolates were mostly transmissions and very few were singletons [24]. This highlights that even in an area that obviously has a high incidence of hospital-acquired infections and patient-to-patient transmissions, including between hospitals, *K. pneumoniae* is not easily transmitted. While most *K. pneumoniae* in this study were not transmissions, they do harbour similar carbapenemases, which may indicate transmission of resistance genesmobile genetic elements, which will be the focus of a follow-up study.

When compared to other African studies, the *K. pneumoniae* population in Southern Nigeria was found to contain four dominant lineages: ST307, ST524, ST15, and ST25, while CR remained low at 8%, with no isolate carrying a combination of carbapenemases, in contrast to our study, where carbapenem-resistance was much higher (41.6% of 84 isolates) and three strains co-harboured NDM and OXA-48 [25]. A multi-centre pilot study in Egypt revealed that ST11, ST147, ST231, ST383, and ST101 were prevalent, which is similar to our study [26]. The observed similarities and difference with the local, regional, and international studies highlights the diversity and dynamic nature of the representative local population of *K. pneumoniae* as well as the potential for regional dissemination of clones in different countries, and the need to support local capacity in robust epidemiological surveillance. The first WGS study on a single ST14 *K. pneumoniae* from Sudan was conducted in 2019 [27]. To the best of our knowledge, this is the first WGS-based study of a large collection of *K. pneumoniae* from Sudan.

Through this study, we aimed to generate robust epidemiological data on *K. pneumoniae*; albeit acknowledging some limitations. We have tried to collect as many *K. pneumoniae* isolates within the indicated timeframes. However, we cannot exclude sampling bias (such as missed samples or misidentification), as the study team was not directly involved in the sampling procedures. Routine sampling of infected patients is not always performed in local hospitals in Sudan, and sampling is frequently sporadic and based on individual doctors' rather than hospital policiesguidelines due to poor microbiological infrastructure in hospitals. The fragmented health service, and the large out-of-pocket contribution by individuals and households, as presented in the introduction, lead to some hospitals also outsourcing microbiology to private laboratories. This study, however, focused on data from hospital-based laboratories only. Furthermore, the study was interrupted during

COVID-19 lockdowns, so isolates between April and July 2020 were not collected. Some associated demographic data were also lacking for some isolates collected in 2018–2019. The sampling strategy therefore cannot exclude missed transmissions. Nevertheless, this study provides a starting point for the understanding of the population structure and diversity of *K. pneumoniae* in Khartoum, Sudan. It furthermore supports national and global efforts in providing robust epidemiological information on important HAI. This study highlights the importance of using WGS in AMR surveillance. Despite the vast possibilities for implementation, major challenges relating to capacity and training, and particularly analysis, of WGS still exist in many LMICs [28].

4. Materials and Methods

A total of 86 non-repetitive isolates were collected within these timeframes: Twenty-four isolates in 2018–2019 from Al-Baraha Hospital (ABH), Al-Saha Hospital (ASH), and Soba University Hospital (SUH); then 60 isolates in March–September 2020 from Fedail Hospital (FH), Ribat University Hospital (RUH), and SUH. Two additional isolates (from 2016) were included from RUH. Descriptive statistics was used to summarise the details of each isolate, including date of isolation, sample type, and hospital (Figure 1 and Table 1). The clinical microbiology laboratories in the named hospitals identify clinical specimens to the genus levels by conventional phenotypic and biochemical methods (growth and colony morphology on various media, and biochemically: indole negative, MR negative, VP positive, citrate positive, oxidase negative, and catalase positive). No specific selection criteria were implemented, as we aimed to collect and characterise any *K. pneumoniae* isolates identified in the hospitals during the study periods. Upon identification of *K. pneumoniae* by the clinical microbiology laboratories, the isolates were stored at 4 °C and collected by the study team within 48 h. It is important to note that the study team cannot guarantee any misidentification or loss of samples that were not submitted to the study.

All acquired isolates were subsequently confirmed phenotypically by conventional culture methods to exclude any contamination, then genotypically by amplification of 16S-23S rDNA internal transcribed spacer (ITS) of *K. pneumoniae*, as described in our preceding study [29]. Further identification confirmation was performed by MALDI-TOF prior to WGS.

Antimicrobial susceptibility (AST) was initially performed by disk-diffusion at the clinical microbiology laboratories and interpreted according to CLSI guidelines [30]. Upon confirmation of species, as described above, the minimum inhibitory concentration (MIC) was determined by MICRONAUT-GN, (Merlin Diagnostika, Germany). This system allows the determination of the MIC for a panel of Gram-negative specific antibiotics: meropenem (MER), gentamicin (GEN), amikacin (AMK), trimethoprim-sulfamethoxazole (SXT), ciprofloxacin (CIP), colistin (COL), Amoxicillin (AMX), AmoxicillunClavulanate (AMC), Cefotaxime (CTX), Ceftazidime (CAZ), Cefuroxim (CXM), Etrapenem (ERT), and Temocillin (TMO), with two concentrations (mgL) based on the EUCAST MIC breakpoints for sensitivity (S) or resistance (R) in a single microtiter plate. Additionally, MIC for imipenem (IMI) was performed by broth microdilution according to the EUCAST guidelines, V2, 2020 [31]. Quality control strains *E. coli* ATCC 13846, ATCC 12241, and *P. aeruginosa* ATCC12903 were used in all AST experiments. Multidrug resistance was defined as resistance to 3 or more antimicrobial classes.

Total DNA was extracted from the bacterial isolates using a MagAttract HMW DNA Kit (Qiagen, Hilden, Germany), according to the manufacturer's instructions. Sequencing libraries were prepared using the Nextera XT library prep kit (Illumina GmbH, Munich, Germany) for a 250-bp paired-end sequencing run on an Illumina MiSeq platform. De novo assembly was constructed using Velvet v1.1.04. Molecular epidemiology of the isolates was investigated by core-genome MLST (cgMLST) using Ridom® SeqSphere^{+} version 8.5.1 [32].

Kleborate v2.0.4 was used to screen genome assemblies for Sequence Types (STs), capsular type, and virulence loci [33]. The resistome for the assembled genomes was identified using Kleborate v2.0.4 and the Resfinder database v3.2 https://cge.cbs.dtu.

dkservicesResFinder (accessed on 10 February 2021) [34]. Sequence alignment and visualisation was performed by Geneious Prime.

All assembled genome sequences were submitted to Genbank under BioPproject ID PRJNA912084.

5. Conclusions

This study aims to provide a population snapshot of *K. pneumoniae* in Sudan. ST101, ST147, and ST437 are epidemic and predominantly present in multiple hospitals. MDR and CR-Kp were also prevalent (80% and 50%, respectively), which is alarming. The detection of multiple virulence genes and the potential emergence of AMR-hypervirulence convergence events must also be considered for local surveillance. We conclude that a diverse population of *K. pneumoniae* is present in hospitals in Khartoum, Sudan. Further genomic investigations, and inclusion of more data from national and regional hospitals, would enable delivery of value on a local, national, regional, and global level in understanding the pathogen dynamics.

Supplementary Materials: The following supporting information can be downloaded at: https://www.mdpi.com/article/10.3390/antibiotics12020233/s1. Figure S1: Minimum spanning tree (MST) of 84 *K. pneumoniae* by hospital information. Ridom SeqSphere+ MST for 84 samples based on 2358 columns, pairwise ignoring missing values, logarithmic scale. Cluster distance threshold: 15. Isolates grouped by colour indicate the different hospitals. Samples were collected from five different hospitals, and 37 different STs were identified, in addition to 14 transmission clusters, represented by shaded nodes and arrows. Numbers between the nodes indicate the number of allelic differences; Table S1: Sequence types, antimicrobial susceptibility and associated resistome; Table S2: Details of ST with virulence genes, K and O loci.

Author Contributions: Conceptualization, E.A.O., H.N.A. and L.A.-H.; methodology, E.A.O., H.N.A. and L.A.-H.; validation, P.G.H., J.W. and L.A.-H.; formal analysis, M.Y. and L.A.-H.; investigation, E.A.O., D.C., J.W. and L.A.-H.; resources, E.A.O., L.A.-H., D.C. and P.G.H.; data curation, M.Y.; writing—original draft preparation, E.A.O. and L.A.-H.; writing—review and editing, E.A.O., M.Y., H.N.A., D.C., J.W., H.S., P.G.H. and L.A.-H.; visualization, M.Y. and P.G.H.; supervision, L.A.-H. and H.N.A.; project administration, E.A.O. and L.A.-H.; funding acquisition, L.A.-H. All authors have read and agreed to the published version of the manuscript.

Funding: This research was funded by the University of Sussex: International Development Challenges Fund, grant number IDCF1-007.

Institutional Review Board Statement: Formal permission was obtained from all hospitals included in this study, and from the Research Ethics Committee, Ministry of Health, Sudan.

Informed Consent Statement: Patient consent was waived by the Sudanese Research Ethics Committee. All patients consent that any microbiological samples collected in the hospital as part of routine clinical work may be used for research purposes. No patient identifying details were collected for this study.

Data Availability Statement: The genome assemblies presented in this study are openly available in NCBI GenBank under BioProject ID PRJNA912084.

Acknowledgments: We would like to acknowledge the medical and technical staff of the microbiology laboratories in the participating hospitals for their help in isolate and data collection. We also want to acknowledge Alaa M. Albasha, Maram M. Elnosh, and Esraa H. Osman for their contribution to part of the isolate collection.

Conflicts of Interest: The authors declare no conflict of interest.

References

1. Boucher, H.W.; Talbot, G.H.; Bradley, J.S.; Edwards, J.E.; Gilbert, D.; Rice, L.B.; Scheld, M.; Spellberg, B.; Bartlett, J. Bad Bugs, No Drugs: No ESKAPE! An Update from the Infectious Diseases Society of America. *Clin. Infect. Dis.* **2009**, *48*, 1–12. [CrossRef] [PubMed]
2. Wyres, K.L.; Lam, M.M.C.; Holt, K.E. Population Genomics of Klebsiella Pneumoniae. *Nat. Rev. Microbiol.* **2020**, *18*, 344–359. [CrossRef] [PubMed]

3. Navon-Venezia, S.; Kondratyeva, K.; Carattoli, A. Klebsiella Pneumoniae: A Major Worldwide Source and Shuttle for Antibiotic Resistance. *FEMS Microbiol. Rev.* **2017**, *41*, 252–275. [CrossRef] [PubMed]
4. Silvester, R.; Madhavan, A.; Kokkat, A.; Parolla, A.; Adarsh, B.M.; Harikrishnan, M.; Abdulla, M.H. Global Surveillance of Antimicrobial Resistance and Hypervirulence in Klebsiella Pneumoniae from LMICs: An in-Silico Approach. *Sci. Total Environ.* **2022**, *802*, 149859. [CrossRef] [PubMed]
5. Murray, C.J.; Ikuta, K.S.; Sharara, F.; Swetschinski, L.; Robles Aguilar, G.; Gray, A.; Han, C.; Bisignano, C.; Rao, P.; Wool, E.; et al. Global Burden of Bacterial Antimicrobial Resistance in 2019: A Systematic Analysis. *Lancet* **2022**, *399*, 629–655. [CrossRef] [PubMed]
6. WHO. *Geneva Global Action Plan on Antimicrobial Resistance*; World Health Organization: Geneva, Switzerland, 2017; pp. 1–28.
7. WHO. *Country Cooperation Strategy for WHO and Sudan*; World Health Organization: Geneva, Switzerland, 2022; ISBN 9789290229698.
8. WHO. *Global Priority List of Antibiotic-Resistant Batceria to Guide Research, Discovery, and Development of New Antibiotics*; WHO: Geneva, Switzerland, 2017; p. 7.
9. Alsammani, M.A.; Ahmed, M.I.; Abdelatif, N.F. Bacterial Uropathogens Isolates and Antibiograms in Children under 5 Years of Age. *Med. Arch.* **2014**, *68*, 239–243. [CrossRef] [PubMed]
10. Albasha, A.M.; Osman, E.H.; Abd-Alhalim, S.; Alshaib, E.F.; Al-Hassan, L.; Altayb, H.N. Detection of Several Carbapenems Resistant and Virulence Genes in Classical and Hyper-Virulent Strains of Klebsiella Pneumoniae Isolated from Hospitalized Neonates and Adults in Khartoum. *BMC Res. Notes* **2020**, *13*, 312. [CrossRef]
11. Elbadawi, H.S.; Elhag, K.M.; Mahgoub, E.; Altayb, H.N.; Abdel Hamid, M.M. Antimicrobial Resistance Surveillance among Gram-Negative Bacterial Isolates from Patients in Hospitals in Khartoum State, Sudan [Version 1; Peer Review: Awaiting Peer Review]. *F1000Research* **2019**, *8*, 156. [CrossRef]
12. Hamdan, H.Z.; Kubbara, E.; Adam, A.M.; Hassan, O.S.; Suliman, S.O.; Adam, I. Urinary Tract Infections and Antimicrobial Sensitivity among Diabetic Patients at Khartoum, Sudan. *Ann. Clin. Microbiol. Antimicrob.* **2015**, *14*, 26. [CrossRef]
13. Abd Alfadil, N.A.; Suliman Mohamed, M.; Ali, M.M.; El Nima, E.A.I. Characterization of Pathogenic Bacteria Isolated from Sudanese Banknotes and Determination of Their Resistance Profile. *Int. J. Microbiol.* **2018**, *2018*, 4375164. [CrossRef]
14. Osman, E.A.; El-Amin, N.E.; Al-Hassan, L.L.; Mukhtar, M. Multiclonal Spread of Klebsiella Pneumoniae across Hospitals in Khartoum, Sudan. *J. Glob. Antimicrob. Resist.* **2021**, *24*, 241–245. [CrossRef] [PubMed]
15. David, S.; Wong, J.L.C.; Sanchez-Garrido, J.; Kwong, H.S.; Low, W.W.; Morecchiato, F.; Giani, T.; Rossolini, G.M.; Brett, S.J.; Clements, A.; et al. Widespread Emergence of OmpK36 Loop 3 Insertions among Multidrug-Resistant Clones of Klebsiella Pneumoniae. *PLoS Pathog.* **2022**, *18*, e1010334. [CrossRef] [PubMed]
16. Wassef, M.; Abdelhaleim, M.; AbdulRahman, E.; Ghaith, D. The Role of OmpK35, OmpK36 Porins, and Production of β-Lactamases on Imipenem Susceptibility in Klebsiella Pneumoniae Clinical Isolates, Cairo, Egypt. *Microb. Drug Resist.* **2015**, *21*, 577–580. [CrossRef]
17. Rodrigues, C.; Desai, S.; Passet, V.; Gajjar, D.; Brisse, S. Genomic Evolution of the Globally Disseminated Multidrug-Resistant Klebsiella Pneumoniae Clonal Group 147. *Microb. Genomics* **2022**, *8*, 000737. [CrossRef]
18. Wong, J.L.C.; Romano, M.; Kerry, L.E.; Kwong, H.S.; Low, W.W.; Brett, S.J.; Clements, A.; Beis, K.; Frankel, G. OmpK36-Mediated Carbapenem Resistance Attenuates ST258 Klebsiella Pneumoniae in Vivo. *Nat. Commun.* **2019**, *10*, 3957. [CrossRef] [PubMed]
19. Papagiannitsis, C.C.; Loli, A.; Tzouvelekis, L.S.; Tzelepi, E.; Arlet, G.; Miriagou, V. SCO-1, a Novel Plasmid-Mediated Class A β-Lactamase with Carbenicillinase Characteristics from Escherichia Coli. *Antimicrob. Agents Chemother.* **2007**, *51*, 2185–2188. [CrossRef]
20. Wyres, K.L.; Wick, R.R.; Judd, L.M.; Froumine, R.; Tokolyi, A.; Gorrie, C.L.; Lam, M.M.; Duchene, S.; Jenney, A.; Holt, E. Distinct Evolutionary Dynamics of Horizontal Gene Transfer in Drug Resistant and Virulent Clones of Klebsiella Pneumoniae. *PLoS Genet.* **2019**, *15*, e100811. [CrossRef]
21. Turton, J.F.; Payne, Z.; Coward, A.; Hopkins, K.L.; Turton, J.A.; Doumith, M.; Woodford, N. Virulence Genes in Isolates of Klebsiella Pneumoniae from the UK during 2016, Including among Carbapenemase Gene-Positive Hypervirulent K1-St23 and 'Non-Hypervirulent' Types ST147, ST15 and ST383. *J. Med. Microbiol.* **2018**, *67*, 118–128. [CrossRef]
22. Edward, E.A.; Mohamed, N.M.; Zakaria, A.S. Whole Genome Characterization of the High-Risk Clone ST383 Klebsiella Pneumoniae with a Simultaneous Carriage of BlaCTX-M-14 on IncLM Plasmid and BlaCTX-M-15 on Convergent IncHI1BIncFIB Plasmid from Egypt. *Microorganisms* **2022**, *10*, 1097. [CrossRef]
23. Adam, M.A.; Elhag, W.I. Prevalence of Metallo-β-Lactamase Acquired Genes among Carbapenems Susceptible and Resistant Gram-Negative Clinical Isolates Using Multiplex PCR, Khartoum Hospitals, Khartoum Sudan. *BMC Infect. Dis.* **2018**, *18*, 4–9. [CrossRef]
24. Al-Hassan, L.; Elbadawi, H.; Osman, E.; Ali, S.; Elhag, K.; Cantillon, D.; Wille, J.; Seifert, H.; Higgins, P.G. Molecular Epidemiology of Carbapenem-Resistant Acinetobacter Baumannii From Khartoum State, Sudan. *Front. Microbiol.* **2021**, *12*, 628736. [CrossRef]
25. Afolayan, A.O.; Oaikhena, A.O.; Aboderin, A.O.; Olabisi, O.F.; Amupitan, A.A.; Abiri, O.V.; Ogunleye, V.O.; Odih, E.E.; Adeyemo, A.T.; Adeyemo, A.T.; et al. Clones and Clusters of Antimicrobial-Resistant Klebsiella from Southwestern Nigeria. *Clin. Infect. Dis.* **2021**, *73*, S308–S315. [CrossRef] [PubMed]
26. Sherif, M.; Palmieri, M.; Mirande, C.; El-Mahallawy, H.; Rashed, H.G.; Abd-El-Reheem, F.; El-Manakhly, A.R.; Abdel-latif, R.A.R.; Aboulela, A.G.; Saeed, L.Y.; et al. Whole-Genome Sequencing of Egyptian Multidrug-Resistant Klebsiella Pneumoniae Isolates: A Multi-Center Pilot Study. *Eur. J. Clin. Microbiol. Infect. Dis.* **2021**, *40*, 1451–1460. [CrossRef] [PubMed]

27. Mohamed, S.B.; Kambal, S.; Munir, A.; Abdalla, N.; Hassan, M.; Hamad, A.; Mohammed, S.; Ahmed, F.; Hamid, O.; Ismail, A.; et al. First Whole-Genome Sequence of a Highly Resistant Klebsiella Pneumoniae Sequence Type 14 Strain Isolated from Sudan. *Microbiol. Resour. Announc.* **2019**, *8*, 14–15. [CrossRef] [PubMed]
28. Osman, E.A.; El-Amin, N.; Adrees, E.A.E.; Al-Hassan, L.; Mukhtar, M. Comparing Conventional, Biochemical and Genotypic Methods for Accurate Identification of Klebsiella Pneumoniae in Sudan. *Access Microbiol.* **2020**, *2*, e000096. [CrossRef] [PubMed]
29. Higgins, P.G.; Prior, K.; Harmsen, D.; Seifert, H. Development and Evaluation of a Core Genome Multilocus Typing Scheme for Whole-Genome Sequence-Based Typing of Acinetobacter Baumannii. *PLoS ONE* **2017**, *12*, e0179228. [CrossRef] [PubMed]
30. Lam, M.M.C.; Wick, R.R.; Watts, S.C.; Cerdeira, L.T.; Wyres, K.L.; Holt, K.E. A Genomic Surveillance Framework and Genotyping Tool for *Klebsiella pneumoniae* and Its Related Species Complex. *Nat. Commun.* **2021**, *12*, 4188. [CrossRef]
31. Zankari, E.; Hasman, H.; Cosentino, S.; Vestergaard, M.; Rasmussen, S.; Lund, O.; Aarestrup, F.M.; Larsen, M.V. Identification of Acquired Antimicrobial Resistance Genes. *J. Antimicrob. Chemother.* **2012**, *67*, 2640–2644. [CrossRef]
32. *Performance Standards for Antimicrobial Susceptibility Testing An Informational Supplement for Global Application Developed through the Clinical and Laboratory Standards Institute*, 26th ed.; Clinical and Laboratory Standards Institute: Wayne, PA, USA, 2016; Volume M100S; ISBN 610.688.0700.
33. EUCAST. *EUCAST Reading Guide for Broth Microdilution*; EUCAST: Växjö, Sweden, 2020.
34. Aanensen, D.M.; Carlos, C.C.; Donado-Godoy, P.; Okeke, I.N.; Ravikumar, K.L.; Abudahab, K.; Abrudan, M.; Argimón, S.; Harste, H.; Kekre, M.; et al. Implementing Whole-Genome Sequencing for Ongoing Surveillance of Antimicrobial Resistance: Exemplifying Insights into Klebsiella Pneumoniae. *Clin. Infect. Dis.* **2021**, *73*, S255–S257. [CrossRef]

antibiotics

Review

Uncovering the Secretion Systems of *Acinetobacter baumannii*: Structures and Functions in Pathogenicity and Antibiotic Resistance

Pu Li [1], Sirui Zhang [2], Jingdan Wang [1], Mona Mohamed Al-Shamiri [2], Bei Han [1], Yanjiong Chen [2], Shaoshan Han [3] and Lei Han [2,*]

[1] School of Public Health, Xi'an Jiaotong University Health Science Center, Xi'an 710061, China
[2] Department of Microbiology and Immunology, School of Basic Medical Sciences, Xi'an Jiaotong University Health Science Center, Xi'an 710061, China
[3] Department of Hepatobiliary Surgery, The First Affiliated Hospital of Xi'an Jiaotong University, Xi'an 710061, China
* Correspondence: lei.han@xjtu.edu.cn

Abstract: Infections led by *Acinetobacter baumannii* strains are of great concern in healthcare environments due to the strong ability of the bacteria to spread through different apparatuses and develop drug resistance. Severe diseases can be caused by *A. baumannii* in critically ill patients, but its biological process and mechanism are not well understood. Secretion systems have recently been demonstrated to be involved in the pathogenic process, and five types of secretion systems out of the currently known six from Gram-negative bacteria have been found in *A. baumannii*. They can promote the fitness and pathogenesis of the bacteria by releasing a variety of effectors. Additionally, antibiotic resistance is found to be related to some types of secretion systems. In this review, we describe the genetic and structural compositions of the five secretion systems that exist in *Acinetobacter*. In addition, the function and molecular mechanism of each secretion system are summarized to explain how they enable these critical pathogens to overcome eukaryotic hosts and prokaryotic competitors to cause diseases.

Keywords: *Acinetobacter baumannii*; secretion systems; pathogenicity; antibiotic resistance

Citation: Li, P.; Zhang, S.; Wang, J.; Al-Shamiri, M.M.; Han, B.; Chen, Y.; Han, S.; Han, L. Uncovering the Secretion Systems of *Acinetobacter baumannii*: Structures and Functions in Pathogenicity and Antibiotic Resistance. *Antibiotics* **2023**, *12*, 195. https://doi.org/10.3390/antibiotics12020195

Academic Editors: Theodoros Karampatakis and Jordi Vila

Received: 22 November 2022
Revised: 6 January 2023
Accepted: 16 January 2023
Published: 17 January 2023

1. Introduction

Acinetobacter baumannii is a strictly aerobic, non-fermenting, Gram-negative coccobacillus with pili and capsule, but no flagella. It is ubiquitous in nature, and used to be considered to be of negligible significance due to its low virulence [1]. However, the rapidly increasing nosocomial infections and high mortality caused by *A. baumannii*, as well as its strong drug resistance, have raised people's attention [2]. Taken together with *Enterococcus faecium*, *Staphylococcus aureus*, *Klebsiella pneumonia*, *Pseudomonas aeruginosa*, *Enterobacter*, *Acinetobacter baumannii* has been enrolled as a member of ESKAPE by the Infectious Diseases Society of America (IDSA) in order to emphasize the importance of these pathogens in causing hospital infections and resisting the effects of a variety of antimicrobial drugs [3,4].

The high frequency of *A. baumannii* nosocomial infections is closely related to its strong environmental persistence. *A. baumannii* can survive in nutrient-limited and desiccation environments, and is capable of resisting disinfections [5]. Moreover, it is able to survive for long periods of time on both biotic and abiotic surfaces [6]. Based on these advantages, *A. baumannii* can be easily transmitted patient to patient by air, water, and contact with medical personnel's hands and equipment, thus colonizing multiple sites and finally leading to a variety of infections, such as pneumonia, septicemia, urinary tract infections, meningitis, and skin and wound infections [7–9].

Antibiotic resistance is another key factor that contributes to *A. baumannii* infections and outbreaks. The increasing rate of infections caused by drug-resistant *A. baumannii* is a significant issue in hospitals all over the world [10]. The continued overuse and misuse of antibiotics enable *A. baumannii* to develop different types of resistance mechanisms, e.g., the acquisition of multiple antibiotic resistance genes to produce degradative enzymes, a decrease in bacterial membrane permeability, the alteration of antibiotic targets, the overexpression of efflux pumps, a change in metabolic status, and the formation of biofilms [11]. Therefore, this bacterium can escape the killing of antibiotics and conquer the stress conditions, further leading to infections. *A. baumannii* has an extraordinary genetic plasticity that results in a high capacity to acquire antimicrobial resistance traits [2], thus producing many multidrug-resistant (MDR), extensively drug-resistant (XDR), and even pan-drug-resistant (PDR) strains, representing a significant challenge for therapy in clinics.

Infections are also dependent on virulence factors. Various genes have been revealed to be involved in the pathogenic procedures of iron acquisition, nutrient uptake, adhesion, biofilm formation, invasion, hemolytic activity, and cytolytic activity [12,13]. Among them, protein secretion systems have received much attention. They can transport the virulence factors produced by bacteria into extracellular environments, meaning that the latter will manipulate the host's defenses and facilitate pathogen infection [14,15]. Until recently, six secretion systems from Gram-negative bacteria have been revealed and studied; namely, type I secretion system (T1SS) to type VI secretion system (T6SS). Some of these have been characterized and reported to have specific roles in the pathophysiology of *A. baumannii*, whereas the gene and protein structures of some secretion systems in *A. baumannii* are still not clear and are being explored. Moreover, the association between secretion systems and drug resistance has been discovered in some bacteria, e.g., the T3SS in *Pseudomonas aeruginosa* correlates with a fluoroquinolone resistance phenotype, and the T4SS in many Gram-negative pathogens mediates antibiotic resistance via conjugation [16–18]. Meanwhile, the contribution of secretion systems to antibiotic resistance in *A. baumannii* is poorly understood. Here, an overview of the progress of the research on the structure, composition, pathogenicity, and relation to antibiotic resistance of the secretion systems in *A. baumannii* is presented.

2. Type I Secretion System (T1SS)

The T1SS is a highly conserved secretion system in pathogenic Gram-negative bacteria. However, it is less reported in *A. baumannii*. In 2017, the T1SS was first identified in the pathogenic *Acinetobacter nosocomialis* strain M2 upon bioinformatic analysis by Harding et al. [19]. Until now, only two reports have described the structure and function of the T1SS in *Acinetobacter* [19,20].

2.1. Gene and Structure

The locus that is homologous to the prototype T1SS of *Escherichia coli* containing the *tolC*, *hlyB*, and *hlyD* genes is found in the M2 chromosome, as well as in *A. baumannii*. In contrast to *E. coli*, these genes are found in three gene clusters, and are most likely in an operon, given that the open reading frame (ORF) for *hlyB* overlaps with both *tolC* and *hlyD* [19] (Figure 1a).

This *tolC-hlyB-hlyD* gene cluster produces three proteins with high molecular weights of 130 kDa, 250 kDa, and 70 kDa. They form a secretion system with the elements of TolC, which is localized in the outer membrane, HlyB, which is anchored in the inner membrane as an ATP-binding cassette transporter, and HlyD as a periplasmic adaptor [19] (Figure 1b).

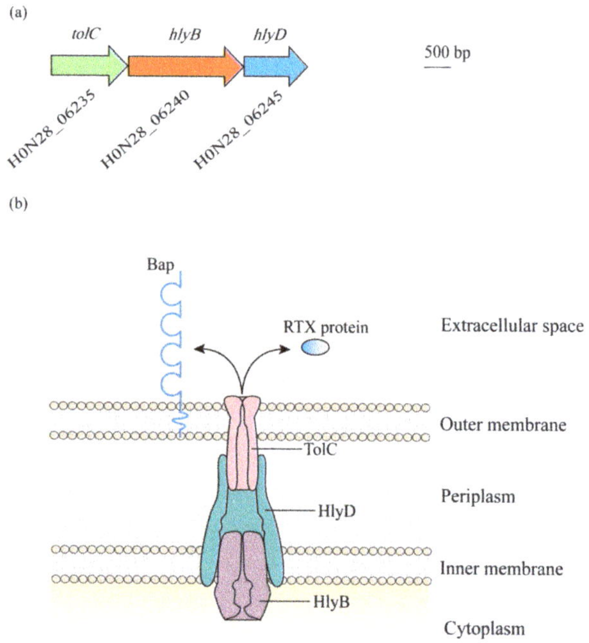

Figure 1. Composition and structure of the type I secretion system (T1SS) in *A. baumannii*: (**a**) Bioinformatic analysis has led to the identification of the T1SS in genomes of *A. baumannii*. Gene locus tags are cited from ATCC 17978. Genes predicted to encode proteins required for the biogenesis of the T1SS are found in three gene clusters, with *hlyB* overlapping with *tolC* and *hlyD*. (**b**) The three components of the T1SS act together to facilitate the secretion of effectors. TolC is a trimeric outer membrane protein with the α-helical barrel forming a tunnel through the periplasm, and it interacts with HlyD. HlyD has a large periplasmic domain linked by a single transmembrane helix, which anchors in the inner membrane. The energy required for the export of specific T1SS substrates is provided by HlyB, which is an ATP-binding protein. Two putative T1SS effectors, namely, Repeats-in-Toxin (RTX)-serralysin-like toxin and biofilm-associated protein (Bap), are involved in the formation and stability of biofilm.

2.2. Function

2.2.1. Secretion of Putative Effectors

The T1SS facilitates the secretion of two putative effectors from the cytoplasm to the extracellular milieu, including Repeats-in-Toxin (RTX)-serralysin-like toxin and the biofilm-associated protein (Bap) [19]. The former belongs to the RTX family, which is a heterogeneous group of proteins translocated out of Gram-negative bacteria by the T1SS. They are commonly involved in bacterial adhesion, pathogenesis, and biofilm formation [21], but the role that RTX-serralysin-like toxin plays in *Acinetobacter* is not yet entirely understood. By contrast, Bap has been well studied. The Bap protein identified from clinical *A. baumannii* isolates that lead to bloodstream infections is homologous to *Staphylococcus aureus*, with nucleotide sequences consistent with cell surface adhesion molecules. It is one of the largest bacterial proteins that localize on the surface of *A. baumannii* and has a remarkably low isoelectric point (pI) at 2.9, making it one of the most acidic bacterial proteins [22,23]. Bap was found to be necessary for mature biofilm formation on medically relevant surfaces, demonstrating the importance of the three-dimensional tower structure and water channel formation. Moreover, it was also involved in the adherence of *A. baumannii* to eukaryotic cells, including human bronchial epithelial cells and neonatal keratinocytes, which is a key step in the biofilm formation of this bacterium in the host [24]. The absence and mutations of Bap diminished both the biovolume and thickness of the biofilm in *A. baumannii* [23]. Interestingly, a stronger biofilm formation was correlated with the overexpression of Bap under the condition of a low iron concentration [25]. As Bap was secreted via the T1SS, Harding et al. further verified that the *Acinetobacter* T1SS was required for biofilm formation [19]. Additionally, the acidic protein Bap may influence the physical and chemical properties of a variety of antibiotics, thus resulting in drug resistance [26].

2.2.2. Cross-Talk with other Secretion Systems

The T1SS has also been revealed to cross-talk with other secretion systems [19]. For example, compared with the wild-type *Acinetobacter* strain, several T2SS-associated proteins, such as CpaA, LipH, a rhombosortase, and a rhombotarget, were found in lower quantities in the T1SS mutant. Moreover, the activity of the T6SS in minimal medium was repressed by the deletion of the T1SS system. This was due to the significantly lower level of the T6SS-associated proteins, VgrG and Hcp, in the T1SS mutant. Specifically, mutations in any component of the T1SS reduced Hcp secretion under nutrient-limited conditions, whereas that in PilD, which is a prepilin peptidase necessary for both T4P and T2SS systems, did not alter Hcp secretion, suggesting a specific association between the T1SS and T6SS. Lastly, two distinct functioning contact-dependent inhibition (CDI) systems were found in pathogenic *A. baumannii* strains. CDI systems are independent from the T1SS and T6SS that facilitate inhibition of the growth of neighboring bacteria and are found to be conserved in medically relevant *Acinetobacter*. However, in the T1SS mutant, a predicted CDI-associated protein was identified at a significantly lower level, indicating the cross-talk between them.

2.2.3. Virulence

The virulence of *A. baumannii* is associated with the T1SS. As observed by Harding et al., T1SS mutants showed attenuated virulence in a *Galleria mellonella* infection model [19]. Moreover, in a recent work, Sycz et al. [20] reported a clinical urinary *A. baumannii* isolate, UPAB1, which was able to replicate in macrophages and escape from them by lysing the host cells. The T1SS was demonstrated to be required for UPAB1 in the process of intracellular replication by secreting two common T1SS-dependent effectors, BapA and RTX2, as well as some additional effectors including proteases, phosphatases, glycosidases, and a putative invasion. Interestingly, the orthologs of this invasion from other bacteria were required to induce bacterial entry and to suppress reactive oxygen species (ROS) generation by the host macrophages [20,27].

3. Type II Secretion System (T2SS)

The T2SS is a multiprotein secretion system that is widely distributed in Gram-negative bacteria, including enterotoxigenic *Escherichia coli*, *Legionella pneumophila*, *Vibrio cholerae*, *Pseudomonas aeruginosa*, and *Klebsiella pneumoniae* [28–34]. It was first reported in *A. baumannii* in 2014 and was subsequently shown to be active in ATCC 17978 by Johnson et al. [35,36]. Further, the T2SS is found in the majority of *A. baumannii* genomes.

3.1. Gene and Structure

In *Acinetobacter* spp., the T2SS is encoded by 12 essential genes, namely, *gspC-M* and *pilD*, and forms an apparatus spanning both the inner membrane and outer membrane [37,38] (Figure 2). In contrast to other Gram-negative pathogens, the core *gsp* genes are not organized in one or two operons, but are grouped into five distinct gene clusters scattered throughout the *Acinetobacter* genome [39] (Figure 2a).

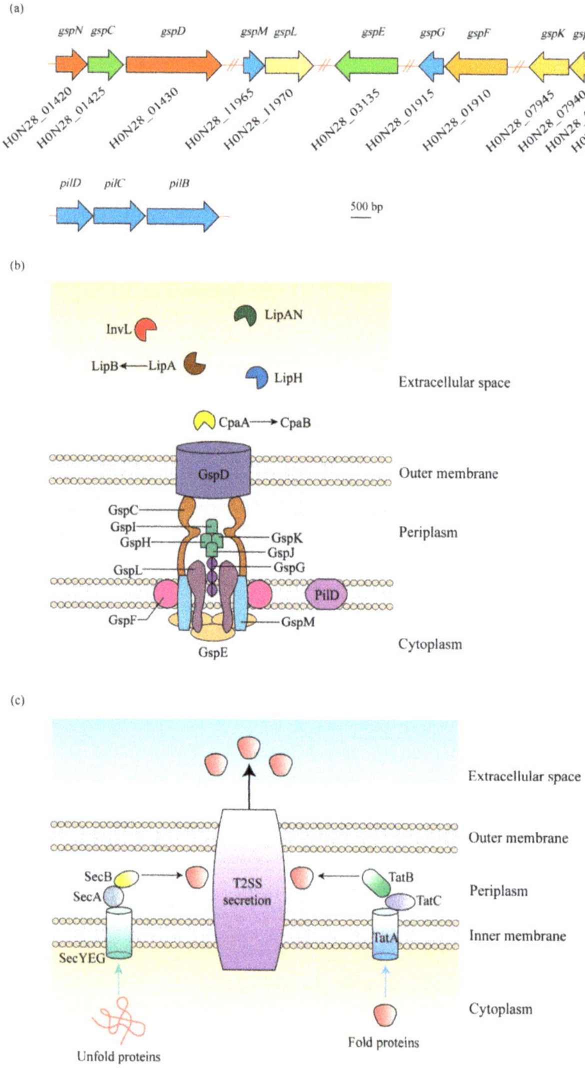

Figure 2. Type II secretion system (T2SS) structure of *A. baumannii* and its protein secretion mechanism: (**a**) As shown in the ATCC 17978 genome, the *gsp* genes required for the T2SS are located in five distant loci, and a single prepilin/pre-pseudopilin peptidase homolog is located in the *pilBCD* cluster. (**b**) The T2SS is composed of an outer membrane (OM) complex (GspD), a periplasmic pseudopilus (GspG, GspH, GspI, GspJ, and GspK), and an inner membrane (IM) platform (GspC, GspF, GspL, and GspM), which relates to the cytoplasmic ATPase GspE. In *A. baumannii*, the T2SS shares a processing protein, PilD, with type IV pili. The T2SS secretes a large number of effectors required for virulence, including the metallopeptidase CpaA (chaperone CpaB), the lipoyl synthases LipA (chaperone LipB), LipH, and LipAN, and a novel lipoprotein, InvL. (**c**) The T2SS-dependent proteins are first exported across the IM to the periplasm via the Sec or Tat pathways in *A. baumannii*. The Sec pathway primarily translocates unfolded proteins, relying on a hydrophobic signal sequence at the N-terminus. On the contrary, the Tat pathway, consisting of TatA, TatB, and TatC, primarily secretes folded proteins. Afterwards, the signal sequence is cleaved, followed by the folding of proteins. Finally, the folded proteins are expelled extracellularly through the OM channel.

In general, the T2SS consists of four parts: (1) an outer membrane (OM) complex; (2) a periplasmic pseudopilus; (3) an inner membrane (IM) complex called the assembly platform (AP); and (4) a cytoplasmic ATPase [40]. The OM complex is composed of GspD, which forms a secretin channel across the outer membrane to transport substrates from the periplasm to the extracellular milieu [41]. The IM platform is composed of GspC, GspF, GspL, and GspM, in which GspC is joined to the periplasmic domains of GspD, thereby connecting the IM platform with the OM complex. In between the OM and IM complexes, the periplasmic pseudopilus, a structure homologous to the type IV pilus, is attached to the IM platform with the composition of major pseudopilin GspG and minor pseudopilins GspH, GspI, GspJ, and GspK. Before the assembly of these subunits, PilD is involved in the cleavage and methylation procedure. Additionally, the cytoplasmic ATPase is formed by a hexamer protein, GspE, to provide ATP to the T2SS for the secretion of effector proteins [40,42] (Figure 2b).

3.2. *Function*

3.2.1. Secretion of Enzymes and Toxins

The T2SS is an important virulence factor that can secrete multiple degradative enzymes and toxins that associate with the fitness of *A. baumannii* in various conditions, including the external environment and mammalian host [43]. The secretion by the T2SS undergoes a two-step process (Figure 2c). Firstly, the substrates of the T2SS, commonly possessing an N-terminal signal peptide, are translocated from the cytoplasm to the periplasm via the general export (Sec) pathway or twin arginine translocation (Tat) pathway. Secondly, after the cleavage of the signal sequence, the proteins fold into a tertiary and/or quaternary structure and exit the bacterial cell through the OM channel [34,43]. After investigation through bioinformatics, proteomics, and mutational analyses, *Acinetobacter* was found to export several substrates through the T2SS, including lipases LipA, LipH, and LipAN, as well as the protease CpaA and lipoprotein InvL [39,44–46].

LipA, after being transferred extracellularly, was reported to be required by *A. baumannii* for utilizing exogenous lipids to obtain nutrients and benefit for colonization in a murine bacteremia model [36]. Its secretion and activity are also related to the chaperone protein LipB, except in the T2SS [34]. Similar to LipA, the secretion of LipH is dependent on a functional T2SS. LipH was discovered to mediate lipase activity as there was residual lipase activity of the culture supernatants in the absence of LipA. Additionally, LipH was found to be secreted by not only the *A. nosocomialis* strain M2, but also a panel of *Acinetobacter clinical strains, including A. baumannii* [39]. LipAN is a newly discovered T2SS-dependent phospholipase from *A. baumannii* ATCC 17978, found in 2016. It locates on the plasmid and contributes to the lung colonization of *A. baumannii*, as investigated in a mouse pneumonia model [44].

CpaA, a zinc-dependent metallo-endopeptidase, was first purified from the culture supernatant of an *A. baumannii* clinical isolate in 2014. It is conserved in most clinical isolates of *A. baumannii*, but it does not exist in ATCC 17978 or ATCC 19606 [45]. Along with LipA and LipH, the potential virulence factor CpaA is also secreted by the T2SS [39]. CpaA is composed of four glycan-binding domains that facilitate this protease to display glycoprotein-targeting activity [47]. It can cleave two *O*-linked glycoproteins, factors V and XII, finally leading to the deregulation of the human blood coagulation system [45,48]. In a recent work, more *O*-linked human glycoproteins were shown that could be cleaved by CpaA, such as CD55 and CD46, that are involved in complement activation, and its activity is unaffected by sialic acid [49]. Similarly to LipA, the chaperone CpaB is essential for the stability and secretion of CpaA. CpaB is a membrane-bound T2SS chaperone that strongly interacts with CpaA in a CpaAB complex with the stoichiometry of 1:1, where the protease (CpaA) surrounds the chaperone (CpaB). However, the proteolytic activity of CpaA is not blocked by the binding of CpaB. This complex structure is a novel model for chaperone–protease interaction [47].

A newly discovered lipoprotein, InvL, is identified as the first effector of the T2SS belonging to the intimin-invasion family. InvL was primarily found in the insoluble fractions of the supernatant of a urinary isolate, UPAB1, and further revealed to be surface-localized. InvL is found to be closely related to international clone (IC) 2 [46]. Its secretion and surface exposure were found to be dependent on the T2SS, since InvL-His_6 expressed in the *gspD*$^+$ strain was readily degraded by proteinase K, whereas the degradation in the Δ*gspD* mutant failed [46].

3.2.2. Pathogenesis

The T2SS plays an important role in infections of *A. baumannii*, primarily by secreting numerous effector substrates, such as the above-mentioned lipases and proteases, to the cell surface or extracellular space. These proteins act on the external environments or target cells, and contribute to the acquisition of nutrients, thus maintaining the survival and colonization of bacteria in the host [50]. Johnson et al. [36] revealed that *A. baumannii* ATCC 17978 mutants lacking either T2SS components GspD and GspE, or its secretion substrate LipA, were unable to grow in vitro when long-chain fatty acids were supplemented as the sole source of carbon. A negative impact on the in vivo fitness of these mutants was also observed in immune-deficient mice.

Moreover, T2SS contributes to the pathogenesis of *A. baumannii* through LipAN, which is a phosphatidylcholine-degrading phospholipase C that displays phospholipase activity and benefits for the improved colonization of *A. baumannii* in the lungs of infected mice [44]. Indeed, by comparing the secretome of *A. baumannii* ATCC 17978 with that of the highly virulent MDR strain 5075, Elhosseiny et al. verified that the T2SS and its secretion substrates provided a colonization advantage to *A. baumannii* 5075 over ATCC 1797, but was more important to the latter for biofilm formation [51]. Similarly, the T2SS-dependent protease CpaA is necessary for the dissemination of *A. nosocomialis* from the initial infection site in the lungs to a distal site in the spleen [52]. The invasion-like adhesin InvL is capable of binding to extracellular matrix (ECM) components, in which fibrinogen shows the highest affinity with it, thus mediating the adhesion to urinary tract cells. Moreover, the *invL* mutant is attenuated in the catheter-associated urinary tract infection (CAUTI) model, verifying that the T2SS plays an important role in the uropathogenesis of *A. baumannii* through InvL [46].

Furthermore, the T2SS and its substrates also participate in the immune escape effect. Waack et al. [43] noticed that the loss of *gspD* resulted in a remarkable reduction in bacterial survival in human serum lacking factor C1q, which is a component of the classical complement pathway; however, it had no such effect in the absence of factor B, which mediates the alternative complement pathway. These findings indicate that the T2SS mediates the outer membrane translocation of an effector protein, thus contributing to in vivo fitness by protecting *A. baumannii* from the human alternative complement pathway [43].

In conclusion, *A. baumannii* mediates the release of various virulent substances through the T2SS and facilitates the adaptation of this organism to the environments and hosts, thus enhancing its ability to cause diseases.

3.2.3. Antibiotic Resistance

Drug resistance is a less mentioned topic in the T2SS field. However, Elhosseiny et al. [51] recently discovered that the T2SS was involved in the resistance of the fluoroquinolone antibiotic ciprofloxacin in AB5075, where an eight-fold increase in the MIC value of ciprofloxacin was detected in the *gspD* loss mutant, and the value was restored upon mutant complementation. The altered expression of outer membrane porins or efflux pumps that are controlled by the T2SS may contribute to antibiotic resistance; however, further research is required to confirm this speculation.

4. Type IV Secretion System (T4SS)

T4SSs are multiprotein nanomachines, widespread in Gram-negative and Gram-positive bacteria, that deliver macromolecules, e.g., DNA and protein, to bacterial recipients or eukaryotic target cells [53]. They are generally divided into three groups; namely, type F and P (IVA), IVB, and GI systems [54,55]. However, T4SSs are less reported in *A. baumannii*. The information can be summarized from five studies, as discussed below. By using the high-density pyrosequencing method, the elements homologous to the Legionella/Coxiella T4S apparatus were first discovered in *A. baumannii* ATCC17978 [56]. Later, in a pathogenic isolate, ACICU, the plasmid pACICU2 was found harboring a complete *tra* locus, which encoded the conjugative apparatus and an F-type T4SS (based on the F-plasmid of *Escherichia coli*) [57]. However, the structure and function of the *A. baumannii* T4SS were not illustrated in these two studies. Furthermore, the plasmid replicase (*rep*) gene *rep*Aci6 from pACICU2 was found widely distributed in *A. baumannii* clinical strains, which carried the T4SS protein TraC coding gene [58,59]. Thus, *rep*Aci6 served as a candidate for screening the F-type T4SS, and the plasmid carried the genes required for the biogenesis of the T4SS, such as *traC*, *traD*, and *traU*, which were identified in clinical carbapenem-resistant *A. baumannii* (CRAB) isolates [60].

4.1. Gene and Structure

The F-type T4SS in *A. baumannii* contains a series of *tra* operon genes, including *traA*, *traB*, *traC*, *traD*, *traE*, *traF*, *traG*, *traH*, *traI*, *traK*, *traL*, *traM*, *traN*, *traU*, *traV*, and *traW*, as well as another two genes, *trbC* and *finO*. Through the alignment of seven F-like *A. baumannii* plasmids, it was observed that the core genes involved in pilus biosynthesis (*traA*, *traB*, *traC*, *traF*, *traH*, *traK*, *traU*, *traV*, *traW*, and *trbC*), nicking (*traI*), the initiation of transfer (*traM* and *traD*), mating aggregate stabilization (*traN* and *traG*), and regulation (*finO*) were highly conserved [60] (Figure 3a).

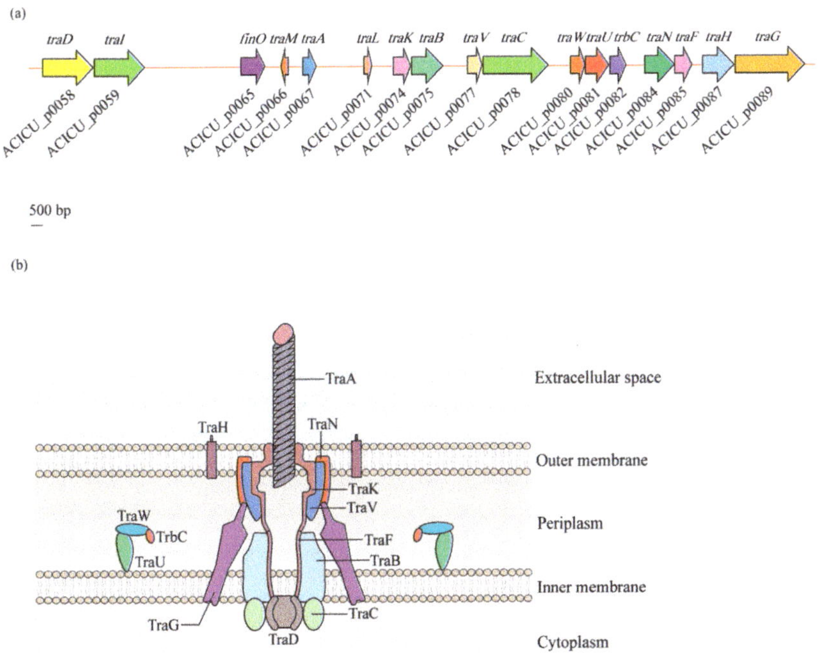

Figure 3. Structural organization of the type IV secretion system (T4SS) in *A. baumannii*: (**a**) Discovered in the *A. baumannii* ACICU plasmid pACICU2, the F-type T4SS contains a series of *tra* operon genes,

and two other genes, *trbC* and *finO*. (**b**) The T4SS is a highly sophisticated nanomachine spanning the entire bacterial cell envelope in *A. baumannii*. The F-like T4SS apparatus is composed of a pilus assembly component (TraA), a core complex (TraK, TraV, TraN, and TraH) embedded in the outer membrane (OM), an inner membrane (IM) platform (TraF, TraB, TraG, TraU, TraW, and TrbC), and components of the cytoplasm (TraC and TraD).

According to the analysis of Liu et al. [60], the T4SS of *A. baumannii* is a symmetrical barrel-shaped structure that is divided into the following units: (1) the pilus assembly component localized in the extracellular space across the OM (TraA); (2) the core complex embedded in the OM (TraK, TraV, TraN, and TraH); (3) the constituents of an IM platform (TraF, TraB, TraG, TraU, TraW, and TrbC); and (4) the components of the cytoplasm (TraC and TraD). This structure is similar to that of the typical VirB/D4 T4SS, which exists on the *Agrobacterium tumefaciens* Ti plasmid, and has gene consistency with *tra* operons as *traB/virB10*, *traC/virB4*, and *traD/virD4* [53,61] (Figure 3b).

4.2. Function

Although there has not been any empirical evidence demonstrating the function of the T4SS in *A. baumannii*, it is rational to speculate that it has similar performances to the universal T4SSs in other bacteria, as they share similar structures [60,62].

4.2.1. DNA Exchange and Antibiotic Resistance

Generally speaking, T4SSs are ancestrally related to bacterial conjugation machines, which mediate the transfer of genes and proteins across membranes [61]. T4SSs can recognize DNA substrates and translocate them to recipient bacterial cells by conjugative transfer. In this way, horizontal gene transfer is performed to disseminate mobile genetic elements, such as antibiotic resistance genes, virulence genes, and other fitness traits, to benefit bacteria by enhancing their survival in various environments and promoting the evolution of infectious pathogens [61]. The spreading of antibiotic resistance genes will lead to the rapid development of drug-resistant bacteria and even cause outbreaks of nosocomial infections. The conservation of most T4SS genes between *A. baumannii* and *E. coli* K-12 indicates that the function of the T4SS could be essential and unique in conjugation-mediated gene transfer [60].

Moreover, instead of connecting with the donor cells, T4SSs can either release naked DNAs to the milieu, or take up DNAs from the extracellular environments, therefore fulfilling the exchange of DNAs with the milieu [62,63].

4.2.2. Virulence

T4SSs can also act as effector translocator systems that deliver bacterial effector proteins across both the membrane of bacteria and eukaryotic target cells, finally contributing to bacterial pathogenicity by assisting the colonization and propagation of bacteria in the eukaryotic host, as well as the activation of pro-inflammation, apoptosis, and cytoskeleton rearrangements of host cells [54,64,65]. In addition, T4SSs are able to deliver a killing toxin to the bacterial neighbors to maintain the advantage of survival [62]. The T4SS may, to a large extent, contribute to the pathogenesis of *A. baumannii*. However, further research needs to be performed to investigate its exact functions.

5. Type V Secretion System (T5SS)

The T5SS, also known as the autotransporter, is a series of simple protein export pathways that are distributed in a large range of Gram-negative bacteria [66]. They are classified into monomeric autotransporters (MA), trimeric autotransporters (TA), and two-partner secretion systems (TPSS), with the composition of a single polypeptide for MA and TA, and separate polypeptide chains for TPSS [67,68]. Depending on the different structural features and domain organization, the T5SS is divided into five known subclasses, so-called types Va to Ve, and possibly another recently identified type, Vf [68]. However, only two types, Vb and Vc, have been identified in *A. baumannii* [34]. Therefore, type Vb and type Vc will be the focused of this review.

5.1. Gene and Structure

In contrast to other types of secretion systems that span the entire cell envelope with a syringe-shape structure, the T5SS only spans the OM. The T5SS consists of three major regions; namely, a signal sequence at the N-terminus, an extracellular secreted passenger, and a β-barrel domain (transporter) at the C-terminal that anchors the protein to the bacterial OM [68,69] (Figure 4a). Being produced in the cytoplasm, the protein is recognized at the N-terminal signal peptide, which targets the Sex complex to mediate the inner-membrane translocation of the protein to the periplasm [34]. Thereafter, the C-terminal transporter domain inserts into the OM and secretes the protein to the external environment through its OM pore. Finally, the passenger domain located between the signal peptide and the β-barrel domain displays the specific effector function extracellularly after proteolytic cleavage [40].

Type Vc is the most popular T5SS in the *A. baumannii* chromosome that belongs to the TA family. Therefore, the protein of type Vc in this bacterium is designated as the *Acinetobacter* trimeric autotransporter (Ata) [70]. Encoded by the *ata* gene, the autotransporter Ata contains a long signal peptide followed by an N-terminus, a surface-exposed passenger domain, and a C-terminal domain encoding four β-strands [70] (Figure 4b).

In contrast to classical autotransporters, type Vb belongs to TPSS, where the passenger and translocator (β-barrel) domains locate in two distinct polypeptide chains that are formed by TpsA and TpsB [67]. TpsA and TpsB are encoded in one operon, and the former connects at the polypeptide transport-associated (POTRA) domain of the latter for secretion through the OM to either be surface-displayed or transported extracellularly [71]. In this way, when releasing the passenger out of the cells after being transported by the β-barrel domain, there is no need for release by proteolytic cleavage [68]. In the *A. baumannii* strain AbH12O-A2, AbFhaB and AbFhaC were found to represent TpsA and TpsB, respectively, due to the highly conserved structure of these proteins [72] (Figure 4c).

Another type of Vb recently observed in *A. baumannii* is the CDI system composed of CdiA and CdiB. CdiA is a large multi-domain protein that forms a filament folded as a β-helix, similarly to TpsA, and has a C-terminal toxin domain. The CdiA protein in the periplasm is released to the cell surface by the OM transporter CdiB, and its β-helix presents the toxin domain to the neighboring bacteria, finally inhibiting their growth [73]. A cytoplasmic immunity protein, CdiI, is also expressed by the CDI operon to protect bacteria from fratricide and auto-inhibition by CdiA toxins [74–76] (Figure 4c).

5.2. Function

T5SSs play crucial roles in the virulence of Gram-negative bacteria. The function of the passenger domain from different bacteria is highly diverse, where it can be enzymatic, proteolytic, toxic, or adhesive, thus contributing to bacterial virulence in colonization, intracellular mobility, nutrient acquisition, immune evasion, the alteration of host cell processes, and biofilm formation [68,77].

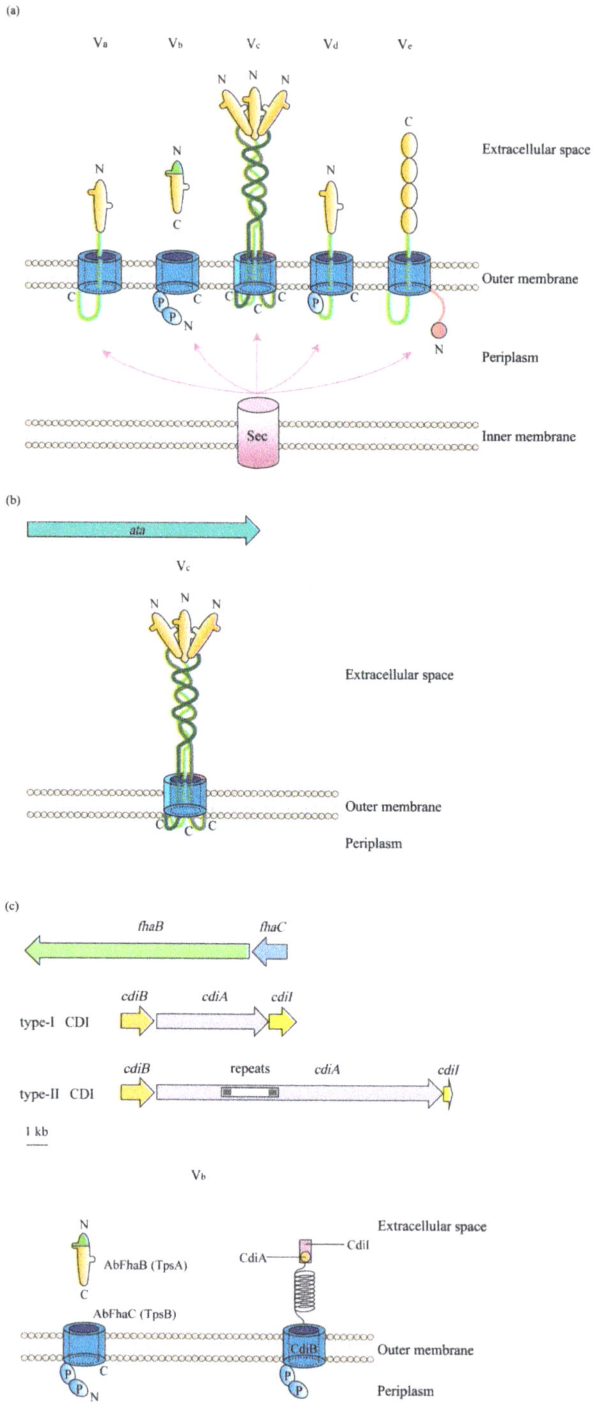

Figure 4. Structure of the type V secretion system (T5SS) in *A. baumannii*: (**a**) There are five types of T5SS in Gram-negative bacteria. They consist of three parts: a signal sequence at the N-terminus, a

secreted passenger in the extracellular milieu, and a transporter at the C-terminal. β-Barrels are displayed in blue; linkers and the two-partner secretion (TPS) domains are in green; passenger regions are in orange; polypeptide transport-associated (POTRA) domains are labeled as P; and the N- and C-termini are indicated. The translocation of substrates for subclasses of T5SS from the cytoplasm to the periplasm relies on the Sec pathway. (**b**) Type Vc is the most frequently identified T5SS in *A. baumannii*. It is formed by a trimeric protein, Ata, which contains a signal peptide at the N-terminus, a surface-exposed passenger domain, and a C-terminal domain. (**c**) Two forms of type Vb are found in *A. baumannii*. The one belonging to the TPSS is constructed of AbFhaB and AbFhaC, which represent TpsA and TpsB in other Gram-negative bacteria, respectively. AbFhaB (TpsA) is the passenger domain that is secreted out of cells through the outer membrane (OM) by AbFhaC (TpsB), which is the translocator domain located in the OM. Another one is the contact-dependent inhibition (CDI) system composed of CdiA and CdiB. Similar to TpsA, the toxin CdiA is released from the periplasm to the cell surface by the OM transporter CdiB.

5.2.1. Function of Type Vc System

In *A. baumannii*, the type Vc autotransporter, Ata, is present in many clinical isolates, and is reported to be produced at the highest level during the very early exponential phase. Ata is critical for biofilm formation, binding to various extracellular/basal matrix proteins, including collagen types I, III, IV, and V and laminin. The biofilm formed on adherent cells was significantly lower in the *ata* deletion mutant. In addition, Ata mediates the virulence of *A. baumannii* by binding to collagen type IV, which promoted the survival of strains in a mouse model of lethal infection [70]. Further study revealed that Ata bound to host glycans with high affinity, including galactose, N-acetylglucosamine, and galactose (β1-3/4) N-acetylglucosamine. This ability was crucial for Ata to recognize human plasma fibronectin during host adherence, as deglycosylated fibronectin had no interaction with Ata [78]. In addition to adhesion, Ata also mediates the in vivo invasion of *A. baumannii* and induces the apoptosis of the host cells [79].

5.2.2. Function of Type Vb Systems

The type Vb systems, including AbFhaB/FhaC and CdiA/CdiB, also play potential roles in the pathobiology of *A. baumannii*. The AbFhaB/FhaC system is involved in the attachment and fibronectin-mediated adherence to host cells. Moreover, these systems participate in the virulence of *A. baumannii*, where higher fertility and survival rates were monitored in *Caenorhabditis elegans* and mouse infection models, respectively, when *fhaC* was absent [72]. In contrast, the CDI system is reported to inhibit the growth of non-immune neighboring cells and, on the other hand, to favor the formation of biofilm structures, thus promoting social interactions between CDI^+ cells to facilitate biofilm formation [80].

6. Type VI Secretion System (T6SS)

The T6SS is a multiprotein transmembrane nanomachine discovered in numerous Gram-negative bacteria, including *Vibrio cholerae*, *Escherichia coli*, *Pseudomonas aeruginosa*, *Klebsiella pneumoniae*, *Francisella tularensis*, and *Yersinia pseudotuberculosis* [81]. It is syringe-shaped and is commonly used by bacteria to inject toxic effectors into competitors or host cells [82]. Several parts of this secretion system are structurally and functionally homologous to the T4 bacteriophage tail, suggesting a common evolutionary origin of this apparatus [83]. In recent years, an increasing number of studies have reported various aspects of the T6SS from *A. baumannii*, including its composition, structure, regulation, and function, confirming it as an important virulence factor.

6.1. Gene and Structure

The T6SS in *A. baumannii* is found in a cluster located in the genome that contains 18 genes, arranged as *asaA-tssBC-hcp(tssD)-tssEFG-asaB-tssM-tagFN-asaC-tssHAKL-asaDE*, while genes of *vgrG*, also known as *tssI*, which are scattered in various numbers throughout the genome [84,85]. In these genes, 12 encode the core T6SS proteins (Tss, Hcp, and VgrG), two encode the TagF and TagN that are associated with the T6SS in other bacteria, and five encode the Asa proteins that only appear in *Acinetobacter* spp. [84] (Figure 5a). Based on the Tss core proteins, the T6SS is composed of three main parts: a membrane complex, a cytoplasmic baseplate, and a contractile tail tube/sheath complex (Figure 5b).

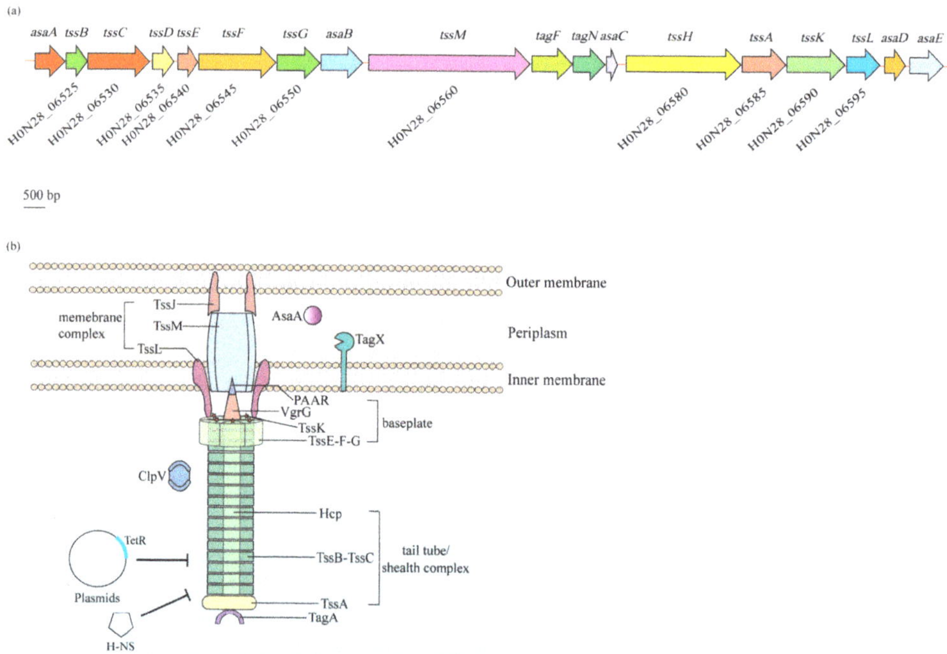

Figure 5. Biogenesis and regulation of the type VI secretion system (T6SS) in *A. baumannii*. The T6SS is a class of macromolecular secretion machines, which translocate proteins into a variety of recipient cells: (**a**) A single gene cluster carries 18 putative genes that are predicted to encode components of the T6SS. Among them, 12 core genes (*tss*) are coded on the chromosome of *A. baumannii* ATCC 17978. (**b**) The T6SS is composed of three main parts: a membrane complex (TssJ, TssL, and TssM), a cytoplasmic baseplate (TssK, TssF, TssG, TssE, VgrG, and PAAR), and a contractile tail tube/sheath complex (Hcp, TssB, TssC, and TssA). The expression of the T6SS is negatively regulated by the TetR-like proteins encoded on the large, conjugative plasmid pAB3 and proteins within the H-NS family.

Normally, in a wide range of bacteria, the membrane complex consists of the TssJ, TssL, and TssM proteins that span the cell envelope, with the complex anchored in the IM and the tip embedded in the OM, but not crossing it [86]. Notably, TssJ, an OM lipoprotein interacting with TssM, is absent in *A. baumannii* [85]. TssM and TssL have strong homology with the T4bSS proteins IcmF and IcmH (or DotU), respectively [87,88]. TssM is a core component of the T6SS that anchors to the IM through three transmembrane segments [88]. Similarly, the cytoplasmic protein TssL is also bound to the IM, but through a single transmembrane helix. Two residues of TssL in *A. baumannii*, Asp98 and Glu99, are strongly conserved among T6SS-encoding Gram-negative bacteria, and remarkably impact the dynamics, expression, and functionality of this protein [89]. TssM and TssL are involved in the recruitment and secretion of Hcp, and are important for the activity of the T6SS [90].

The baseplate complex is a central piece of the T6SS machinery that consists of six $(TssK)_6$-$(TssF)_2$-$(TssG)_1$-$(TssE)_1$ wedges around a central $(VgrG)_3$-PAAR spike. It connects the tail to the membrane complex and initiates the polymerization of the tail tube/sheath complex [91]. TssG is the core component of a baseplate wedge, where its C-terminal domain acts as an adaptor to interact with both TssF and TssK. VgrG, which binds to the PAAR-repeat protein at its distal extremity, is essential for the assembly of the Hcp tube, thus significantly contributing to the structure of the T6SS in various bacteria, including *A. baumannii* [92–94].

The tail tube/sheath complex is a contractile structure formed by the Hcp tube, TssBC sheath, and TssA cap. Although VgrG locates in the center of the baseplate complex, it is identified as an extension of the Hcp tube, as the central density of the latter is uniform from the first ring docked on top of the $(VgrG)_3$-PAAR spike [93]. Normally, the inner Hcp tube assembles onto the base of VgrG and extends into the cytoplasm. Simultaneously, the TssBC helical sheath polymerizes around the Hcp tube in an extended, high-energy "primed" conformation [95]. Additionally, its proximal ring has been suggested to interact with the TssK-TssF-TssG complex [92]. After contraction, the sheath is disassembled by the AAA^+ ATPase ClpV for a new assembly cycle of an extended sheath [96]. Lastly, TssA is involved in the assembly of Hcp-TssBC, and caps the distal end of this structure [95,97].

In addition to the core components, additional auxiliaries are required for the *A. baumannii* T6SS to ensure the correct assembly and full activity. For example, TagF and TagN were identified to negatively regulate the activity of the T6SS, where the absence of these two proteins increased the secretion of Hcp [98]. Moreover, AsaA was demonstrated to localize in the periplasmic space and affect the assembly or stability of the T6SS by interacting with TssM [99]. Additionally, a novel peptidoglycan hydrolase, TagX, was proposed to be required for the transit of the T6SS machinery across the peptidoglycan layer, thus finally allowing the assembly of the T6SS [98].

6.2. Function

6.2.1. Virulence

Similarly to other Gram-negative bacteria, the T6SS is a multifunctional apparatus in *A. baumannii*. Bacterial virulence is the most concerning effect raised by the T6SS. First of all, outcompeting other bacterial competitors is one critical factor to evaluate the virulence. As observed by Kim et al., clinical *A. baumannii* isolates causing bacteremia were shown to outcompete *E. coli* in a T6SS-dependent manner [100]. This is primarily due to the injection of toxic effectors into target cells, which may have bacteriostatic or bactericidal activities [84]. The T6SS effectors that have been characterized to date include $NAD(P)^+$ glycohydrolase, ADP-ribosylating toxins, (p)ppApp synthetase, and Rhs [101–104]. Current research has found a synergistic effect of D-lysine on the peptidoglycanase activity of the T6SS effector Tse4 in *A. baumannii*. Additionally, the T6SS-mediated killing effect on Gram-positive bacteria was also seen with the lethal combination of D-lysine and Tse4 [105]. Moreover, the significantly enhanced efficacy of the T6SS can be induced by toxins that not only kill, but also quickly lyse, competitor bacteria [82]. The injection procedure was revealed to be reliant upon TssB, Hcp, and TssM [84,106].

Secondly, the T6SS is involved in pathogenicity in eukaryotic hosts. Repizo et al. reported that the T6SS was required for the host colonization of *A. baumannii* in the *G. mellonella* model [106]. Furthermore, T6SS-active clinical strains were found to survive better in the presence of human serum and were more frequently detected in patients with a catheter-related bloodstream infection, hematopoietic stem cell transplantation, and immunosuppressive agent therapy [100]. Higher *hcp* expression was found in invasive *A. baumannii* isolates under the status of respiratory infection and could be triggered by the acid environment [107]. VgrG is also involved in the virulence of *A. baumannii*. It was found to be beneficial to cellular adherence and systemic infection in hosts. The number of *vgrG* deletion mutants adhering to human lung epithelial cells was much lower than that of the wild-type strain, and the lethal ability of this mutant to mice was also decreased [108].

6.2.2. Antibiotic Resistance

Antibiotic resistance is also found to be closely related to the T6SS. Dong et al. revealed that the presence of the T6SS in XDR *A. baumannii* isolates was significantly higher than that in MDR strains, followed by that in sensitive isolates [109]. Compared with the wild-type *A. baumannii* strain, an increased resistance to ampicillin/sulbactam and a decreased resistance to chloramphenicol were detected in a *vgrG*-lacking mutant [108]. The T6SS may be a contributor to inter-species horizontal gene transfer (HGT), which is one of the critical drug resistance mechanisms in bacteria. It was revealed to promote HGT by lysing the neighboring *E. coli* cells, whose genes were subsequently gained by *Acinetobacter* [110]. Indeed, when *hcp* was missing from T6SS+ *A. baumannii* A152, the ability of this strain to acquire antimicrobial resistance plasmids from *E. coli* was reduced remarkably [109]. Nevertheless, the function of the T6SS in *A. baumannii* is instead repressed during the intra-species dissemination of multidrug-resistant plasmids. As discovered by Weber et al., the T6SS was negatively regulated by a large resistance plasmid containing TetR-like regulators, which presented in a wide range of *A. baumannii* [111]. Additionally, the dissemination and conjugation of MDR plasmids among *A. baumannii* isolates relied on their distinctive ability to repress the function of the T6SS [112].

7. Conclusions and Future Perspectives

Although six types of secretion systems have been well characterized in numerous Gram-negative bacteria over the past decades, the study in this field is in the initial stage for *A. baumannii*. This bacterium is commonly regarded as a low-grade pathogen, whereas, based on the current knowledge, five types of secretion systems have been discovered in this species, including the T1SS, T2SS, T4SS, T5SS, and T6SS. Among them, the T2SS, T6SS, and Ata from the T5SS are the most frequently reported secretion systems in *A. baumannii*, and can coexist in the majority of isolates. They share high similarity with the composition and structure of the secretion systems in other bacteria and are involved in different aspects, such as pathogenicity and antibiotic resistance.

Virulence, mediated by this mechanism, is mainly achieved via the secretion of various effectors, which not only promote the fitness of *A. baumannii* in different environments, but also alter the physiology of its hosts. Further research to investigate additional effectors, as well as their enzymatic activity, will provide more information on the contribution of secretion systems during human infections.

Secretion systems are normally considered to be involved in bacterial virulence, but less is known about their relationship with antibiotic resistance. A few studies have begun to reveal the changes in antibiotic susceptibility in the mutants of some types of secretion systems, such as the T2SS and T6SS in *A. baumannii*. However, the mechanism for drug resistance mediated by secretion systems is not fully understood. Additional work is necessary to confirm the role that various secretion systems play in the antibiotic resistance in *A. baumannii*. Additionally, more functional components and effectors should be genetically investigated to facilitate a deeper understanding of the mechanism.

Treatment and prevention strategies based on secretion systems can be considered. Currently, sulbactam, carbapenems, aminoglycosides, polymyxins, tigecycline, and tetracycline are recommended for the therapy of *Acinetobacter* infections [113]. Among them, carbapenems are known as the "last line of defense" against Gram-negative bacteria and have been the preferred remedy choice for MDR *A. baumannii* infections in the past few decades. However, an increased incidence of resistance towards this sort of antibiotic has been reported in recent years, and carbapenem-resistant *A. baumannii* (CRAB) has become a global threat to human health, as it is commonly associated with a broad range of co-resistance to other antibiotic classes [114–116]. Consequently, the World Health Organization (WHO) lists CRAB in the critical group for the research and development of new antibiotics [115]. In addition, its resistance to other antimicrobials also continues to increase. Therefore, new drugs are in urgent need, and those that target secretion systems could be a good choice.

The T2SS and T6SS are the most frequently discovered secretion systems in *A. baumannii*, demonstrating their key roles in this bacterium. Some medicines, such as Orlistat, have been reported to be able to prevent the growth of *A. baumannii* by inhibiting LipA [36], while blocking the function of GspD or the pathways of Sec and Tat can contribute to the inhibition of virulence mediated by the T2SS [117]. Conserved sequences of VgrG were identified to be antigenic in various strains of *A. baumannii*; thus, it could be used to develop multivalent vaccines [118,119]. In addition, the T5SS protein, Ata, has been reported to be a potential candidate vaccine by many researchers because of its outstanding protection effect against the lethal challenges of various *A. baumannii* strains [120–122]. Additionally, further investigations on more effectors of the T2SS and T6SS, as well as other less reported secretion systems, may offer further opportunities to control the infections caused by drug-resistant *A. baumannii*.

Author Contributions: Conceptualization, L.H.; writing—original draft preparation, P.L. and L.H.; writing—review and editing, L.H., P.L., S.Z., J.W., M.M.A.-S., B.H., Y.C. and S.H.; visualization, P.L.; supervision, L.H.; funding acquisition, L.H. All authors have read and agreed to the published version of the manuscript.

Funding: This work was funded by the National Natural Science Foundation of China (81702043), the Fundamental Research Funds for the Central Universities (xzy012020050), and the "Basic-Clinical" Fusion Innovation Project of Xi'an Jiaotong University (YXJLRH2022020).

Institutional Review Board Statement: Not applicable.

Informed Consent Statement: Not applicable.

Data Availability Statement: Not applicable.

Conflicts of Interest: The authors declare no conflict of interest.

References

1. Giamarellou, H.; Antoniadou, A.; Kanellakopoulou, K. Acinetobacter baumannii: A universal threat to public health? *Int. J. Antimicrob. Agents* **2008**, *32*, 106–119. [CrossRef] [PubMed]
2. Ramirez, M.S.; Bonomo, R.A.; Tolmasky, M.E. Carbapenemases: Transforming Acinetobacter baumannii into a Yet More Dangerous Menace. *Biomolecules* **2020**, *10*, 720. [CrossRef] [PubMed]
3. Boucher, H.W.; Talbot, G.H.; Bradley, J.S.; Edwards, J.E.; Gilbert, D.; Rice, L.B.; Scheld, M.; Spellberg, B.; Bartlett, J. Bad bugs, no drugs: No ESKAPE! An update from the Infectious Diseases Society of America. *Clin. Infect. Dis.* **2009**, *48*, 1–12. [CrossRef] [PubMed]
4. Pogue, J.M.; Kaye, K.S.; Cohen, D.A.; Marchaim, D. Appropriate antimicrobial therapy in the era of multidrug-resistant human pathogens. *Clin. Microbiol. Infect.* **2015**, *21*, 302–312. [CrossRef]
5. Espinal, P.; Marti, S.; Vila, J. Effect of biofilm formation on the survival of Acinetobacter baumannii on dry surfaces. *J. Hosp. Infect.* **2012**, *80*, 56–60. [CrossRef]
6. Longo, F.; Vuotto, C.; Donelli, G. Biofilm formation in Acinetobacter baumannii. *New Microbiol.* **2014**, *37*, 119–127.
7. Harris, A.D.; Johnson, J.K.; Pineles, L.; O'Hara, L.M.; Bonomo, R.A.; Thom, K.A. Patient-to-Patient Transmission of Acinetobacter baumannii Gastrointestinal Colonization in the Intensive Care Unit. *Antimicrob. Agents Chemother.* **2019**, *63*, e00392-19. [CrossRef]
8. Bayuga, S.; Zeana, C.; Sahni, J.; Della-Latta, P.; El-Sadr, W.; Larson, E. Prevalence and antimicrobial patterns of Acinetobacter baumannii on hands and nares of hospital personnel and patients: The iceberg phenomenon again. *Heart Lung* **2002**, *31*, 382–390. [CrossRef]
9. Nasr, P. Genetics, epidemiology, and clinical manifestations of multidrug-resistant Acinetobacter baumannii. *J. Hosp. Infect.* **2020**, *104*, 4–11. [CrossRef]
10. Nowak, P.; Paluchowska, P. Acinetobacter baumannii: Biology and drug resistance—Role of carbapenemases. *Folia Histochem. Cytobiol.* **2016**, *54*, 61–74. [CrossRef]
11. Lee, C.R.; Lee, J.H.; Park, M.; Park, K.S.; Bae, I.K.; Kim, Y.B.; Cha, C.J.; Jeong, B.C.; Lee, S.H. Biology of Acinetobacter baumannii: Pathogenesis, Antibiotic Resistance Mechanisms, and Prospective Treatment Options. *Front. Cell. Infect. Microbiol.* **2017**, *7*, 55. [CrossRef]
12. Gheorghe, I.; Barbu, I.C.; Surleac, M.; Sarbu, I.; Popa, L.I.; Paraschiv, S.; Feng, Y.; Lazar, V.; Chifiriuc, M.C.; Otelea, D.; et al. Subtypes, resistance and virulence platforms in extended-drug resistant Acinetobacter baumannii Romanian isolates. *Sci. Rep.* **2021**, *11*, 13288. [CrossRef] [PubMed]
13. Smiline Girija, A.S.; Ganesh, P.S. Virulence of Acinetobacter baumannii in proteins moonlighting. *Arch. Microbiol.* **2021**, *204*, 96. [CrossRef] [PubMed]
14. Wong, D.; Chao, J.D.; Av-Gay, Y. Mycobacterium tuberculosis-secreted phosphatases: From pathogenesis to targets for TB drug development. *Trends Microbiol.* **2013**, *21*, 100–109. [CrossRef]

15. Gerlach, R.G.; Hensel, M. Protein secretion systems and adhesins: The molecular armory of Gram-negative pathogens. *Int. J. Med. Microbiol.* **2007**, *297*, 401–415. [CrossRef]
16. Sawa, T.; Shimizu, M.; Moriyama, K.; Wiener-Kronish, J.P. Association between Pseudomonas aeruginosa type III secretion, antibiotic resistance, and clinical outcome: A review. *Crit. Care* **2014**, *18*, 668. [CrossRef]
17. Boudaher, E.; Shaffer, C.L. Inhibiting bacterial secretion systems in the fight against antibiotic resistance. *Medchemcomm* **2019**, *10*, 682–692. [CrossRef]
18. Ding, M.; Ye, Z.; Liu, L.; Wang, W.; Chen, Q.; Zhang, F.; Wang, Y.; Sjoling, A.; Martin-Rodriguez, A.J.; Hu, R.; et al. Subinhibitory antibiotic concentrations promote the horizontal transfer of plasmid-borne resistance genes from Klebsiellae pneumoniae to Escherichia coli. *Front. Microbiol.* **2022**, *13*, 1017092. [CrossRef] [PubMed]
19. Harding, C.M.; Pulido, M.R.; Di Venanzio, G.; Kinsella, R.L.; Webb, A.I.; Scott, N.E.; Pachon, J.; Feldman, M.F. Pathogenic Acinetobacter species have a functional type I secretion system and contact-dependent inhibition systems. *J. Biol. Chem.* **2017**, *292*, 9075–9087. [CrossRef]
20. Sycz, G.; Di Venanzio, G.; Distel, J.S.; Sartorio, M.G.; Le, N.H.; Scott, N.E.; Beatty, W.L.; Feldman, M.F. Modern Acinetobacter baumannii clinical isolates replicate inside spacious vacuoles and egress from macrophages. *PLoS Pathog.* **2021**, *17*, e1009802. [CrossRef]
21. Satchell, K.J. Structure and function of MARTX toxins and other large repetitive RTX proteins. *Annu. Rev. Microbiol.* **2011**, *65*, 71–90. [CrossRef] [PubMed]
22. Rahbar, M.R.; Rasooli, I.; Mousavi Gargari, S.L.; Amani, J.; Fattahian, Y. In silico analysis of antibody triggering biofilm associated protein in Acinetobacter baumannii. *J. Theor. Biol.* **2010**, *266*, 275–290. [CrossRef] [PubMed]
23. Loehfelm, T.W.; Luke, N.R.; Campagnari, A.A. Identification and characterization of an Acinetobacter baumannii biofilm-associated protein. *J. Bacteriol.* **2008**, *190*, 1036–1044. [CrossRef]
24. Brossard, K.A.; Campagnari, A.A. The Acinetobacter baumannii biofilm-associated protein plays a role in adherence to human epithelial cells. *Infect. Immun.* **2012**, *80*, 228–233. [CrossRef]
25. Azizi, O.; Shahcheraghi, F.; Salimizand, H.; Modarresi, F.; Shakibaie, M.R.; Mansouri, S.; Ramazanzadeh, R.; Badmasti, F.; Nikbin, V. Molecular Analysis and Expression of bap Gene in Biofilm-Forming Multi-Drug-Resistant Acinetobacter baumannii. *Rep. Biochem. Mol. Biol.* **2016**, *5*, 62–72.
26. Yang, C.; Huang, Q. Regulation of *Acinetobacter baumannii* biofilm formation. *Chin. J. Infect. Control.* **2012**, *11*, 228–235.
27. Teymournejad, O.; Rikihisa, Y. Ehrlichia chaffeensis Uses an Invasin To Suppress Reactive Oxygen Species Generation by Macrophages via CD147-Dependent Inhibition of Vav1 To Block Rac1 Activation. *mBio* **2020**, *11*, e00267-20. [CrossRef]
28. Ho, T.D.; Davis, B.M.; Ritchie, J.M.; Waldor, M.K. Type 2 secretion promotes enterohemorrhagic Escherichia coli adherence and intestinal colonization. *Infect. Immun.* **2008**, *76*, 1858–1865. [CrossRef]
29. Baldi, D.L.; Higginson, E.E.; Hocking, D.M.; Praszkier, J.; Cavaliere, R.; James, C.E.; Bennett-Wood, V.; Azzopardi, K.I.; Turnbull, L.; Lithgow, T.; et al. The type II secretion system and its ubiquitous lipoprotein substrate, SslE, are required for biofilm formation and virulence of enteropathogenic Escherichia coli. *Infect. Immun.* **2012**, *80*, 2042–2052. [CrossRef] [PubMed]
30. McCoy-Simandle, K.; Stewart, C.R.; Dao, J.; DebRoy, S.; Rossier, O.; Bryce, P.J.; Cianciotto, N.P. Legionella pneumophila type II secretion dampens the cytokine response of infected macrophages and epithelia. *Infect. Immun.* **2011**, *79*, 1984–1997. [CrossRef]
31. Sikora, A.E.; Zielke, R.A.; Lawrence, D.A.; Andrews, P.C.; Sandkvist, M. Proteomic analysis of the Vibrio cholerae type II secretome reveals new proteins, including three related serine proteases. *J. Biol. Chem.* **2011**, *286*, 16555–16566. [CrossRef] [PubMed]
32. Jyot, J.; Balloy, V.; Jouvion, G.; Verma, A.; Touqui, L.; Huerre, M.; Chignard, M.; Ramphal, R. Type II secretion system of Pseudomonas aeruginosa: In vivo evidence of a significant role in death due to lung infection. *J. Infect. Dis.* **2011**, *203*, 1369–1377. [CrossRef] [PubMed]
33. Tomas, A.; Lery, L.; Regueiro, V.; Perez-Gutierrez, C.; Martinez, V.; Moranta, D.; Llobet, E.; Gonzalez-Nicolau, M.; Insua, J.L.; Tomas, J.M.; et al. Functional Genomic Screen Identifies Klebsiella pneumoniae Factors Implicated in Blocking Nuclear Factor kappaB (NF-kappaB) Signaling. *J. Biol. Chem.* **2015**, *290*, 16678–16697. [CrossRef] [PubMed]
34. Elhosseiny, N.M.; Attia, A.S. Acinetobacter: An emerging pathogen with a versatile secretome. *Emerg. Microbes. Infect.* **2018**, *7*, 33. [CrossRef]
35. Eijkelkamp, B.A.; Stroeher, U.H.; Hassan, K.A.; Paulsen, I.T.; Brown, M.H. Comparative analysis of surface-exposed virulence factors of Acinetobacter baumannii. *BMC Genomics* **2014**, *15*, 1020. [CrossRef]
36. Johnson, T.L.; Waack, U.; Smith, S.; Mobley, H.; Sandkvist, M. Acinetobacter baumannii Is Dependent on the Type II Secretion System and Its Substrate LipA for Lipid Utilization and In Vivo Fitness. *J. Bacteriol.* **2015**, *198*, 711–719. [CrossRef]
37. Korotkov, K.V.; Sandkvist, M.; Hol, W.G. The type II secretion system: Biogenesis, molecular architecture and mechanism. *Nat. Rev. Microbiol.* **2012**, *10*, 336–351. [CrossRef]
38. Thomassin, J.L.; Santos Moreno, J.; Guilvout, I.; Tran Van Nhieu, G.; Francetic, O. The trans-envelope architecture and function of the type 2 secretion system: New insights raising new questions. *Mol. Microbiol.* **2017**, *105*, 211–226. [CrossRef]
39. Harding, C.M.; Kinsella, R.L.; Palmer, L.D.; Skaar, E.P.; Feldman, M.F. Medically Relevant Acinetobacter Species Require a Type II Secretion System and Specific Membrane-Associated Chaperones for the Export of Multiple Substrates and Full Virulence. *PLoS Pathog.* **2016**, *12*, e1005391. [CrossRef]

40. Costa, T.R.; Felisberto-Rodrigues, C.; Meir, A.; Prevost, M.S.; Redzej, A.; Trokter, M.; Waksman, G. Secretion systems in Gram-negative bacteria: Structural and mechanistic insights. *Nat. Rev. Microbiol.* **2015**, *13*, 343–359. [CrossRef]
41. Yan, Z.; Yin, M.; Xu, D.; Zhu, Y.; Li, X. Structural insights into the secretin translocation channel in the type II secretion system. *Nat. Struct. Mol. Biol.* **2017**, *24*, 177–183. [CrossRef]
42. Naskar, S.; Hohl, M.; Tassinari, M.; Low, H.H. The structure and mechanism of the bacterial type II secretion system. *Mol. Microbiol.* **2021**, *115*, 412–424. [CrossRef] [PubMed]
43. Waack, U.; Johnson, T.L.; Chedid, K.; Xi, C.; Simmons, L.A.; Mobley, H.L.T.; Sandkvist, M. Targeting the Type II Secretion System: Development, Optimization, and Validation of a High-Throughput Screen for the Identification of Small Molecule Inhibitors. *Front. Cell. Infect. Microbiol.* **2017**, *7*, 380. [CrossRef] [PubMed]
44. Elhosseiny, N.M.; El-Tayeb, O.M.; Yassin, A.S.; Lory, S.; Attia, A.S. The secretome of Acinetobacter baumannii ATCC 17978 type II secretion system reveals a novel plasmid encoded phospholipase that could be implicated in lung colonization. *Int. J. Med. Microbiol.* **2016**, *306*, 633–641. [CrossRef] [PubMed]
45. Tilley, D.; Law, R.; Warren, S.; Samis, J.A.; Kumar, A. CpaA a novel protease from Acinetobacter baumannii clinical isolates deregulates blood coagulation. *FEMS Microbiol. Lett.* **2014**, *356*, 53–61. [CrossRef]
46. Jackson-Litteken, C.D.; Di Venanzio, G.; Le, N.H.; Scott, N.E.; Djahanschiri, B.; Distel, J.S.; Pardue, E.J.; Ebersberger, I.; Feldman, M.F. InvL, an Invasin-Like Adhesin, Is a Type II Secretion System Substrate Required for Acinetobacter baumannii Uropathogenesis. *mBio* **2022**, *13*, e00258-22. [CrossRef] [PubMed]
47. Urusova, D.V.; Kinsella, R.L.; Salinas, N.D.; Haurat, M.F.; Feldman, M.F.; Tolia, N.H. The structure of Acinetobacter-secreted protease CpaA complexed with its chaperone CpaB reveals a novel mode of a T2SS chaperone-substrate interaction. *J. Biol. Chem.* **2019**, *294*, 13344–13354. [CrossRef]
48. Waack, U.; Warnock, M.; Yee, A.; Huttinger, Z.; Smith, S.; Kumar, A.; Deroux, A.; Ginsburg, D.; Mobley, H.L.T.; Lawrence, D.A.; et al. CpaA Is a Glycan-Specific Adamalysin-like Protease Secreted by Acinetobacter baumannii That Inactivates Coagulation Factor XII. *mBio* **2018**, *9*, e01606-18. [CrossRef]
49. Haurat, M.F.; Scott, N.E.; Di Venanzio, G.; Lopez, J.; Pluvinage, B.; Boraston, A.B.; Ferracane, M.J.; Feldman, M.F. The Glycoprotease CpaA Secreted by Medically Relevant Acinetobacter Species Targets Multiple O-Linked Host Glycoproteins. *mBio* **2020**, *11*, e02033-20. [CrossRef]
50. Korotkov, K.V.; Sandkvist, M. Architecture, Function, and Substrates of the Type II Secretion System. *EcoSal Plus* **2019**, *8*. [CrossRef]
51. Elhosseiny, N.M.; Elhezawy, N.B.; Attia, A.S. Comparative proteomics analyses of Acinetobacter baumannii strains ATCC 17978 and AB5075 reveal the differential role of type II secretion system secretomes in lung colonization and ciprofloxacin resistance. *Microb. Pathog.* **2019**, *128*, 20–27. [CrossRef] [PubMed]
52. Kinsella, R.L.; Lopez, J.; Palmer, L.D.; Salinas, N.D.; Skaar, E.P.; Tolia, N.H.; Feldman, M.F. Defining the interaction of the protease CpaA with its type II secretion chaperone CpaB and its contribution to virulence in Acinetobacter species. *J. Biol. Chem.* **2017**, *292*, 19628–19638. [CrossRef] [PubMed]
53. Costa, T.R.D.; Harb, L.; Khara, P.; Zeng, L.; Hu, B.; Christie, P.J. Type IV secretion systems: Advances in structure, function, and activation. *Mol. Microbiol.* **2021**, *115*, 436–452. [CrossRef] [PubMed]
54. Christie, P.J.; Atmakuri, K.; Krishnamoorthy, V.; Jakubowski, S.; Cascales, E. Biogenesis, architecture, and function of bacterial type IV secretion systems. *Annu. Rev. Microbiol.* **2005**, *59*, 451–485. [CrossRef]
55. Juhas, M.; Crook, D.W.; Dimopoulou, I.D.; Lunter, G.; Harding, R.M.; Ferguson, D.J.; Hood, D.W. Novel type IV secretion system involved in propagation of genomic islands. *J. Bacteriol.* **2007**, *189*, 761–771. [CrossRef]
56. Smith, M.G.; Gianoulis, T.A.; Pukatzki, S.; Mekalanos, J.J.; Ornston, L.N.; Gerstein, M.; Snyder, M. New insights into Acinetobacter baumannii pathogenesis revealed by high-density pyrosequencing and transposon mutagenesis. *Genes Dev.* **2007**, *21*, 601–614. [CrossRef]
57. Iacono, M.; Villa, L.; Fortini, D.; Bordoni, R.; Imperi, F.; Bonnal, R.J.; Sicheritz-Ponten, T.; De Bellis, G.; Visca, P.; Cassone, A.; et al. Whole-genome pyrosequencing of an epidemic multidrug-resistant Acinetobacter baumannii strain belonging to the European clone II group. *Antimicrob. Agents Chemother.* **2008**, *52*, 2616–2625. [CrossRef]
58. Povilonis, J.; Seputiene, V.; Krasauskas, R.; Juskaite, R.; Miskinyte, M.; Suziedelis, K.; Suziedeliene, E. Spread of carbapenem-resistant Acinetobacter baumannii carrying a plasmid with two genes encoding OXA-72 carbapenemase in Lithuanian hospitals. *J. Antimicrob. Chemother.* **2013**, *68*, 1000–1006. [CrossRef]
59. Towner, K.J.; Evans, B.; Villa, L.; Levi, K.; Hamouda, A.; Amyes, S.G.; Carattoli, A. Distribution of intrinsic plasmid replicase genes and their association with carbapenem-hydrolyzing class D beta-lactamase genes in European clinical isolates of Acinetobacter baumannii. *Antimicrob. Agents Chemother.* **2011**, *55*, 2154–2159. [CrossRef]
60. Liu, C.C.; Kuo, H.Y.; Tang, C.Y.; Chang, K.C.; Liou, M.L. Prevalence and mapping of a plasmid encoding a type IV secretion system in Acinetobacter baumannii. *Genomics* **2014**, *104*, 215–223. [CrossRef]
61. Juhas, M.; Crook, D.W.; Hood, D.W. Type IV secretion systems: Tools of bacterial horizontal gene transfer and virulence. *Cell. Microbiol.* **2008**, *10*, 2377–2386. [CrossRef] [PubMed]
62. Grohmann, E.; Christie, P.J.; Waksman, G.; Backert, S. Type IV secretion in Gram-negative and Gram-positive bacteria. *Mol. Microbiol.* **2018**, *107*, 455–471. [CrossRef] [PubMed]

63. Hamilton, H.L.; Dominguez, N.M.; Schwartz, K.J.; Hackett, K.T.; Dillard, J.P. Neisseria gonorrhoeae secretes chromosomal DNA via a novel type IV secretion system. *Mol. Microbiol.* **2005**, *55*, 1704–1721. [CrossRef] [PubMed]
64. Schulein, R.; Guye, P.; Rhomberg, T.A.; Schmid, M.C.; Schroder, G.; Vergunst, A.C.; Carena, I.; Dehio, C. A bipartite signal mediates the transfer of type IV secretion substrates of Bartonella henselae into human cells. *Proc. Natl. Acad. Sci. USA* **2005**, *102*, 856–861. [CrossRef]
65. Backert, S.; Meyer, T.F. Type IV secretion systems and their effectors in bacterial pathogenesis. *Curr. Opin. Microbiol.* **2006**, *9*, 207–217. [CrossRef] [PubMed]
66. Bernstein, H.D. Type V Secretion in Gram-Negative Bacteria. *EcoSal Plus* **2019**, *8*. [CrossRef]
67. Leo, J.C.; Grin, I.; Linke, D. Type V secretion: Mechanism(s) of autotransport through the bacterial outer membrane. *Philos. Trans. R. Soc. Lond B Biol. Sci.* **2012**, *367*, 1088–1101. [CrossRef]
68. Meuskens, I.; Saragliadis, A.; Leo, J.C.; Linke, D. Type V Secretion Systems: An Overview of Passenger Domain Functions. *Front. Microbiol.* **2019**, *10*, 1163. [CrossRef]
69. Jose, J.; Jahnig, F.; Meyer, T.F. Common structural features of IgA1 protease-like outer membrane protein autotransporters. *Mol. Microbiol.* **1995**, *18*, 378–380. [CrossRef]
70. Bentancor, L.V.; Camacho-Peiro, A.; Bozkurt-Guzel, C.; Pier, G.B.; Maira-Litran, T. Identification of Ata, a multifunctional trimeric autotransporter of Acinetobacter baumannii. *J. Bacteriol.* **2012**, *194*, 3950–3960. [CrossRef]
71. Thanassi, D.G.; Stathopoulos, C.; Karkal, A.; Li, H. Protein secretion in the absence of ATP: The autotransporter, two-partner secretion and chaperone/usher pathways of gram-negative bacteria (review). *Mol. Membr. Biol.* **2005**, *22*, 63–72. [CrossRef] [PubMed]
72. Perez, A.; Merino, M.; Rumbo-Feal, S.; Alvarez-Fraga, L.; Vallejo, J.A.; Beceiro, A.; Ohneck, E.J.; Mateos, J.; Fernandez-Puente, P.; Actis, L.A.; et al. The FhaB/FhaC two-partner secretion system is involved in adhesion of Acinetobacter baumannii AbH12O-A2 strain. *Virulence* **2017**, *8*, 959–974. [CrossRef] [PubMed]
73. Guerin, J.; Botos, I.; Zhang, Z.; Lundquist, K.; Gumbart, J.C.; Buchanan, S.K. Structural insight into toxin secretion by contact-dependent growth inhibition transporters. *Elife* **2020**, *9*, e58100. [CrossRef] [PubMed]
74. Aoki, S.K.; Pamma, R.; Hernday, A.D.; Bickham, J.E.; Braaten, B.A.; Low, D.A. Contact-dependent inhibition of growth in Escherichia coli. *Science* **2005**, *309*, 1245–1248. [CrossRef]
75. Ruhe, Z.C.; Subramanian, P.; Song, K.; Nguyen, J.Y.; Stevens, T.A.; Low, D.A.; Jensen, G.J.; Hayes, C.S. Programmed Secretion Arrest and Receptor-Triggered Toxin Export during Antibacterial Contact-Dependent Growth Inhibition. *Cell* **2018**, *175*, 921–933. [CrossRef]
76. Ruhe, Z.C.; Nguyen, J.Y.; Xiong, J.; Koskiniemi, S.; Beck, C.M.; Perkins, B.R.; Low, D.A.; Hayes, C.S. CdiA Effectors Use Modular Receptor-Binding Domains To Recognize Target Bacteria. *mBio* **2017**, *8*, e00290-17. [CrossRef]
77. van Ulsen, P.; Rahman, S.; Jong, W.S.; Daleke-Schermerhorn, M.H.; Luirink, J. Type V secretion: From biogenesis to biotechnology. *Biochim. Biophys. Acta.* **2014**, *1843*, 1592–1611. [CrossRef]
78. Tram, G.; Poole, J.; Adams, F.G.; Jennings, M.P.; Eijkelkamp, B.A.; Atack, J.M. The Acinetobacter baumannii Autotransporter Adhesin Ata Recognizes Host Glycans as High-Affinity Receptors. *ACS Infect. Dis.* **2021**, *7*, 2352–2361. [CrossRef]
79. Weidensdorfer, M.; Ishikawa, M.; Hori, K.; Linke, D.; Djahanschiri, B.; Iruegas, R.; Ebersberger, I.; Riedel-Christ, S.; Enders, G.; Leukert, L.; et al. The Acinetobacter trimeric autotransporter adhesin Ata controls key virulence traits of Acinetobacter baumannii. *Virulence* **2019**, *10*, 68–81. [CrossRef]
80. De Gregorio, E.; Esposito, E.P.; Zarrilli, R.; Di Nocera, P.P. Contact-Dependent Growth Inhibition Proteins in Acinetobacter baylyi ADP1. *Curr. Microbiol.* **2018**, *75*, 1434–1440. [CrossRef]
81. Monjaras Feria, J.; Valvano, M.A. An Overview of Anti-Eukaryotic T6SS Effectors. *Front. Cell. Infect. Microbiol.* **2020**, *10*, 584751. [CrossRef] [PubMed]
82. Hofer, U. T6SS: Shoot and scrub. *Nat. Rev. Microbiol.* **2020**, *18*, 412–413. [CrossRef] [PubMed]
83. Leiman, P.G.; Basler, M.; Ramagopal, U.A.; Bonanno, J.B.; Sauder, J.M.; Pukatzki, S.; Burley, S.K.; Almo, S.C.; Mekalanos, J.J. Type VI secretion apparatus and phage tail-associated protein complexes share a common evolutionary origin. *Proc. Natl. Acad. Sci. USA* **2009**, *106*, 4154–4159. [CrossRef] [PubMed]
84. Carruthers, M.D.; Nicholson, P.A.; Tracy, E.N.; Munson, R.S., Jr. Acinetobacter baumannii utilizes a type VI secretion system for bacterial competition. *PLoS ONE* **2013**, *8*, e59388. [CrossRef]
85. Weber, B.S.; Miyata, S.T.; Iwashkiw, J.A.; Mortensen, B.L.; Skaar, E.P.; Pukatzki, S.; Feldman, M.F. Genomic and functional analysis of the type VI secretion system in Acinetobacter. *PLoS ONE* **2013**, *8*, e55142. [CrossRef] [PubMed]
86. Rapisarda, C.; Cherrak, Y.; Kooger, R.; Schmidt, V.; Pellarin, R.; Logger, L.; Cascales, E.; Pilhofer, M.; Durand, E.; Fronzes, R. In Situ and high-resolution cryo-EM structure of a bacterial type VI secretion system membrane complex. *EMBO J.* **2019**, *38*, e100886. [CrossRef]
87. Nguyen, V.S.; Douzi, B.; Durand, E.; Roussel, A.; Cascales, E.; Cambillau, C. Towards a complete structural deciphering of Type VI secretion system. *Curr. Opin. Struct. Biol.* **2018**, *49*, 77–84. [CrossRef] [PubMed]
88. Silverman, J.M.; Brunet, Y.R.; Cascales, E.; Mougous, J.D. Structure and regulation of the type VI secretion system. *Annu. Rev. Microbiol.* **2012**, *66*, 453–472. [CrossRef]
89. Ruiz, F.M.; Lopez, J.; Ferrara, C.G.; Santillana, E.; Espinosa, Y.R.; Feldman, M.F.; Romero, A. Structural Characterization of TssL from Acinetobacter baumannii: A Key Component of the Type VI Secretion System. *J. Bacteriol.* **2020**, *202*, e00210-20. [CrossRef]

90. Ma, L.S.; Narberhaus, F.; Lai, E.M. IcmF family protein TssM exhibits ATPase activity and energizes type VI secretion. *J. Biol. Chem.* **2012**, *287*, 15610–15621. [CrossRef]
91. Cherrak, Y.; Rapisarda, C.; Pellarin, R.; Bouvier, G.; Bardiaux, B.; Allain, F.; Malosse, C.; Rey, M.; Chamot-Rooke, J.; Cascales, E.; et al. Biogenesis and structure of a type VI secretion baseplate. *Nat. Microbiol.* **2018**, *3*, 1404–1416. [CrossRef] [PubMed]
92. Park, Y.J.; Lacourse, K.D.; Cambillau, C.; DiMaio, F.; Mougous, J.D.; Veesler, D. Structure of the type VI secretion system TssK-TssF-TssG baseplate subcomplex revealed by cryo-electron microscopy. *Nat. Commun.* **2018**, *9*, 5385. [CrossRef] [PubMed]
93. Nazarov, S.; Schneider, J.P.; Brackmann, M.; Goldie, K.N.; Stahlberg, H.; Basler, M. Cryo-EM reconstruction of Type VI secretion system baseplate and sheath distal end. *EMBO J.* **2018**, *37*, e97103. [CrossRef] [PubMed]
94. Lopez, J.; Ly, P.M.; Feldman, M.F. The Tip of the VgrG Spike Is Essential to Functional Type VI Secretion System Assembly in Acinetobacter baumannii. *mBio* **2020**, *11*, e02761-19. [CrossRef] [PubMed]
95. Coulthurst, S. The Type VI secretion system: A versatile bacterial weapon. *Microbiology* **2019**, *165*, 503–515. [CrossRef]
96. Forster, A.; Planamente, S.; Manoli, E.; Lossi, N.S.; Freemont, P.S.; Filloux, A. Coevolution of the ATPase ClpV, the sheath proteins TssB and TssC, and the accessory protein TagJ/HsiE1 distinguishes type VI secretion classes. *J. Biol. Chem.* **2014**, *289*, 33032–33043. [CrossRef]
97. Dix, S.R.; Owen, H.J.; Sun, R.; Ahmad, A.; Shastri, S.; Spiewak, H.L.; Mosby, D.J.; Harris, M.J.; Batters, S.L.; Brooker, T.A.; et al. Structural insights into the function of type VI secretion system TssA subunits. *Nat. Commun.* **2018**, *9*, 4765. [CrossRef]
98. Weber, B.S.; Hennon, S.W.; Wright, M.S.; Scott, N.E.; de Berardinis, V.; Foster, L.J.; Ayala, J.A.; Adams, M.D.; Feldman, M.F. Genetic Dissection of the Type VI Secretion System in Acinetobacter and Identification of a Novel Peptidoglycan Hydrolase, TagX, Required for Its Biogenesis. *mBio* **2016**, *7*, e01253-16. [CrossRef]
99. Li, L.; Wang, Y.N.; Jia, H.B.; Wang, P.; Dong, J.F.; Deng, J.; Lu, F.M.; Zou, Q.H. The type VI secretion system protein AsaA in Acinetobacter baumannii is a periplasmic protein physically interacting with TssM and required for T6SS assembly. *Sci. Rep.* **2019**, *9*, 9438. [CrossRef]
100. Kim, J.; Lee, J.Y.; Lee, H.; Choi, J.Y.; Kim, D.H.; Wi, Y.M.; Peck, K.R.; Ko, K.S. Microbiological features and clinical impact of the type VI secretion system (T6SS) in Acinetobacter baumannii isolates causing bacteremia. *Virulence* **2017**, *8*, 1378–1389. [CrossRef]
101. Whitney, J.C.; Quentin, D.; Sawai, S.; LeRoux, M.; Harding, B.N.; Ledvina, H.E.; Tran, B.Q.; Robinson, H.; Goo, Y.A.; Goodlett, D.R.; et al. An interbacterial NAD(P)(+) glycohydrolase toxin requires elongation factor Tu for delivery to target cells. *Cell* **2015**, *163*, 607–619. [CrossRef] [PubMed]
102. Ting, S.Y.; Bosch, D.E.; Mangiameli, S.M.; Radey, M.C.; Huang, S.; Park, Y.J.; Kelly, K.A.; Filip, S.K.; Goo, Y.A.; Eng, J.K.; et al. Bifunctional Immunity Proteins Protect Bacteria against FtsZ-Targeting ADP-Ribosylating Toxins. *Cell* **2018**, *175*, 1380–1392.e4. [CrossRef] [PubMed]
103. Ahmad, S.; Wang, B.; Walker, M.D.; Tran, H.R.; Stogios, P.J.; Savchenko, A.; Grant, R.A.; McArthur, A.G.; Laub, M.T.; Whitney, J.C. An interbacterial toxin inhibits target cell growth by synthesizing (p)ppApp. *Nature* **2019**, *575*, 674–678. [CrossRef] [PubMed]
104. Alcoforado Diniz, J.; Coulthurst, S.J. Intraspecies Competition in Serratia marcescens Is Mediated by Type VI-Secreted Rhs Effectors and a Conserved Effector-Associated Accessory Protein. *J. Bacteriol.* **2015**, *197*, 2350–2360. [CrossRef] [PubMed]
105. Le, N.H.; Pinedo, V.; Lopez, J.; Cava, F.; Feldman, M.F. Killing of Gram-negative and Gram-positive bacteria by a bifunctional cell wall-targeting T6SS effector. *Proc. Natl. Acad. Sci. USA* **2021**, *118*, e2106555118. [CrossRef]
106. Repizo, G.D.; Gagne, S.; Foucault-Grunenwald, M.L.; Borges, V.; Charpentier, X.; Limansky, A.S.; Gomes, J.P.; Viale, A.M.; Salcedo, S.P. Differential Role of the T6SS in Acinetobacter baumannii Virulence. *PLoS ONE* **2015**, *10*, e0138265. [CrossRef]
107. Hu, Y.Y.; Liu, C.X.; Liu, P.; Wu, Z.Y.; Zhang, Y.D.; Xiong, X.S.; Li, X.Y. Regulation of gene expression of hcp, a core gene of the type VI secretion system in Acinetobacter baumannii causing respiratory tract infection. *J. Med. Microbiol.* **2018**, *67*, 945–951. [CrossRef]
108. Wang, J.; Zhou, Z.; He, F.; Ruan, Z.; Jiang, Y.; Hua, X.; Yu, Y. The role of the type VI secretion system vgrG gene in the virulence and antimicrobial resistance of Acinetobacter baumannii ATCC 19606. *PLoS ONE* **2018**, *13*, e0192288. [CrossRef]
109. Dong, J.F.; Liu, C.W.; Wang, P.; Li, L.; Zou, Q.H. The type VI secretion system in Acinetobacter baumannii clinical isolates and its roles in antimicrobial resistance acquisition. *Microb. Pathog.* **2022**, *169*, 105668. [CrossRef]
110. Cooper, R.M.; Tsimring, L.; Hasty, J. Inter-species population dynamics enhance microbial horizontal gene transfer and spread of antibiotic resistance. *eLife* **2017**, *6*, e25950. [CrossRef]
111. Weber, B.S.; Ly, P.M.; Irwin, J.N.; Pukatzki, S.; Feldman, M.F. A multidrug resistance plasmid contains the molecular switch for type VI secretion in Acinetobacter baumannii. *Proc. Natl. Acad. Sci. USA* **2015**, *112*, 9442–9447. [CrossRef] [PubMed]
112. Di Venanzio, G.; Moon, K.H.; Weber, B.S.; Lopez, J.; Ly, P.M.; Potter, R.F.; Dantas, G.; Feldman, M.F. Multidrug-resistant plasmids repress chromosomally encoded T6SS to enable their dissemination. *Proc. Natl. Acad. Sci. USA* **2019**, *116*, 1378–1383. [CrossRef] [PubMed]
113. Fishbain, J.; Peleg, A.Y. Treatment of Acinetobacter infections. *Clin. Infect. Dis.* **2010**, *51*, 79–84. [CrossRef] [PubMed]
114. Nordmann, P.; Poirel, L. Epidemiology and Diagnostics of Carbapenem Resistance in Gram-negative Bacteria. *Clin. Infect. Dis.* **2019**, *69*, S521–S528. [CrossRef] [PubMed]
115. Tacconelli, E.; Carrara, E.; Savoldi, A.; Harbarth, S.; Mendelson, M.; Monnet, D.L.; Pulcini, C.; Kahlmeter, G.; Kluytmans, J.; Carmeli, Y.; et al. Discovery, research, and development of new antibiotics: The WHO priority list of antibiotic-resistant bacteria and tuberculosis. *Lancet Infect. Dis.* **2018**, *18*, 318–327. [CrossRef] [PubMed]
116. Kyriakidis, I.; Vasileiou, E.; Pana, Z.D.; Tragiannidis, A. Acinetobacter baumannii Antibiotic Resistance Mechanisms. *Pathogens* **2021**, *10*, 373. [CrossRef]

117. Ji, S.; Li, G. Advances in type II secretion system of *Acinetobacter baumannii*. *Prog. Microbiol. Immunol.* **2021**, *49*, 81–86. [CrossRef]
118. Alipouri, S.; Rasooli, I.; Ghaini, M.H.; Jahangiri, A.; Darvish Alipour Astaneh, S.; Ramezanalizadeh, F. Immunity induced by valine-glycine repeat protein G imparts histoprotection of vital body organs against Acinetobacter baumannii. *J. Genet. Eng. Biotechnol.* **2022**, *20*, 42. [CrossRef]
119. Pazoki, M.; Darvish Alipour Astaneh, S.; Ramezanalizadeh, F.; Jahangiri, A.; Rasooli, I. Immunoprotectivity of Valine-glycine repeat protein G, a potent mediator of pathogenicity, against Acinetobacter baumannii. *Mol. Immunol.* **2021**, *135*, 276–284. [CrossRef]
120. Hatefi Oskuei, R.; Darvish Alipour Astaneh, S.; Rasooli, I. A conserved region of Acinetobacter trimeric autotransporter adhesion, Ata, provokes suppression of Acinetobacter baumannii virulence. *Arch. Microbiol.* **2021**, *203*, 3483–3493. [CrossRef]
121. Sun, P.; Li, X.; Pan, C.; Liu, Z.; Wu, J.; Wang, H.; Zhu, L. A Short Peptide of Autotransporter Ata Is a Promising Protective Antigen for Vaccination Against Acinetobacter baumannii. *Front. Immunol.* **2022**, *13*, 884555. [CrossRef] [PubMed]
122. Ren, S.; Guan, L.; Dong, Y.; Wang, C.; Feng, L.; Xie, Y. Design and evaluation of a multi-epitope assembly peptide vaccine against Acinetobacter baumannii infection in mice. *Swiss Med. Wkly.* **2019**, *149*, w20052. [CrossRef] [PubMed]

antibiotics

MDPI

Article

Molecular Characteristics and Quantitative Proteomic Analysis of *Klebsiella pneumoniae* Strains with Carbapenem and Colistin Resistance

Ling Hao [1], Xiao Yang [2], Huiling Chen [2], Zexun Mo [1], Yujun Li [1], Shuquan Wei [1] and Ziwen Zhao [1,*]

[1] Department of Pulmonary and Critical Care Medicine, Guangzhou First People's Hospital, Guangzhou 510180, China
[2] Department of Laboratory Medicine, Guangzhou First People's Hospital, Guangzhou 510180, China
* Correspondence: eyzhaoziwen@scut.edu.cn

Abstract: Carbapenem-resistant *Klebsiella pneumoniae* (CRKP) are usually multidrug resistant (MDR) and cause serious therapeutic problems. Colistin is a critical last-resort therapeutic option for MDR bacterial infections. However, increasing colistin use has led to the emergence of extensively drug-resistant (XDR) strains, raising a significant challenge for healthcare. In order to gain insight into the antibiotic resistance mechanisms of CRKP and identify potential drug targets, we compared the molecular characteristics and the proteomes among drug-sensitive (DS), MDR, and XDR *K. pneumoniae* strains. All drug-resistant isolates belonged to ST11, harboring bla_{KPC} and hypervirulent genes. None of the plasmid-encoded *mcr* genes were detected in the colistin-resistant XDR strains. Through a tandem mass tag (TMT)-labeled proteomic technique, a total of 3531 proteins were identified in the current study. Compared to the DS strains, there were 247 differentially expressed proteins (DEPs) in the MDR strains and 346 DEPs in the XDR strains, respectively. Gene Ontology (GO) and Kyoto Encyclopedia of Genes and Genomes (KEGG) enrichment analysis revealed that a majority of the DEPs were involved in various metabolic pathways, which were beneficial to the evolution of drug resistance in *K. pneumoniae*. In addition, a total of 67 DEPs were identified between the MDR and XDR strains. KEGG enrichment and protein–protein interaction network analysis showed their participation in cationic antimicrobial peptide resistance and two-component systems. In conclusion, our results highlight the emergence of colistin-resistant and hypervirulent CRKP, which is a noticeable superbug. The DEPs identified in our study are of great significance for the exploration of effective control strategies against infections of CRKP.

Keywords: *Klebsiella pneumoniae*; multidrug-resistant; extensively drug-resistant; colistin; carbapenem; quantitative proteomics; tandem mass tag

check for updates

Citation: Hao, L.; Yang, X.; Chen, H.; Mo, Z.; Li, Y.; Wei, S.; Zhao, Z. Molecular Characteristics and Quantitative Proteomic Analysis of *Klebsiella pneumoniae* Strains with Carbapenem and Colistin Resistance. *Antibiotics* **2022**, *11*, 1341. https://doi.org/10.3390/antibiotics11101341

Academic Editors: Theodoros Karampatakis and Marc Maresca

Received: 7 September 2022
Accepted: 29 September 2022
Published: 30 September 2022

1. Introduction

Carbapenem-resistant *Klebsiella pneumoniae* (CRKP) strains, which account for 70–90% of clinical carbapenem resistant Enterobacteriaceae (CRE) infections [1,2], pose urgent threats to global public health because they are usually multidrug resistant (MDR), and few antibiotics retain activity against them [3]. Infections caused by CRKP are associated with greater disease severity, more complications, and higher mortality [4]. Especially for severely ill patients, mortality could exceed 50% even with adequate antibiotic treatment [5]. Colistin is a critical last-resort antimicrobial for serious infections caused by CRKP in clinics [6]. Although several novel drugs and combinations have been considered in the therapy of MDR Gram-negative bacteria, colistin will still be considered as a fundamental companion drug for the treatment of CRE [7]. However, the increasing and suboptimal use of colistin has led to the emergence of colistin-resistant CRKP, which are identified as extensively drug-resistant (XDR) strains [8], raising a significant challenge for healthcare.

Hence, additional knowledge about the mechanism of drug resistance in CRKP is urgently needed to aid in the development of more effective anti-infectious methodologies.

The precise mechanism of action of colistin remains unclear, but a number of models have been proposed. Studies have shown that lipid A, a lipid component of lipopolysaccharide (LPS) in the bacterial outer membrane (OM), is the primary target of colistin [6]. Colistin is a cationic antimicrobial peptide that binds lipid A and causes cell membrane leakage [6]. The interaction between colistin and lipid A is affected when lipid A is modified with 4-amino-4-deoxy-l-arabinose (L-Ara4N) or phosphoethanolamine (PEtN), thus resulting in colistin resistance [6,9]. There are two main mechanisms of modification of lipid A: chromosome-mediated regulation pathways and plasmid-mediated mobile-resistance genes (*mcr-1* to *mcr-10*) [10–13]. In chromosomes, modifications of lipid A with L-Ara4N or PEtN are achieved by the activities of the *arnBCADTEF* and *pmrCAB* operons, respectively [10,11]. Mobile *mcr* genes can facilitate resistant gene transfer from one strain to another and spread rapidly [12,13]. In addition, it has been reported that capsular polysaccharide and multidrug efflux pumps can also confer tolerance to colistin, although their impact is limited [11,14,15]. Currently, the steps toward the development new drugs cannot keep up with the pace at which resistance is acquired by *Klebsiella pneumoniae* (*K. pneumoniae*). Significant improvements in the therapeutic strategies for XDR bacterial infections are still very poor.

Proteomic analyses based on mass spectrometry and bioinformatics have provided an opportunity to combat antibiotic resistance and reveal unknown functional relationships [16]. In this context, the mining of specific proteins or proteome sets from clinically useful drug-resistant (DR) *K. pneumoniae* strains is very important. The tandem mass tag (TMT)-labeled quantitative proteomic approach is an in vitro labeling technology that can detect differentially abundant proteins from up to 16 different samples at the same time, so inconsistent conditions for the identification and separation of all samples can be avoided and more accurate quantitative results can be obtained [17]. In order to gain insight into the antibiotic resistance mechanisms of *K. pneumoniae* and identify potential drug targets, the TMT-labeled proteomic technique was used to compare the proteomes of drug-sensitive (DS), MDR, and XDR strains. Meanwhile, we also investigated the presence of *mcr* genes in colistin-resistant XDR strains. Such strains were also examined for the presence of carbapenemases and hypervirulent genes, sequence types (STs), hypermucoviscosity phenotypes, and their ability to form biofilms.

2. Results

2.1. Antimicrobial Susceptibility Profile and Molecular Characteristics of K. pneumoniae Isolates

The antimicrobial susceptibility profile showed that all DR isolates included in this study were resistant to 13 out of the 16 tested antibiotics. The MIC values for colistin of the XDR strains ranged from 32 to >64 μg/mL. The antibiotic susceptibility results for the nine isolates are summarized in Table 1.

Table 1. Antimicrobial susceptibility of the DS, MDR, and XDR isolates.

Antibiotics	MIC of Strains (μg/mL)								
	DS-1	DS-2	DS-3	MDR-1	MDR-2	MDR-3	XDR-1	XDR-2	XDR-3
CST	≤0.5	≤0.5	≤0.5	≤0.5	≤0.5	≤0.5	**32**	**>64**	**>64**
AMK	≤2	≤2	≤2	**≥64**	**≥64**	**≥64**	**≥64**	**≥64**	**≥64**
GEN	≤1	≤1	≤1	**≥16**	**≥16**	**≥16**	**≥16**	**≥16**	**≥16**
TOB	≤1	≤1	≤1	**≥16**	**≥16**	**≥16**	**≥16**	**≥16**	**≥16**
IPM	≤1	≤1	≤1	**≥16**	**≥16**	**≥16**	**≥16**	**≥16**	**≥16**
MEM	≤0.25	≤0.25	≤0.25	**≥16**	**≥16**	**≥16**	**≥16**	**≥16**	**≥16**

Table 1. *Cont.*

Antibiotics	MIC of Strains (μg/mL)								
	DS-1	DS-2	DS-3	MDR-1	MDR-2	MDR-3	XDR-1	XDR-2	XDR-3
CAZ	≤1	≤1	≤1	**≥64**	**≥64**	**≥64**	**≥64**	**≥64**	**≥64**
CRO	≤1	≤1	≤1	**≥64**	**≥64**	**≥64**	**≥64**	**≥64**	**≥64**
FEP	≤1	≤1	≤1	**≥32**	**≥32**	**≥32**	**≥32**	**≥32**	**≥32**
TZP	≤4	≤4	≤4	**≥128**	**≥128**	**≥128**	**≥128**	**≥128**	**≥128**
AZM	≤1	≤1	≤1	**≥64**	**≥64**	**≥64**	**≥64**	**≥64**	**≥64**
CIP	≤0.25	≤0.25	≤0.25	**≥4**	**≥4**	**≥4**	**≥4**	**≥4**	**≥4**
LVX	≤0.25	≤0.25	≤0.25	**≥8**	**≥8**	**≥8**	**≥8**	**≥8**	**≥8**
SXT	≤20	≤20	≤20	40	≤20	≤20	≤20	≤20	≤20
SFP	≤16	≤16	≤16	**≥64**	**≥64**	**≥64**	**≥64**	**≥64**	**≥64**
TGC	2	≤1	≤1	2	≤1	≤1	2	≤1	2

CST, Colistin; AMK, Amikacin; GEN, Gentamicin; TOB, Tobramycin; IPM, Imipenem; MEM, Meropenem; CAZ, Ceftazidime; CRO, Ceftriaxone; FEP, Cefepime; TZP, Piperacillin/Tazobactam; AZM, Aztreonam; CIP, Ciprofloxacin; LVX, Levofloxacin; SXT, Trimethoprim/Sulfa; SFP, Cefoperazone/Sulbactam; TGC, Tigecycline. Resistance is emphasized in bold.

All DR isolates were found to be positive for bla_{KPC}. However, none of the plasmid-encoded *mcr* genes were detected in the colistin-resistant XDR strains. According to Multilocus Sequence Typing (MLST), all DR strains belonged to ST11 (Table 2). Hypervirulent genes *peg-344*, *iroB*, *iucA*, $_{p}rmpA$, and $_{p}rmpA2$ were reported to be the biomarkers for defining hypervirulent *K. pneumoniae* (hvKP) (diagnostic accuracy >0.95), which is an evolving pathotype with more virulence than classical *K. pneumoniae*, and it infects healthy individuals in communities [18]. In our study, all DR isolates carried at least two hypervirulent genes, indicating the cooccurrence of drug resistance and hypervirulence in *K. pneumoniae* (Table 2). The MDR2 strain from ST11 harboring all five hypervirulent biomarkers might have the highest virulence potential. Interestingly, a string test showed that all MDR strains were positive and considered to have hypermucoviscosity, while all XDR strains were negative. In addition, the mean OD590 nm values obtained with the quantitative biofilm production assay are listed in Table 2. The XDR1 strain was a strong biofilm producer. The DS1, DS3, and MDR1 strains were weak biofilm producers, and the MDR3 strain showed no biofilm capabilities. The rest of the strains showed moderate biofilm-forming capacities. The results showed that the colistin-resistant XDR strains showed higher biofilm-forming capacities (0.59467 ± 0.14465) than the colistin-sensitive MDR strains (0.26778 ± 0.08842).

Table 2. MLST and molecular characteristics of the DS, MDR, and XDR isolates.

Isolate	Housekeeping Genes							ST	Carbapenemases	Virulence Genes	String Test	Biofilm Formation (OD590 nm)
	gapA	*infB*	*mdh*	*pgi*	*phoE*	*rpoB*	*tonB*					
DS1	2	2	1	1	3	3	3	ST5	-	*iucA*, *iroB*, *peg-344*	-	0.2264
DS2	18	15	26	108	32	37	51	ST1304	-	*iucA*, *peg-344*	-	0.5713
DS3	2	1	1	1	9	4	12	ST23	-	*iucA*, *iroB*, *peg-344*, $_{p}rmpA$, $_{p}rmpA2$	+	0.2334
MDR1	3	3	1	1	1	1	4	ST11	KPC	*iucA*, *peg-344*, $_{p}rmpA2$	+	0.28933
MDR2	3	3	1	1	1	1	4	ST11	KPC	*iucA*, *iroB*, *peg-344*, $_{p}rmpA$, $_{p}rmpA2$	+	0.36367

Table 2. *Cont.*

Isolate	Housekeeping Genes							ST	Carbapenemases	Virulence Genes	String Test	Biofilm Formation (OD590 nm)
	gapA	*infB*	*mdh*	*pgi*	*phoE*	*rpoB*	*tonB*					
MDR3	3	3	1	1	1	1	4	ST11	KPC	*iucA*, *peg-344*, *prmpA2*	+	0.15033
XDR1	3	3	1	1	1	1	4	ST11	KPC	*iucA*, *iroB*, *peg-344*, *prmpA2*	-	0.76367
XDR2	3	3	1	1	1	1	4	ST11	KPC	*iucA*, *peg-344*, *prmpA2*	-	0.41033
XDR3	3	3	1	1	1	1	4	ST11	KPC	*iucA*, *peg-344*	-	0.61

2.2. Statistics of the Spectrum Data of the Proteome

In total, there were three groups in our experimental design, and TMT-labeling was employed for protein quantification (Figure 1). LC-MS/MS analysis of the whole-cell-lysate proteins (extracted at 4 °C) from nine *K. pneumoniae* isolates generated a total of 513,635 raw spectra. With an FDR of <1% for protein and peptide identification, we obtained 96,134 qualified spectra, including 95,529 unique spectra, which were associated with 34,739 peptides (34,548 unique peptides). These peptides were mapped to 3531 proteins with at least one unique peptide per protein; there were at least two unique peptides for 3032 (85.9%) of the 3531 proteins identified (Figure 2A). The distribution of the peptide length is shown in Figure 2B. A total of 87.4% of the peptides had 5 to 20 amino acids after enzymatic hydrolysis, which indicated that the enzymatic hydrolysis was sufficient and the identification results were reliable. The molecular masses of most proteins were from 10 to 60 kDa, and 2% of the proteins with molecular masses >100 kDa were identified (Figure 2C). The identification was more convincing with the expansion of the peptide coverage distribution. In this study, 82.2% of the identified proteins had a peptide coverage of more than 10% (Figure 2D).

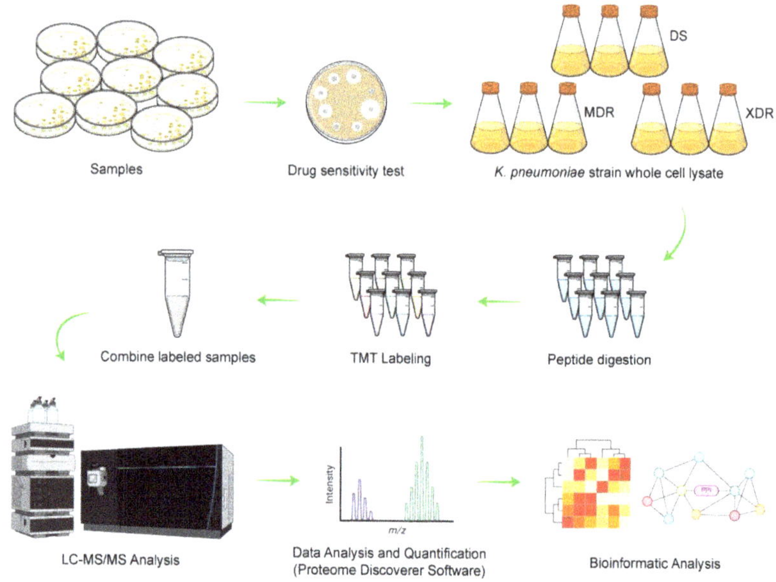

Figure 1. Experimental workflow of the quantitative proteomic analysis of DEPs in *K. pneumoniae*.

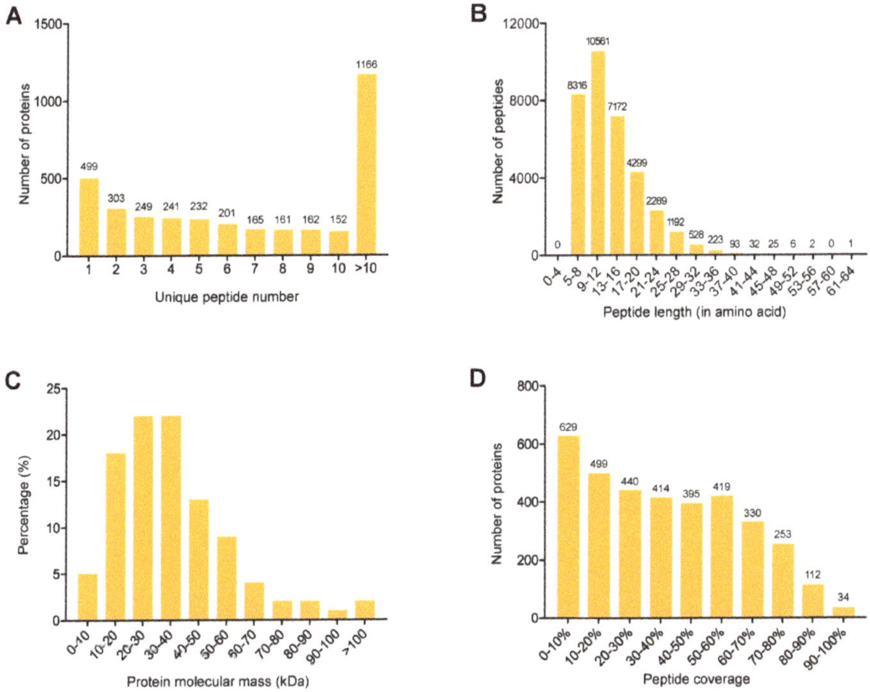

Figure 2. Profiles of the whole-cell-lysate proteins identified in nine *K. pneumoniae* isolates. (**A**) Quantitative distribution of unique peptides identified for individual proteins. (**B**) Length distribution of the identified peptides. (**C**) Distribution of molecular mass among all identified proteins. (**D**) Proportional distribution of the amino acid sequence of each identified protein covered by peptides.

2.3. Identification of Differentially Expressed Proteins (DEPs) for Each Compared Group

Based on the quantitative spectrum data, significantly, the DEPs between each compared group were determined with a fold change of >2 and a p-value < 0.05 (Figure 3). When compared to the DS strains, there were a total of 247 DEPs, including 194 up-regulated and 53 down-regulated proteins, in the MDR strains (Figure 3A). A total of 346 DEPs were identified between the XDR and DS strains, among which 270 proteins were significantly increased, while 76 proteins were decreased in the XDR strains (Figure 3B). In addition, there were a total of 67 DEPs, including 49 up-regulated and 18 down-regulated proteins, between the XDR and MDR strains (Figure 3C). The details of the DEPs in the MDR vs. DS strains, the XDR vs. DS strains, and the MDR vs. XDR strains are shown in Tables S1–S3.

2.4. Prediction of the Protein Functions of the DEPs between the MDR and DS Strains

The Gene Ontology (GO) classification of the 247 DEPs between the MDR and DS strains resulted in three broad categories (biological processes, cellular components, and molecular functions) and 24 GO terms (Figure 4A). In biological process category, 69 DEPs were associated with metabolic processes, 62 DEPs with cellular processes, and 24 DEPs with localization. In the cellular component category, 35 DEPs were detected in the membrane, 32 DEPs in the membrane part, and 29 DEPs in the cell part. In addition, the molecular functions associated with the DEPs were as follows: 111 DEPs were involved in catalytic activity, 74 DEPs were involved in binding, and 17 DEPs were detected mainly in transporter activity (Table S4). The functional enrichment analysis further determined the significantly enriched GO classification of the DEPs according to the P-values. The results indicated that the biological processes of the DEPs were related to transport, localization, and catabolic processes (Figure 4B). The molecular functions of the DEPs were concentrated on oxidoreductase activity, DNA binding, and endonuclease activity (Figure 4C).

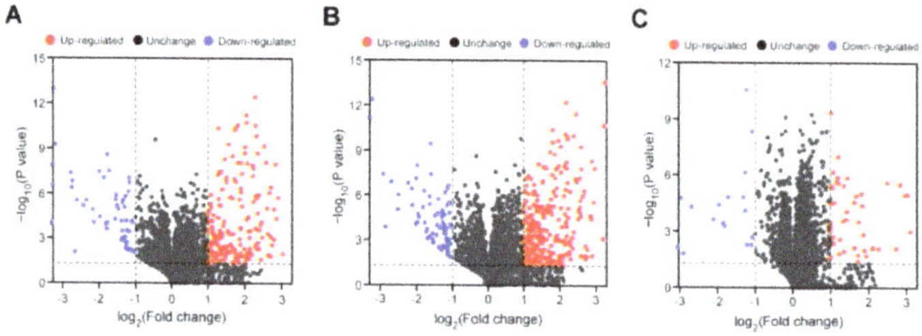

Figure 3. Volcano plots of the DEPs for each compared group. (**A–C**) Volcano plots for the screening of the DEPs in the MDR vs. DS strains (**A**), in the XDR vs. DS strains (**B**), and in the XDR vs. MDR strains (**C**). The dashed line represents the applied threshold (p-value < 0.05, fold change >2). The up-regulated and down-regulated proteins are shown in red and blue, respectively.

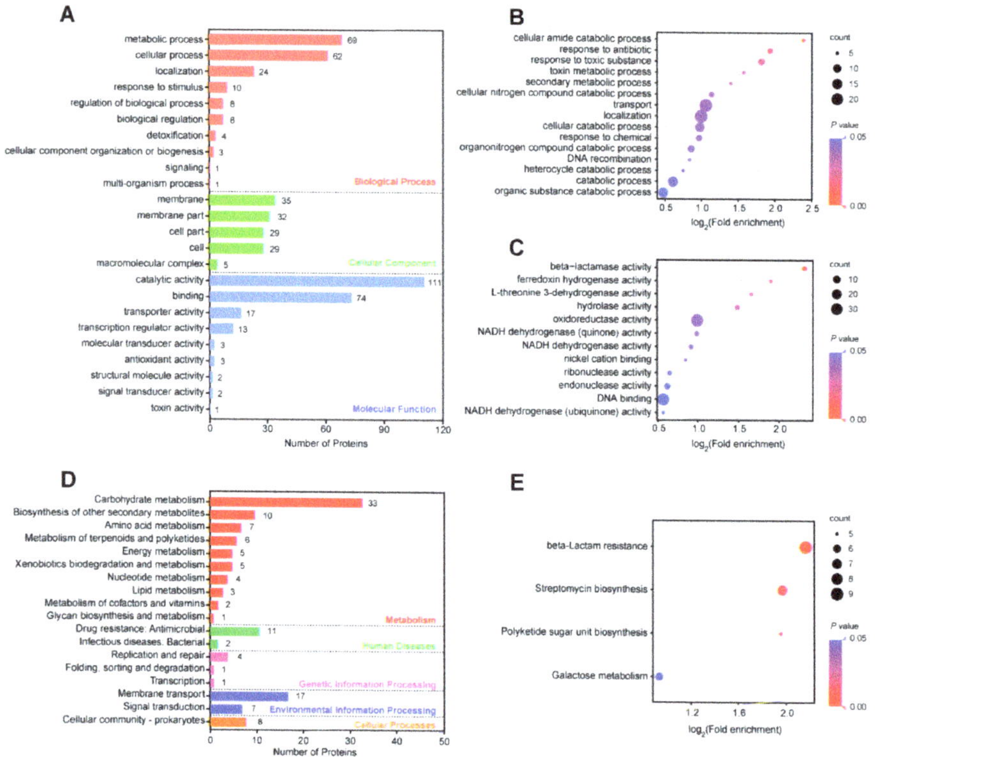

Figure 4. Functional and pathway enrichment analysis of the DEPs between the MDR and DS strains. (**A**) Numbers of the 247 DEPs in the biological process, cellular component, and molecular function categories revealed by the GO annotations. (**B**,**C**) GO functional enrichment analysis of biological processes (**B**) and molecular functions (**C**). (**D**) The KEGG pathway analysis of the 247 DEPs. (**E**) The enrichment analysis of the KEGG pathway.

Based on the Kyoto Encyclopedia of Genes and Genomes (KEGG) analysis, the DEPs were classified and annotated into metabolism, human diseases, genetic information processing, environmental information processing, and cellular processes (Figure 4D). Carbohydrate metabolism, biosynthesis of other secondary metabolites, and amino acid metabolism were the most common annotations of DEPs in the metabolism category, with 33, 10, and 7 proteins, respectively. Importantly, 11 DEPs, including AcrB, WecH, Bm3R1, SapB, OppC, OppA, OppF, and 4 β-lactamases (KPC-2, CTX-M-14, SHV-11, TEM-1), were enriched in drug resistance (Table S5). The pathway enrichment analysis showed that the DEPs were linked to beta-lactam resistance, streptomycin biosynthesis, polyketide sugar unit biosynthesis, and galactose metabolism (Figure 4E).

2.5. Prediction of the Protein Functions of the DEPs between the XDR and DS Strains

The effects of the 346 DEPs between the XDR and DS strains on cellular processes were clarified using GO annotation and KEGG pathway analysis (Figure 5). The GO annotation of the 346 DEPs resulted in 27 GO terms (Figure 5A). The numbers of DEPs involved in biological process were as follows: 110 proteins were associated with metabolic processes, 100 proteins with cellular processes, and 36 proteins with localization. The DEPs in cellular components were mainly concentrated in the membrane (66 proteins), cell part (58 proteins), and cell (58 proteins). In addition, the top three items of molecular functions associated with DEPs were catalytic activity (176 proteins), binding (100 proteins), and transporter activity (28 proteins) (Table S6). The enrichment analysis of the biological processes showed that the DEPs were related to transport, localization, and establishment of localization (Figure 5B). In the molecular functions, the DEPs were associated with oxidoreductase activity and lyase activity (Figure 5C).

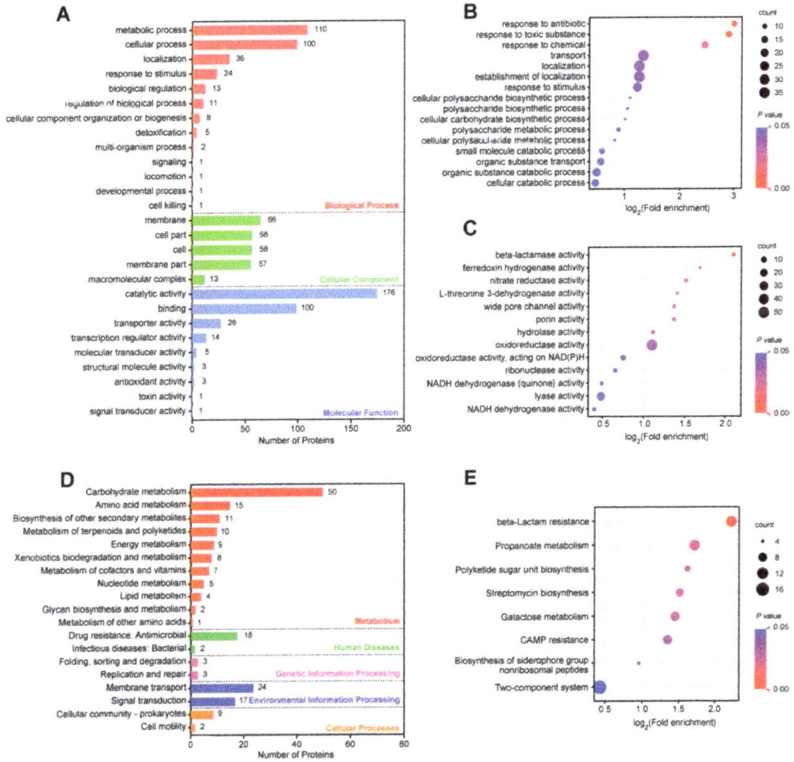

Figure 5. Functional and pathway enrichment analysis of the DEPs between the XDR and DS strains. (**A**) Numbers of the 346 DEPs in the biological process, cellular component, and molecular function categories revealed by the GO annotations. (**B**,**C**) GO functional enrichment analysis of biological processes (**B**) and molecular functions (**C**). (**D**) The KEGG pathway analysis of the 346 DEPs. (**E**) The enrichment analysis of the KEGG pathway.

The KEGG pathway analysis for the DEPs between the XDR and DS strains showed that carbohydrate metabolism, amino acid metabolism, and biosynthesis of other secondary metabolites were still the most commonly annotated DEPs in the metabolism category (Figure 5D). Importantly, 18 DEPs, including AcrA, AcrB, WecH, Bm3R1, OppC, OppA, OppF, DegP, PmrD, 4 β-lactamases (KPC-2, CTX-M-14, SHV-11, TEM-1), and five enzymes that are responsible for lipid A modification (ArnB, ArnC, ArnA, ArnD, ArnT), were enriched in drug resistance (Table S7). The pathway enrichment analysis showed that the DEPs were linked to two-component systems (TCSs), beta-lactam resistance, propanoate metabolism, galactose metabolism, and cationic antimicrobial peptide (CAMP) resistance (Figure 5E).

2.6. *Prediction of the Protein Functions of the DEPs between the MDR and XDR Strains*

We further analyzed the effects of the DEPs between the XDR and MDR strains, which may offer an opportunity to find the potential mechanism of colistin resistance in *K. pneumoniae*. In the GO analysis of the 67 DEPs, there were 10, 5, and 6 enriched GO terms in the biological process, cellular component, and molecular function categories, respectively (Figure 6A). The most abundant GO terms included metabolic processes (17 proteins) in the biological process category, membrane (25 proteins) and membrane part (23 proteins) in the cellular component category, and catalytic activity (30 proteins) in the molecular function category (Table S8). The enrichment analysis of the biological process showed that the DEPs were related to transport, localization, and establishment of localization (Figure 6B). In molecular function, the DEPs were associated with transporter activity and transmembrane transporter activity (Figure 6C).

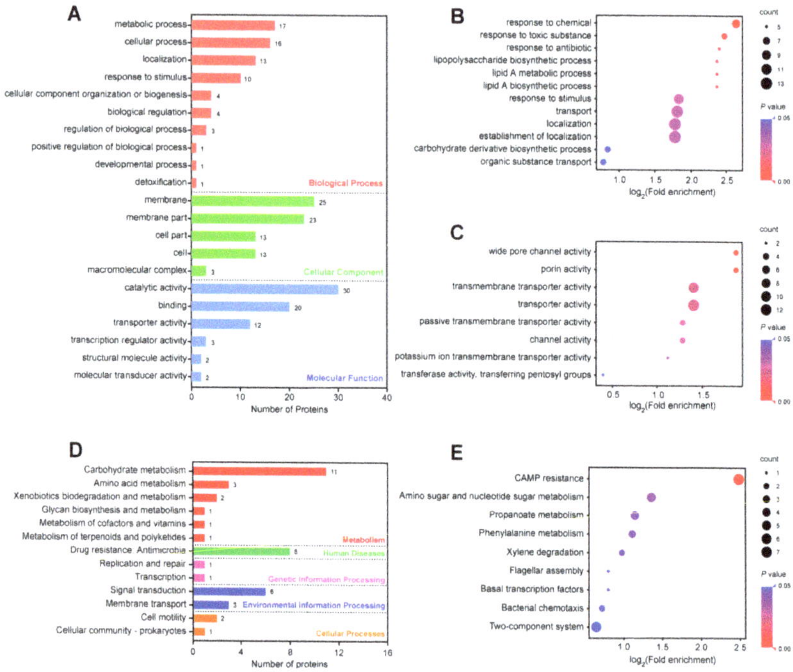

Figure 6. Functional and pathway enrichment analysis of the DEPs between the MDR and XDR strains. (**A**) Numbers of the 67 DEPs in the biological process, cellular component, and molecular function categories revealed by the GO annotations. (**B**,**C**) GO functional enrichment analysis of biological processes (**B**) and molecular functions (**C**). (**D**) The KEGG pathway analysis of the 67 DEPs. (**E**) The enrichment analysis of the KEGG pathway.

The KEGG pathway analysis for the DEPs between the MDR and XDR strains showed that most DEPs were annotated in carbohydrate metabolism in the metabolism category. In particular, eight proteins (ArnT, ArnD, ArnA, ArnC, ArnB, PmrD, YddW, OmpK36) were enriched in drug resistance (Table S9). The pathway enrichment analysis showed that a majority of the DEPs were involved in TCS, CAMP resistance, amino sugar and nucleotide sugar metabolism, propanoate metabolism, and phenylalanine metabolism.

The 67 DEPs were mapped onto the STRING database, and 50 of them were retained to build a protein–protein interaction (PPI) network (Figure 7). Proteins that did not play roles in the construction of the network were filtered out. In particular, PfeA, OmpN, and YfeY, which had the largest number of interacting proteins, were the hubs of the network. The closely interacting DEPs were mainly distributed in four modules, namely, propanoate metabolism, phosphotransferase system (PTS), TCSs, and CAMP resistance.

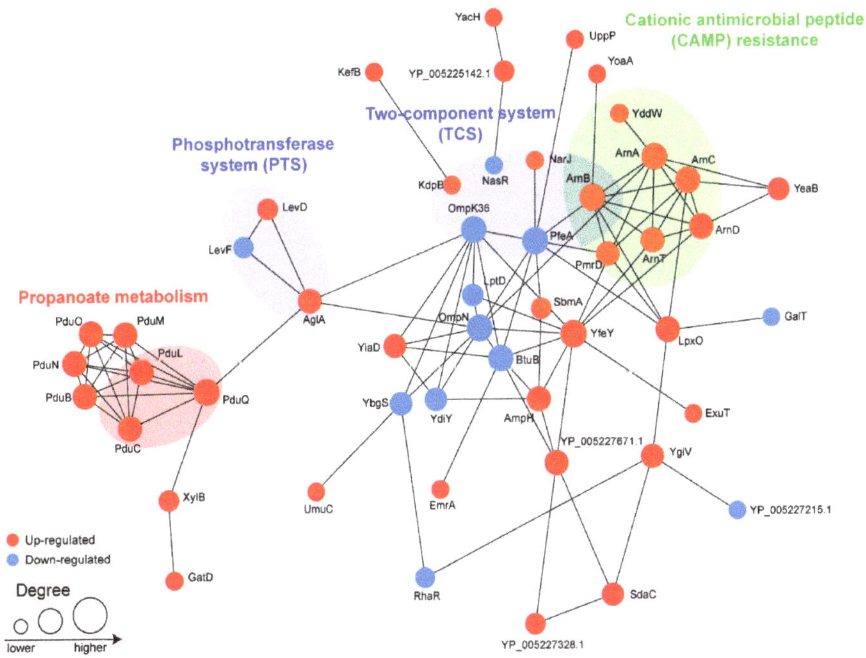

Figure 7. The interaction network analysis of the DEPs between the MDR and XDR strains. The size of the node represents the number of interacting proteins. The up-regulated proteins are marked in red, while the down-regulated proteins are marked in blue.

3. Discussion

Nowadays, *K. pneumoniae* has become a significant clinical concern in nosocomial environments worldwide because of its remarkable abilities for drug resistance [1,2]. Worse still, a combination of drug-resistant and hypervirulent genes in *K. pneumoniae* strains could exacerbate the scarcity of effective treatments, resulting in related fatal hospital outbreaks [19,20]. A comparative analysis of molecular characteristics and proteomic changes from clinical *K. pneumoniae* isolates allowed us to reveal the unknown mechanism of antibiotic resistance of this bacterial pathogen. In the present study, antibiotic resistance, virulence-associated genes, STs, hypermucoviscosity phenotypes, biofilm-forming ability, and the proteomes among DS, MDR, and XDR *K. pneumoniae* strains were investigated. All DR strains belonged to ST11, harboring bla_{KPC} and hypervirulent genes. None of the plasmid-encoded *mcr* genes were detected in the colistin-resistant XDR strains. The string test showed that all MDR strains were positive and considered to display hypermucoviscosity, while the

XDR strains were negative. The biofilm formation capacity of the colistin-resistant XDR strains was higher than that of the colistin-sensitive MDR strains. Through the TMT-labeled quantitative proteomic technique, we identified a total of 34,548 unique peptides associated with 3531 proteins among the DS, MDR, and XDR strains. Compared to the DS strains, there were 247 DEPs in the MDR strains and 346 DEPs in the XDR strains. The GO and KEGG enrichment analysis revealed that various metabolic pathways in which the DEPs participated were conducive to the evolution of DS strains into MDR and XDR strains. In addition, a total of 67 DEPs were identified between the MDR and XDR strains, of which 49 were up-regulated and 18 were down-regulated. Notably, the DEPs related to drug resistance were ArnT, ArnD, ArnA, ArnC, ArnB, PmrD, YddW, and OmpK36. Bioinformatic studies such as KEGG enrichment and PPI network analysis showed their participation in CAMP resistance and TCSs. Therefore, these DEPs are of great significance for exploring effective control strategies against infections of CRKP.

Under antibiotic pressure, *K. pneumoniae* can rapidly evolve and develop resistance, showing alterations in protein expressions, responsive signals, and cellular processes [21]. In our report, there were 247 DEPs identified in the MDR strains and 346 DEPs identified in the XDR strains when compared to the DS strains. Notably, four β-lactamases, namely, KPC-2, CTX-M-14, SHV-11, and TEM-1, were detected in all DR strains, and these comprise the major mechanism of the resistance of *K. pneumoniae* to most β-lactam antibiotics. The expression of efflux pumps in order to pump out an antibiotic is another common resistance mechanism for many antibiotics. In our study, KexD, AcrA, and AcrB were up-regulated in the DR strains compared to the DS strains. KexD is a component of an energy-dependent efflux pump that belongs to the resistance nodulation division (RND) superfamily [22]. It could interact with other RND efflux components (e.g., the periplasmic protein AcrA and the porin TolC) to efflux antibiotics, such as novobiocin or erythromycin, and it was induced by *crrB* mutation to result in colistin resistance [23]. AcrA and ArcB are the components of an archetypal RND multidrug efflux pump, AcrAB-TolC, which has a critical role in resistance to multiple antibiotics, especially fluoroquinolones, tetracycline, and beta-lactam antibiotics, in MDR strains of *K. pneumoniae* [24,25]. In addition, WecH, Bm3R1, OppC, OppA, and OppF were the same DEPs in the MDR and XDR strains compared to the DS strains, and they were enriched in drug resistance. WecH is an O-acetyltransferase that is responsible for the incorporation of O-acetyl groups into the enterobacterial common antigen (ECA) trisaccharide repeat units [26]. Bm3R1 is a TetR family transcriptional regulator that acts as a chemical sensor to monitor the cellular environmental dynamics, which mediate an adaptive response to toxic fatty acids in *Bacillus megaterium* [27,28]. OppC, OppA, and OppF are oligopeptide permeases that have been proved to play a variety of important roles in nutrition and virulence in several bacteria [29]. Low expression of OppC, OppA, and OppF was associated with β-lactam resistance in *Streptococcus agalactiae* [30]. Importantly, bacterial metabolism plays a crucial role in mediating the cellular responses to antibiotic treatment [31]. Through the GO and KEGG pathway analysis, a majority of the DEPs in the DR vs. DS strains were involved in several metabolic processes, such as carbohydrate metabolism, amino acid metabolism, metabolism of terpenoids and polyketides, and galactose metabolism. Studies have shown that metabolism of carbohydrates, starches, sucrose, arginine, and its derivative, proline, played an important role in biofilm formation and maintenance [32–34]. Biofilms are communities of microorganisms that adhere to a surface, and they play a significant role in persistent bacterial infections [31]. Compared with planktonic bacteria, the resistance of bacteria within a biofilm to antibiotics is several orders of magnitude higher. Biofilm-forming ability has been associated with antibiotic resistance in *K. pneumoniae* [35,36]. Consistently with these studies [35,36], we found that the colistin-resistant XDR strains could have a strong biofilm-forming ability, which might make colonized *K. pneumoniae* more difficult to treat.

More importantly, we further identified 67 DEPs between the MDR and XDR strains and found that ArnT, ArnD, ArnA, ArnC, ArnB, PmrD, YddW, and OmpK36 were enriched

in drug resistance. It is widely approved that colistin destroys the bacterial OM by binding to lipid A of LPS, leading to osmotic imbalance and cell death [6]. Alterations in LPS helped *K. pneumoniae* survive under colistin [6]. ArnT, ArnD, ArnA, ArnC, and ArnB, the gene products of the *arnBCADTEF* operon, are the enzymes responsible for the biosynthesis and addition of L-Ara4N to lipid A, which reduces the negative charge carried by LPS, causing a decreased ability of colistin to combine with bacteria [6]. Constitutive expression of the *arnBCADTEF* operon is achieved by the mutations in TCSs, mainly PmrA/PmrB and PhoP/PhoQ, leading to colistin resistance [10,11,23]. PmrD is a connector protein that conveys feedback between the PmrA/PmrB and PhoP/PhoQ TCSs [11,37]. Expression of *pmrD* in *K. pneumoniae* was dependent on PhoP; the activated PmrD would then bind to PmrA to inhibit its dephosphorylation and, eventually, turn on the expression of the *arnBCADTEF* operon for the resistance to colistin [11]. YddW is a divisome-localized glycosyl hydrolase that cleaves peptide-free peptidoglycan glycans in the OM layer, promoting OM constriction during cell division [38]. The overexpression of *yddW* alleviates the phenotypes of *Escherichia coli* (*E. coli*) in a defective Tol-Pal system, a multiprotein system in the OM of Gram-negative bacteria [38]. Inactivation of this system compromises the OM layer, resulting in hypersensitivity to many antibiotics [38]. In addition, porin loss could contribute to phenotypes that are resistant against antibiotics such as β-lactams [39]. Studies have shown that defects in *K. pneumoniae* porins OmpK35 and OmpK36 (the *E. coli* homologs are OmpF and OmpC, respectively) could lead to reduced sensitivity to carbapenems [40,41]. Substitutions in the *ompC* gene mediated carbapenem resistance among *Enterobacteriaceae* and were partly involved in colistin resistance [42]. In line with these studies [39–41], our proteomic data also showed that OmpK36 was decreased in the XDR strains compared to the MDR strains. In addition, we found that OmpN was also down-regulated in the colistin-resistant XDR strains. The overexpression of OmpN enhanced the sensitivity of *K. pneumoniae* to several antibiotics, such as cefuroxime, cefuroxime axetil, nitrofurantoin, and ampicillin/sulbactam [43], indicating that the decrease in OmpN might be responsible for drug resistance. However, the regulatory effect of OmpN on colistin resistance is unknown. Moreover, the KEGG pathway analysis and the PPI network showed that the DEPs between the MDR and XDR strains were mainly enriched in CAMP resistance and TCSs. In the CAMP resistance pathway, ArnBCADT, PmrD, and YddW were highly expressed in the colistin-resistant XDR strains, which indicated that lipid A modification is still the main mechanism of colistin resistance in *K. pneumoniae*. TCSs, which are typically composed of two proteins (a sensor kinase and a response regulator), are crucial for bacteria to maintain homeostasis, and they effectively sense and respond to environmental changes, such as nutrition, osmotic pressure, and antibiotic exposure [44]. In the current study, KdpB, OmpK36, PfeA, NasR, NarJ, and ArnB were involved in the TCS pathway. KdpB is a subunit of potassium-transporting ATPase [45]. The KdpATPase is an inducible high-affinity K^+ transporter that maintains the desired concentration of internal K^+ in bacterial cells [45]. The porin protein OmpK36 is regulated by the EnvZ/OmpR TCS and plays an important role in iron homeostasis [46]. PfeA is a ferric enterobactin receptor inducible by enterobactin under the conditions of iron limitation, which is regulated by a TCS PfeR/PfeS [47]. NasR and a TCS NARL/NARP regulate the synthesis of enzymes for nitrate/nitrite respiration and assimilation [48]. NarJ is a system-specific chaperone required for the formation of the respiratory nitrate reductase complex in *Escherichia coli* [49]. ArnB is the promoter region of the *arn* operon, which binds to the PhoP/PhoQ response regulator and regulates the modification of lipid A [50]. Studies have shown that multiple pathways might operate together to cause drug resistance, such as a cross-talk between the PhoP/PhoQ TCS and the Rcs system (regulator capsule synthesis) [51], as well as a cross-regulatory interaction between a specific CroS/CroR TCS and the HPr protein of PTS [52]. Future work will focus on understanding the detailed roles of these pathways in colistin resistance in order to make the bacteria sensitive.

In conclusion, our data highlight that the co-existence of drug-resistant and hypervirulent genes, as well as the strong biofilm-forming ability, will cause colistin-resistant CRKP

to emerge as super-bugs. The proteomic results obtained the DEPs of DR strains, providing insights for understanding the mechanism of resistance to colistin in *K. pneumoniae*. Functional and pathway enrichment analyses of these DEPs laid a foundation for establishing effective control strategies against infections of CRKP.

4. Materials and Methods

4.1. Bacterial Strains and Culture

Nine strains of *K. pneumoniae* were isolated from the clinical samples of inpatients in our hospital. Among these strains, three (DS1, DS2, and DS3) were sensitive to all tested drugs, three (MDR1, MDR2, and MDR3) were CRKP strains that were susceptible to colistin, and the other three (XDR1, XDR2, and XDR3) were colistin-resistant CRKP strains. The XDR strains were matched with the MDR strains from the same patient obtained before the initiation of colistin treatment. All strains were identified using biochemical tests and the IVD MALDI Biotype system. Bacterial cells were cultured at 37 °C overnight on Luri-aBertani (LB) plates (Difco, Sparks, MD). Then, a single colony was selected and inoculated under different culture conditions: The DS strains were inoculated in LB broth with no antibiotic, while the DR strains were cultured in LB broth supplemented with meropenem (16 μg/mL) and imipenem (16 μg/mL) and with or without colistin (4 μg/mL). The cultures were grown at 37 °C with shaking at 180 rpm until the OD_{600} reached 04–06 (mid-log growth phase), washed twice with cold PBS, and harvested for proteomic analysis.

4.2. Drug Sensitivity Test of Bacterial Strains

Routine antimicrobial susceptibility testing was performed with the Vitek-2 Compact automatic microbiology system (bioMerieux, Inc., Durham, NC, USA). The minimum inhibitory concentrations (MICs) were interpreted according to the breakpoints defined by the Clinical and Laboratory Standards Institute (CLSI), as described previously [53]. Susceptibility testing for colistin was further confirmed for all strains with the microdilution method outlined by the CLSI [54]. *E. coli* ATCC 25922 and *Pseudomonas aeruginosa* ATCC 27853 were used as quality control organisms for the antimicrobial susceptibility testing.

4.3. Identification of mcr, Carbapenemases, and Hypervirulent Genes

The genomic DNA of each isolate was extracted using a DNA extraction kit (Tiangen, Beijing, China) according to the manufacturer's protocol. The presence of antibiotic resistance genes, including *mcr* genes (*mcr-1* to *mcr-8*) and carbapenemases (bla_{KPC}, bla_{NDM}, bla_{VIM}, bla_{IMP}, and $bla_{OXA\text{-}48}$ type), was analyzed with conventional PCR using validated primers and methods, and the amplified products were confirmed through sequencing [55,56]. The frequency of five hypervirulent genes, namely, *iucA*, *iroB*, *peg-344*, $_{p}rmpA$, and $_{p}rmpA2$, was determined through PCR with established primers and methods [18]. All analyses were performed with corresponding positive controls. All primer sequences are listed in Table S10.

4.4. Multilocus Sequence Typing (MLST)

MLST was performed on all *K. pneumoniae* isolates by amplifying and sequencing the seven standard housekeeping genes (*gapA*, *infB*, *mdh*, *pgi*, *phoE*, *rpoB*, and *tonB*), as described previously [57]. STs were assigned using the online database on the Pasteur Institute MLST website (http://bigsdb.pasteur.fr/klebsiella/klebsiella.html, accessed on 28 July 2022).

4.5. Detection of the Hypermucoviscosity Phenotype

A string test was used to detect the hypermucoviscosity phenotype, as described previously [58]. Briefly, an inoculation loop was used to stretch the bacterial colonies of *K. pneumoniae* isolates on Columbia blood agar plates (Crmicrobio, Jiangmen, China) from an overnight culture. Strains with the formation of viscous strings that were >5 mm in length were considered to be positive for the string test.

4.6. Biofilm Assay

Biofilm formation was determined by using the crystal violet staining method from the previous literature with little modification [59,60]. Briefy, 100 µL of overnight-grown culture with an optical density of OD_{600} = 0.1 was added into a 96-well plate. Sterile LB broth served as a negative control. After static incubation at 37 °C for 24 h, each well was washed gently with sterile PBS three times to remove the planktonic cells. After air drying, biofilms were stained with 1% (*w*/*v*) crystal violet for 15 min and then washed again with sterile PBS three times. Finally, the dye bound to adherent biomass was dissolved with 200 µL of absolute ethanol. Absorbance was measured using an automated microplate reader (Thermo Scientific, Waltham, MA, USA) at 590 nm. The tests were performed in triplicate, and the results were averaged. The strains were classified into non-biofilm producers, weak biofilm producers, moderate biofilm producers, and strong biofilm producers according to a previous report [35].

4.7. Whole-Cell-Lysate Protein Extraction from Bacterial Strains

The cell pellets were suspended in 500 µL of lysis buffer (4% SDS, 7 M urea, 2 M thiourea, 20 mM Tris-HCl, proteinase inhibitor cocktail, PH = 8.0) and lysed through ultrasonic decomposition in an ice bath at a speed of 60 Hz for 2 min. The supernatants were collected by centrifugation at 25,000× *g* at 4 °C for 15 min, reacted with 10 mM DTT (1,4-Dithiothreitol) for 30 min at 37 °C, and alkylated with 55 mM IAA (iodoacetamide) for another 45 min at room temperature in the dark. Then, the soluble proteins were precipitated for 2 h at −20 °C by adding cold acetone to the protein solution in excess (1:5 in volume). The precipitated proteins were collected by centrifugation (25,000× *g*, 4 °C, 20 min), air-dried, and then dissolved in an appropriate volume of lysis buffer (without SDS). Samples were further treated with ultrasound (60 Hz, 2 min) in an ice bath to promote protein dissolution. The collected whole-cell-lysate proteins were purified through ultracentrifugation at 25,000× *g* for 15 min at 4 °C. The protein concentrations were determined using the Bradford assay. The lysate proteins were identified with 12% SDS-PAGE.

4.8. Peptide Digestion and TMT Labeling

Peptide digestion was performed by following a previously published protocol [61]. Briefly, 100 µg of each lysate protein sample was first diluted in an appropriate volume of 0.5 M triethylamine borane (TEAB, Sigma-Aldrich, St Louis, MO, USA) to make a final concentration of <2 M urea, and then digested with 5 µg of trypsin at 37 °C for 4 h. The obtained peptide digests were desalted using a Strata-X column (Phenomenex, Torrance, CA, United States) and freeze-dried after salt removal. Then, 100 µg of dried peptide fragments were dissolved in 0.1 M TEAB (Sigma) to a final concentration of 3.74 µg/µL and mixed with different TMT isotopes at room temperature for 2 h (Table S11). The labeled peptides were pooled together for LC-MS/MS analysis.

4.9. LC-MS/MS Analysis of the Labeled Peptides

A total of 20 µg of pooled peptides were dried and then dissolved in 2 mL of Buffer A (5% acetonitrile, pH 9.8). Fractionation was carried out in an HPLC system (LC-20AD, Shimadzu, Kyoto, Japan) using a Gemini C18 column containing 5 µm particles (Phenomenex, Torrance, CA, USA). A total of 20 fractions were collected at a flow rate of 1 mL/min with Buffer B (95% acetonitrile, pH 9.8) with a multi-step gradient system as follows: 5% for 10 min, 5–35% for 40 min, 35–95% for 1 min, 95% for 3 min, and 5% for 10 min. Then, the fractions were freeze-dried, re-dissolved in Solvent A (2% acetonitrile, 0.1% formic acid), and centrifuged at 20,000× *g* for 10 min. The supernatant was applied to an EASY-nLC^{TM} 1200 system (Thermo Fisher Scientific, San Jose, CA, USA) equipped with a 1.9 µm C18 column. Samples were separated using a 3 min gradient from 5 to 8%, 42 min from 8 to 44%, 5 min from 44 to 60%, 7 min from 60 to 100%, and 7 min maintained at 80% of Solvent B (80% acetonitrile, 0.1% formic acid) at 200 nL/min. Separated peptides were analyzed

using a tandem mass spectrometer (Orbitrap Exploris 480, Thermo Fisher Scientific, San Jose, CA, USA) with its parameters set as a data-dependent acquisition mode with a full MS scan (350–1600 *m/z* range) at a resolution of 60,000, followed by a full MS2 scan (100 *m/z*) at a resolution of 15,000, charge of 2+ to 7+, and dynamic exclusion setting of 30 s.

4.10. TMT Protein Identification and Quantification

Raw MS/MS spectrum data were processed with Thermo Proteome Discoverer 1.4.1.14 (Thermo Fisher Scientific, Waltham, MA, USA) and searched with the MASCOT search engine (Matrix Science, London, UK; version 2.3.02) against the *K. pneumoniae* database (NCBI: txid573). The peptide identification was based on a mass tolerance of 20 ppm for intact peptide masses and a mass tolerance of 0.05 Da for fragmented ions. Carbamidomethyl (C), TMT10plex (N-term), and TMT10plex (K) represent fixed modifications, and Oxidation (M), Deamidated (NQ), and TMT10plex (Y) represent variable modifications. All unique peptides (at least one unique spectrum) were required for protein identification and quantification. Then, labeled peptides were quantitatively analyzed with the IQuant software [62] using the Mascot Percolator algorithm [63] in both the peptide spectrum match (PSM) and protein- level with the chosen protein false discovery rate (FDR) strategy (FDR = 1%) [64]. The significant DEPs between the comparison groups were identified by fold change of >2 and p-values < 0.05.

4.11. Bioinformatics and Statistical Analysis

The GO and KEGG databases were used for functional and pathway annotations of all identified proteins. Enrichment analyses for the GO and KEGG pathways were carried out by using the hypergeometric distribution method to explore the biological functions and enrichment pathways of the DEPs between the compared samples. Entries with p-values < 0.05 were considered significant. The DEPs identified from the proteomic results were mapped onto the Search Tool for the Retrieval of Interaction Genes/Proteins (STRING) database (http://string-db.org/, accessed on 1 May 2022) to develop a PPI network (confidence score > 0.7). The Cytoscape software was used to visualize the PPI network.

Supplementary Materials: The following are available online at https://www.mdpi.com/article/10.3390/antibiotics11101341/s1, Table S1:The details of the DEPs between the MDR and DS strains; Table S2: The details of the DEPs between the XDR and DS strains; Table S3: The details of the DEPs between the MDR and XDR strains; Table S4: GO function analysis of the DEPs between the MDR and DS strains; Table S5: KEGG pathway analysis of the DEPs between the MDR and DS strains; Table S6: GO function analysis of the DEPs between the XDR and DS strains; Table S7: KEGG pathway analysis of the DEPs between the XDR and DS strains; Table S8: GO function analysis of the DEPs between the MDR and XDR strains; Table S9: KEGG pathway analysis of the DEPs between the MDR and XDR strains; Table S10: The sequences of primers used in this study [18,55,56]; Table S11: TMT labeling information of the DS, MDR, and XDR strains.

Author Contributions: Z.Z. and L.H. designed this project. L.H. performed the experiments. L.H., X.Y., H.C., Z.M., Y.L. and S.W. analyzed the proteomic data. L.H. wrote the manuscript with support from Z.Z. The final manuscript was thoroughly revised by Z.Z. All authors have read and agreed to the published version of the manuscript.

Funding: This work was financially supported by Science and Technology Projects in Guangzhou (No.202201010177, 202102010018).

Institutional Review Board Statement: Not applicable.

Informed Consent Statement: Not applicable.

Data Availability Statement: The data used to support the findings of this study are available from the corresponding author upon request.

Conflicts of Interest: The authors declare that the research was conducted in the absence of any commercial or financial relationships that could be construed as a potential conflict of interest.

49. Bay, D.C.; Chan, C.S.; Turner, R.J. NarJ subfamily system specific chaperone diversity and evolution is directed by respiratory enzyme associations. *BMC Evol. Biol.* **2015**, *15*, 110. [CrossRef] [PubMed]
50. Kang, K.N.; Klein, D.R.; Kazi, M.I.; Guérin, F.; Cattoir, V.; Brodbelt, J.S.; Boll, J.M. Colistin heteroresistance in *Enterobacter cloacae* is regulated by PhoPQ-dependent 4-amino-4-deoxy-l-arabinose addition to lipid A. *Mol. Microbiol.* **2019**, *111*, 1604–1616. [CrossRef]
51. Llobet, E.; Campos, M.A.; Giménez, P.; Moranta, D.; Bengoechea, J.A. Analysis of the networks controlling the antimicrobial-peptide-dependent induction of *Klebsiella pneumoniae* virulence factors. *Infect. Immun.* **2011**, *79*, 3718–3732. [CrossRef]
52. Snyder, H.; Kellogg, S.L.; Skarda, L.M.; Little, J.L.; Kristich, C.J. Nutritional control of antibiotic resistance via an interface between the phosphotransferase system and a two-component signaling system. *Antimicrob. Agents Chemother.* **2014**, *58*, 957–965. [CrossRef] [PubMed]
53. Humphries, R.; Bobenchik, A.M.; Hindler, J.A.; Schuetz, A.N. Overview of Changes to the Clinical and Laboratory Standards Institute Performance Standards for Antimicrobial Susceptibility Testing, M100, 31st Edition. *J. Clin. Microbiol.* **2021**, *59*, e0021321. [CrossRef]
54. Satlin, M.J.; Lewis, J.S.; Weinstein, M.P.; Patel, J.; Humphries, R.M.; Kahlmeter, G.; Giske, C.G.; Turnidge, J. Clinical and Laboratory Standards Institute and European Committee on Antimicrobial Susceptibility Testing Position Statements on Polymyxin B and Colistin Clinical Breakpoints. *Clin. Infect. Dis.* **2020**, *71*, e523–e529. [CrossRef] [PubMed]
55. Wang, X.; Wang, Y.; Zhou, Y.; Li, J.; Yin, W.; Wang, S.; Zhang, S.; Shen, J.; Shen, Z.; Wang, Y. Emergence of a novel mobile colistin resistance gene, mcr-8, in NDM-producing *Klebsiella pneumoniae. Emerg. Microbes Infect.* **2018**, *7*, 122. [CrossRef]
56. Poirel, L.; Walsh, T.R.; Cuvillier, V.; Nordmann, P. Multiplex PCR for detection of acquired carbapenemase genes. *Diagn. Microbiol. Infect. Dis.* **2011**, *70*, 119–123. [CrossRef] [PubMed]
57. Diancourt, L.; Passet, V.; Verhoef, J.; Grimont, P.A.; Brisse, S. Multilocus sequence typing of *Klebsiella pneumoniae* nosocomial isolates. *J. Clin. Microbiol.* **2005**, *43*, 4178–4182. [CrossRef] [PubMed]
58. Zhan, L.; Wang, S.; Guo, Y.; Jin, Y.; Duan, J.; Hao, Z.; Lv, J.; Qi, X.; Hu, L.; Chen, L.; et al. Outbreak by Hypermucoviscous *Klebsiella pneumoniae* ST11 Isolates with Carbapenem Resistance in a Tertiary Hospital in China. *Front. Cell. Infect. Microbiol.* **2017**, *7*, 182. [CrossRef] [PubMed]
59. Liu, X.; Wu, Y.; Zhu, Y.; Jia, P.; Li, X.; Jia, X.; Yu, W.; Cui, Y.; Yang, R.; Xia, W.; et al. Emergence of colistin-resistant hypervirulent Klebsiella pneumoniae (CoR-HvKp) in China. *Emerg. Microbes Infect.* **2022**, *11*, 648–661. [CrossRef]
60. Mirzaie, A.; Ranjbar, R. Antibiotic resistance, virulence-associated genes analysis and molecular typing of *Klebsiella pneumoniae* strains recovered from clinical samples. *AMB Express* **2021**, *11*, 122. [CrossRef] [PubMed]
61. Yan, J.; Xia, Y.; Yang, M.; Zou, J.; Chen, Y.; Zhang, D.; Ma, L. Quantitative Proteomics Analysis of Membrane Proteins in Enterococcus faecalis With Low-Level Linezolid-Resistance. *Front. Microbiol.* **2018**, *9*, 1698. [CrossRef]
62. Wen, B.; Zhou, R.; Feng, Q.; Wang, Q.; Wang, J.; Liu, S. IQuant: An automated pipeline for quantitative proteomics based upon isobaric tags. *Proteomics* **2014**, *14*, 2280–2285. [CrossRef]
63. Brosch, M.; Yu, L.; Hubbard, T.; Choudhary, J. Accurate and sensitive peptide identification with Mascot Percolator. *J. Proteome Res.* **2009**, *8*, 3176–3181. [CrossRef]
64. Savitski, M.M.; Wilhelm, M.; Hahne, H.; Kuster, B.; Bantscheff, M. A Scalable Approach for Protein False Discovery Rate Estimation in Large Proteomic Data Sets. *Mol. Cell. Proteom.* **2015**, *14*, 2394–2404. [CrossRef] [PubMed]

24. Shi, X.; Chen, M.; Yu, Z.; Bell, J.M.; Wang, H.; Forrester, I.; Villarreal, H.; Jakana, J.; Du, D.; Luisi, B.F.; et al. In situ structure and assembly of the multidrug efflux pump AcrAB-TolC. *Nat. Commun.* **2019**, *10*, 2635. [CrossRef] [PubMed]
25. Shao, L.; Yao, B.; Yang, J.; Li, X.; Ye, K.; Zhang, Y.; Wang, C. Characterization of a multidrug-resistant *Klebsiella pneumoniae* ST3330 clone responsible for a nosocomial outbreak in a neonatal intensive care unit. *Ann. Palliat. Med.* **2020**, *9*, 1092–1102. [CrossRef]
26. Kajimura, J.; Rahman, A.; Hsu, J.; Evans, M.R.; Gardner, K.H.; Rick, P.D. O acetylation of the enterobacterial common antigen polysaccharide is catalyzed by the product of the yiaH gene of *Escherichia coli* K-12. *J. Bacteriol.* **2006**, *188*, 7542–7550. [CrossRef] [PubMed]
27. Deng, W.; Li, C.; Xie, J. The underling mechanism of bacterial TetR/AcrR family transcriptional repressors. *Cell. Signal.* **2013**, *25*, 1608–1613. [CrossRef] [PubMed]
28. Palmer, C.N.; Axen, E.; Hughes, V.; Wolf, C.R. The repressor protein, Bm3R1, mediates an adaptive response to toxic fatty acids in *Bacillus megaterium*. *J. Biol. Chem.* **1998**, *273*, 18109–18116. [CrossRef] [PubMed]
29. Liu, W.; Huang, L.; Su, Y.; Qin, Y.; Zhao, L.; Yan, Q. Contributions of the oligopeptide permeases in multistep of *Vibrio alginolyticus* pathogenesis. *Microbiologyopen* **2017**, *6*, e00511. [CrossRef] [PubMed]
30. Liu, Y.; Li, L.; Huang, T.; Wu, W.; Liang, W.; Chen, M. The Interaction between Phagocytes and *Streptococcus agalactiae* (GBS) Mediated by the Activated Complement System is the Key to GBS Inducing Acute Bacterial Meningitis of Tilapia. *Animals* **2019**, *9*, 818. [CrossRef] [PubMed]
31. Keasey, S.L.; Suh, M.J.; Das, S.; Blancett, C.D.; Zeng, X.; Andresson, T.; Sun, M.G.; Ulrich, R.G. Decreased Antibiotic Susceptibility Driven by Global Remodeling of the *Klebsiella pneumoniae* Proteome. *Mol. Cell. Proteom.* **2019**, *18*, 657–668. [CrossRef] [PubMed]
32. Sheppard, D.C.; Howell, P.L. Biofilm Exopolysaccharides of Pathogenic Fungi: Lessons from Bacteria. *J. Biol. Chem.* **2016**, *291*, 12529–12537. [CrossRef] [PubMed]
33. Mitchell, K.F.; Zarnowski, R.; Sanchez, H.; Edward, J.A.; Reinicke, E.L.; Nett, J.E.; Mitchell, A.P.; Andes, D.R. Community participation in biofilm matrix assembly and function. *Proc. Natl. Acad. Sci. USA* **2015**, *112*, 4092–4097. [CrossRef] [PubMed]
34. Shen, F.; Ge, C.; Yuan, P. Metabolomics Study Reveals Inhibition and Metabolic Dysregulation in *Staphylococcus aureus* Planktonic Cells and Biofilms Induced by Carnosol. *Front. Microbiol.* **2020**, *11*, 538572. [CrossRef]
35. Vuotto, C.; Longo, F.; Pascolini, C.; Donelli, G.; Balice, M.P.; Libori, M.F.; Tiracchia, V.; Salvia, A.; Varaldo, P.E. Biofilm formation and antibiotic resistance in *Klebsiella pneumoniae* urinary strains. *J. Appl. Microbiol.* **2017**, *123*, 1003–1018. [CrossRef] [PubMed]
36. Chen, X.; Tian, J.; Luo, C.; Wang, X.; Li, X.; Wang, M. Cell Membrane Remodeling Mediates Polymyxin B Resistance in *Klebsiella pneumoniae*: An Integrated Proteomics and Metabolomics Study. *Front. Microbiol.* **2022**, *13*, 810403. [CrossRef] [PubMed]
37. Luo, S.C.; Lou, Y.C.; Rajasekaran, M.; Chang, Y.W.; Hsiao, C.D.; Chen, C. Structural basis of a physical blockage mechanism for the interaction of response regulator PmrA with connector protein PmrD from *Klebsiella pneumoniae*. *J. Biol. Chem.* **2013**, *288*, 25551–25561. [CrossRef] [PubMed]
38. Yakhnina, A.A.; Bernhardt, T.G. The Tol-Pal system is required for peptidoglycan-cleaving enzymes to complete bacterial cell division. *Proc. Natl. Acad. Sci. USA* **2020**, *117*, 6777–6783. [CrossRef]
39. Rocker, A.; Lacey, J.A.; Belousoff, M.J.; Wilksch, J.J.; Strugnell, R.A.; Davies, M.R.; Lithgow, T. Global Trends in Proteome Remodeling of the Outer Membrane Modulate Antimicrobial Permeability in *Klebsiella pneumoniae*. *mBio* **2020**, *11*, e00603-20. [CrossRef]
40. Hamzaoui, Z.; Ocampo-Sosa, A.; Fernandez Martinez, M.; Landolsi, S.; Ferjani, S.; Maamar, E.; Saidani, M.; Slim, A.; Martinez-Martinez, L.; Boutiba-Ben Boubaker, I. Role of association of OmpK35 and OmpK36 alteration and blaESBL and/or blaAmpC genes in conferring carbapenem resistance among non-carbapenemase-producing *Klebsiella pneumoniae*. *Int. J. Antimicrob. Agents.* **2018**, *52*, 898–905. [CrossRef] [PubMed]
41. Tsai, Y.K.; Fung, C.P.; Lin, J.C.; Chen, J.H.; Chang, F.Y.; Chen, T.L.; Siu, L.K. Klebsiella pneumoniae outer membrane porins OmpK35 and OmpK36 play roles in both antimicrobial resistance and virulence. *Antimicrob. Agents Chemother.* **2011**, *55*, 1485–1493. [CrossRef] [PubMed]
42. Ngbede, E.O.; Adekanmbi, F.; Poudel, A.; Kalalah, A.; Kelly, P.; Yang, Y.; Adamu, A.M.; Daniel, S.T.; Adikwu, A.A.; Akwuobu, C.A.; et al. Concurrent Resistance to Carbapenem and Colistin Among Enterobacteriaceae Recovered from Human and Animal Sources in Nigeria Is Associated with Multiple Genetic Mechanisms. *Front. Microbiol.* **2021**, *12*, 740348. [CrossRef]
43. Cai, R.; Deng, H.; Song, J.; Zhang, L.; Zhao, R.; Guo, Z.; Zhang, X.; Zhang, H.; Tian, T.; Ji, Y.; et al. Phage resistance mutation triggered by OmpC deficiency in *Klebsiella pneumoniae* induced limited fitness costs. *Microb. Pathog.* **2022**, *167*, 105556. [CrossRef] [PubMed]
44. Bhagirath, A.Y.; Li, Y.; Patidar, R.; Yerex, K.; Ma, X.; Kumar, A.; Duan, K. Two Component Regulatory Systems and Antibiotic Resistance in Gram-Negative Pathogens. *Int. J. Mol. Sci.* **2019**, *20*, 1781. [CrossRef] [PubMed]
45. Ballal, A.; Basu, B.; Apte, S.K. The Kdp-ATPase system and its regulation. *J. Biosci.* **2007**, *32*, 559–568. [CrossRef]
46. Gerken, H.; Vuong, P.; Soparkar, K.; Misra, R. Roles of the EnvZ/OmpR Two-Component System and Porins in Iron Acquisition in *Escherichia coli*. *mBio* **2020**, *11*, e01192-20. [CrossRef]
47. Dean, C.R.; Poole, K. Expression of the ferric enterobactin receptor (PfeA) of Pseudomonas aeruginosa: Involvement of a two-component regulatory system. *Mol. Microbiol.* **1993**, *8*, 1095–1103. [CrossRef] [PubMed]
48. Boudes, M.; Lazar, N.; Graille, M.; Durand, D.; Gaidenko, T.A.; Stewart, V.; van Tilbeurgh, H. The structure of the NasR transcription antiterminator reveals a one-component system with a NIT nitrate receptor coupled to an ANTAR RNA-binding effector. *Mol. Microbiol.* **2012**, *85*, 431–444. [CrossRef] [PubMed]

References

1. Grundmann, H.; Glasner, C.; Albiger, B.; Aanensen, D.M.; Tomlinson, C.T.; Andrasević, A.T.; Cantón, R.; Carmeli, Y.; Friedrich, A.W.; Giske, C.G.; et al. Occurrence of carbapenemase-producing Klebsiella pneumoniae and *Escherichia coli* in the European survey of carbapenemase-producing *Enterobacteriaceae* (EuSCAPE): A prospective, multinational study. *Lancet Infect. Dis.* **2017**, *17*, 153–163. [CrossRef]
2. Zhang, R.; Liu, L.; Zhou, H.; Chan, E.W.; Li, J.; Fang, Y.; Li, Y.; Liao, K.; Chen, S. Nationwide Surveillance of Clinical Carbapenem-resistant *Enterobacteriaceae* (CRE) Strains in China. *EBioMedicine* **2017**, *19*, 98–106. [CrossRef]
3. Perez, F.; Bonomo, R.A. Carbapenem-resistant *Enterobacteriaceae*: Global action required. *Lancet Infect. Dis.* **2019**, *19*, 561–562. [CrossRef]
4. Cienfuegos-Gallet, A.V.; Ocampo de Los Ríos, A.M.; Sierra Viana, P.; Ramirez Brinez, F.; Restrepo Castro, C.; Roncancio Villamil, G.; Del Corral Londoño, H.; Jiménez, J.N. Risk factors and survival of patients infected with carbapenem-resistant *Klebsiella pneumoniae* in a KPC endemic setting: A case-control and cohort study. *BMC Infect. Dis.* **2019**, *19*, 830. [CrossRef] [PubMed]
5. Cassini, A.; Högberg, L.D.; Plachouras, D.; Quattrocchi, A.; Hoxha, A.; Simonsen, G.S.; Colomb-Cotinat, M.; Kretzschmar, M.E.; Devleesschauwer, B.; Cecchini, M.; et al. Attributable deaths and disability-adjusted life-years caused by infections with antibiotic-resistant bacteria in the EU and the European Economic Area in 2015: A population-level modelling analysis. *Lancet Infect. Dis.* **2019**, *19*, 56–66. [CrossRef]
6. Li, J.; Nation, R.L.; Turnidge, J.D.; Milne, R.W.; Coulthard, K.; Rayner, C.R.; Paterson, D.L. Colistin: The re-emerging antibiotic for multidrug-resistant Gram-negative bacterial infections. *Lancet Infect. Dis.* **2006**, *6*, 589–601. [CrossRef]
7. Karaiskos, I.; Lagou, S.; Pontikis, K.; Rapti, V.; Poulakou, G. The "Old" and the "New" Antibiotics for MDR Gram-Negative Pathogens: For Whom, When, and How. *Front. Public Health* **2019**, *7*, 151. [CrossRef] [PubMed]
8. Jousset, A.B.; Bonnin, R.A.; Rosinski-Chupin, I.; Girlich, D.; Cuzon, G.; Cabanel, N.; Frech, H.; Farfour, E.; Dortet, L.; Glaser, P.; et al. A 4.5-Year Within-Patient Evolution of a Colistin-Resistant *Klebsiella pneumoniae* Carbapenemase-Producing *K. pneumoniae* Sequence Type 258. *Clin. Infect. Dis.* **2018**, *67*, 1388–1394. [CrossRef] [PubMed]
9. Jeannot, K.; Bolard, A.; Plésiat, P. Resistance to polymyxins in Gram-negative organisms. *Int. J. Antimicrob. Agents.* **2017**, *49*, 526–535. [CrossRef]
10. Cannatelli, A.; D'Andrea, M.M.; Giani, T.; Di Pilato, V.; Arena, F.; Ambretti, S.; Gaibani, P.; Rossolini, G.M. In vivo emergence of colistin resistance in *Klebsiella pneumoniae* producing KPC-type carbapenemases mediated by insertional inactivation of the PhoQ/PhoP mgrB regulator. *Antimicrob. Agents Chemother.* **2013**, *57*, 5521–5526. [CrossRef]
11. Cheng, H.Y.; Chen, Y.F.; Peng, H.L. Molecular characterization of the PhoPQ-PmrD-PmrAB mediated pathway regulating polymyxin B resistance in *Klebsiella pneumoniae* CG43. *J. Biomed. Sci.* **2010**, *17*, 60. [CrossRef] [PubMed]
12. Liu, Y.Y.; Wang, Y.; Walsh, T.R.; Yi, L.X.; Zhang, R.; Spencer, J.; Doi, Y.; Tian, G.; Dong, B.; Huang, X.; et al. Emergence of plasmid-mediated colistin resistance mechanism MCR-1 in animals and human beings in China: A microbiological and molecular biological study. *Lancet Infect. Dis.* **2016**, *16*, 161–168. [CrossRef]
13. Chen, F.J.; Lauderdale, T.L.; Huang, W.C.; Shiau, Y.R.; Wang, H.Y.; Kuo, S.C. Emergence of mcr-1, mcr-3 and mcr-8 in clinical *Klebsiella pneumoniae* isolates in Taiwan. *Clin. Microbiol. Infect.* **2021**, *27*, 305–307. [CrossRef]
14. Rosenblum, R.; Khan, E.; Gonzalez, G.; Hasan, R.; Schneiders, T. Genetic regulation of the ramA locus and its expression in clinical isolates of *Klebsiella pneumoniae*. *Int. J. Antimicrob. Agents.* **2011**, *38*, 39–45. [CrossRef] [PubMed]
15. Ni, W.; Li, Y.; Guan, J.; Zhao, J.; Cui, J.; Wang, R.; Liu, Y. Effects of Efflux Pump Inhibitors on Colistin Resistance in Multidrug-Resistant Gram-Negative Bacteria. *Antimicrob. Agents Chemother.* **2016**, *60*, 3215–3218. [CrossRef]
16. Chen, B.; Zhang, D.; Wang, X.; Ma, W.; Deng, S.; Zhang, P.; Zhu, H.; Xu, N.; Liang, S. Proteomics progresses in microbial physiology and clinical antimicrobial therapy. *Eur. J. Clin. Microbiol. Infect. Dis.* **2017**, *36*, 403–413. [CrossRef]
17. Moulder, R.; Bhosale, S.D.; Goodlett, D.R.; Lahesmaa, R. Analysis of the plasma proteome using iTRAQ and TMT-based Isobaric labeling. *Mass. Spectrom. Rev.* **2018**, *37*, 583–606. [CrossRef] [PubMed]
18. Russo, T.A.; Olson, R.; Fang, C.T.; Stoesser, N.; Miller, M.; MacDonald, U.; Hutson, A.; Barker, J.H.; La Hoz, R.M.; Johnson, J.R. Identification of Biomarkers for Differentiation of Hypervirulent *Klebsiella pneumoniae* from Classical *K. pneumoniae*. *J. Clin. Microbiol.* **2018**, *56*, e00776-18. [CrossRef] [PubMed]
19. Gu, D.; Dong, N.; Zheng, Z.; Lin, D.; Huang, M.; Wang, L.; Chan, E.W.; Shu, L.; Yu, J.; Zhang, R.; et al. A fatal outbreak of ST11 carbapenem-resistant hypervirulent *Klebsiella pneumoniae* in a Chinese hospital: A molecular epidemiological study. *Lancet Infect. Dis.* **2018**, *18*, 37–46. [CrossRef]
20. Huang, Y.H.; Chou, S.H.; Liang, S.W.; Ni, C.E.; Lin, Y.T.; Huang, Y.W.; Yang, T.C. Emergence of an XDR and carbapenemase-producing hypervirulent *Klebsiella pneumoniae* strain in Taiwan. *J. Antimicrob. Chemother.* **2018**, *73*, 2039–2046. [CrossRef]
21. Navon-Venezia, S.; Kondratyeva, K.; Carattoli, A. Klebsiella pneumoniae: A major worldwide source and shuttle for antibiotic resistance. *FEMS Microbiol. Rev.* **2017**, *41*, 252–275. [CrossRef] [PubMed]
22. Ogawa, W.; Onishi, M.; Ni, R.; Tsuchiya, T.; Kuroda, T. Functional study of the novel multidrug efflux pump KexD from *Klebsiella pneumoniae*. *Gene* **2012**, *498*, 177–182. [CrossRef] [PubMed]
23. Sun, L.; Rasmussen, P.K.; Bai, Y.; Chen, X.; Cai, T.; Wang, J.; Guo, X.; Xie, Z.; Ding, X.; Niu, L.; et al. Proteomic Changes of *Klebsiella pneumoniae* in Response to Colistin Treatment and crrB Mutation-Mediated Colistin Resistance. *Antimicrob. Agents Chemother.* **2020**, *64*, e02200-19. [CrossRef] [PubMed]

www.ingramcontent.com/pod-product-compliance
Lightning Source LLC
Chambersburg PA
CBHW061509210926
51850CB00008B/1862
* 9 7 8 3 7 2 5 8 1 4 1 8 3 *

MDPI AG
Grosspeteranlage 5
4052 Basel
Switzerland
Tel.: +41 61 683 77 34

Antibiotics Editorial Office
E-mail: antibiotics@mdpi.com
www.mdpi.com/journal/antibiotics